Accountability in Social Research

Accountability in Social Research

Issues and Debates

Norma R. A. Romm

University of Hull
Hull, United Kingdom

Kluwer Academic / Plenum Publishers
New York • Boston • Dordrecht • London • Moscow

Library of Congress Cataloging-in-Publication Data

Romm, Norma R. A.
Accountability in social research: issues and debates/Norma R. A. Romm.
p. cm.
Includes bibliographical references and index.
ISBN 0-306-46564-7
1. Social sciences—Research—Moral and ethical aspects. 2. Social scientists—Professional ethics. 3. Science—Philosophy. I. Title.

H62 .R644 2001
300.7′2—dc21

2001018618

ISBN 0-306-46564-7

233 Spring Street, New York, New York 10013

http://www.wkap.nl/

10 9 8 7 6 5 4 3 2 1

A C.I.P. record for this book is available from the Library of Congress

Printed in the United States of America

To Peter
who contributed so much toward this book

Preface

In this book I have concentrated on drawing attention to various conceptions of accountability that might be brought to bear in judging the practice of social research. Much of the book is organized around making explicit the assumptions that influence what counts as "proper" research in society, including assumptions about how social inquirers might be held accountable. My focus is on reviewing discourses around the practice of "professional" inquiry, with a view to reconsidering the way in which people create expectations for accountable social inquiry. My focus hereon is related to my concern that the manner in which judgments about researchers' accountability are made, is not without social consequences for our way of living in society.

I have approached the issues by beginning with a discussion of tenets of the position called "positivism" (so named by certain proponents), and by considering the view on accountability that is implied by adherence to these tenets. Briefly expressed, positivist argumentation suggests that researchers are required to "do science" in a manner that warrants their being considered, indeed, scientists. I use my discussion of accountability as seen within positivist argumentation to explicate ways in which alternative positions have arisen as ways of treating accountability issues. Through my way of comparing the various positions, I hope to provide some indication of the complexity of ethical and accountability issues in social inquiry.

The book elucidates six positions that offer alternatives to positivist conceptions of researcher accountability. These positions are introduced in Chapter 1. Five of them are spelled out in detail in Chapter 2 — where I show how they offer differing ways of approaching positivist argumentation. Responses to positivism are discussed in terms of: non-foundationalism; scientific realism; interpretivism; critical theory; and anti-foundationalist feminism. I also consider in detail their various views on accountability in relation to one another (and in relation to what I call discursively-oriented constructivism). In Chapter 3, I go on to provide an elaboration of constructivist argumentation in relation to positivist and other views on researcher accountability. In doing so, I extend constructivist thinking toward what I name a "trusting constructivist" position. This position focuses on ways in which trust earning and trust awarding in the context of social inquiry can proceed without researchers having to justify themselves as striving to gain access to knowledge as representation of reality. Through my development of the trusting constructivist position, I show how other ways of creating trust through processes of social discourse can be explored.

I have attempted to structure my theoretical discussion of the positions in a way that will be accessible to readers who are not altogether familiar with differing arguments about the proper practice of social science. I have engaged with debates about what is properly involved in social research by explicating alternative ways in which the tenets of positivism can be approached and by showing implications hereof for proposals for researcher accountability. I then carry through my discussion of the various positions as explored in Chapters 1-3, by linking these to a detailed assessment of some actual research examples (in Chapters 4-7). And in Chapter 8, I summarize the various arguments that I have developed throughout the book, bringing together the discussions in Chapters 1-3 with those in Chapters 4-7.

It is possible that some readers will not wish to read the book in terms of the order of presentation given. They may wish to read it in terms of a different way of entering into the discussions. For instance, after reading my introductory chapter (Chapter 1), they may wish to move on to my account of one or more of the research examples (from Chapters 4-7), before delving into my detailed examination and comparison of differing arguments as developed in Chapters 2-3. And the reading of any of the material could perhaps be facilitated by looking again at the outline of the arguments as given in Chapter 1.

In whatever way readers decide to approach the discussions in the book, I hope that they will find it a valuable source to extend their considerations of the issues that I have raised in it.

I wish to make the following acknowledgments. To start with, I wish to thank the University of Hull for affording me the time to create the book (through my position as senior researcher in the Centre for Systems Studies). I have also benefited from many interesting discussions with members of the Centre. I would like to thank Ken Derham of Kluwer Academic/Plenum publishers, for his advice (in terms of the structuring of the book and its editing). I wish also to thank my dear friend Veronica McKay, for her warm and insightful energies. And Susan Weil offered valuable commentary. (Indeed the idea of the Preface sprang from her input.) I am exceedingly thankful to Ernst Onkenhout (my longstanding partner) for his careful proofreading. And I wish to express my deep gratitude to Peter Adman, my friend and colleague, for the tremendous care, energy, and intelligence that he put into helping me to design the various chapters (including ways of expressing arguments), and for his meticulous and well thought-out designs for all the figures that appear in Chapters 1-7.

Norma Romm
November 2000

Contents

Chapter 4: Exploring Experimentation

Chapter 5: Exploring Survey Research

Chapter 6: Exploring the Ethnographic Study of Lives

LIST OF TABLES

LIST OF FIGURES

Accountability in Social Research

1
Introduction

The focus of this book is on the way in which accountability in the practice of social research can be approached. As is spelled out in the book, the terms "accountability" and "social research" depend for their meanings on the way they are fleshed out in specific contexts of use. Usage of these terms invokes discussion around the question of what might be achieved through the practice of what is called "scientific" or "professional" social inquiry in society. Answers that are provided are bound up with differing conceptions of the character of science, and specifically, of *social* science (or social inquiry) as an enterprise in society. My discussion of various arguments as explored through the book is directed at highlighting ways in which criteria of accountability have become determined in definitions of the character of social science. My aim in foregrounding the various criteria, is to extend our considerations of the terms on which researchers' accountabilities might be judged.

Much argumentation in the literature on the character and achievements of science can be clarified by considering how these arguments relate to a view of science called "positivism". Bacon's work (in the late 16th and early 17th centuries), concerning the development of an empirical scientific approach grounded in observation and experimentation as the basis for the study of nature, is often considered as a forerunner of what later was named (by Comte) the positivist philosophy.[1] Crotty notes that Comte popularized the word positivism in his *Cours de philosophie positive,* which appeared between 1830 and 1842 (Crotty, 1998, p. 19). In his writings, he also pleaded for the extension of natural scientific procedures to the social sciences, and he coined both the terms "sociology" and "social physics". His argument was couched in terms of the need for the development of a "positive" (scientific) approach to the study of society.

Delanty suggests that an exploration of positivist argumentation provides a way of opening a discussion on the meaning of the social scientific enterprise in society, for, as he suggests, "philosophical debates on the methodology and self-understanding of social science have been, for the greater part, shaped by the positivist dispute" (1997, p. 11). Putting aside considerations of to what extent "positivist" thinking may (arguably) be influential in the self-understanding of those regarding themselves as social inquirers, and/or in the minds of philosophers of social science, and/or in the minds of others in society who have developed expectations concerning the practice of social

---

[1]Comte himself refers favorably to Bacon's influence on "positive" philosophy. He suggests that the positive philosophy is a system of thought that "has been growing steadily since Bacon's time" (1970, p. 20).

inquiry, I suggest that it can at least be considered as one platform from which to explore debates about the accountability of social researchers. For this reason I begin my discussion in this book with a categorization of arguments labeled as "positivist".

In Chapter 2, I organize my discussion of various viewpoints concerning what is, and/or should be, involved in the practice of social inquiry by positioning them in terms of their relationship with positivist argumentation, before going on to discuss their relationship with one another. The chapter presents some detail on certain "tenets" associated with positivism, as located by Delanty (1997). Delanty cautions that not all authors who might be labeled (by themselves and/or others) as positivist in orientation need uphold all these tenets. Decisions can be made in specific circumstances as to whether or not to apply the label to some argument advanced by an author. The categories that I present in this book are meant to be considered as enablers of discursive engagement around the issues brought to the fore through use of the categories. This still allows people to reconsider the manner in which their own and others' arguments are being placed if they feel that the debate around questions of concern can be furthered in this way. (This issue is also discussed in Chapter 8, Section 8.8.)

The tenets identified by Delanty (1997, pp. 12-13) as defining the position of positivism are: *scientism; phenomenalism; empiricism; value freedom; and instrumental knowledge*.

- *Scientism* suggests that the organization of investigations in the natural sciences can be taken as exemplary for all scientific inquiry.
- *Phenomenalism* asserts that "observable units" of reality can be isolated and can be observed without prejudice (or bias).
- *Empiricism* asserts that empirical experience (observation via the senses) forms the foundation on which science as an enterprise rests: scientific reasoning must be firmly grounded in such experience.
- *Value freedom* suggests that the personal values of scientists should not affect their readiness to employ a scientific approach to make advancements in knowledge.
- *Instrumental knowledge* refers to the capacity of science to furnish some information concerning the relationship between conditions and ensuing outcomes in reality: those seeking goals in society are thus given the instruments to act in terms of scientifically informed understanding.

Summarizing the substance of these tenets, the first three could be argued to refer to the *processes* involved in activities that warrant the label of "science" being attached to them; the last two tenets can be argued to refer to the status of the *product* (the development of knowledge) produced through the scientific process. The first three tenets suggest that the processes involved in natural and social science are not different in kind; that science requires developing observation techniques to get to grips with factual phenomena; and that by relating observation of phenomena through scientific logic to more general theoretical statements about reality, knowledge can be sought. The fourth tenet suggests that a product that reflects reality (rather than being a reflection of scientists' particular values and preferences) is likely to emerge through this process. And this product is seen as having instrumental utility (in terms of the fifth tenet) precisely because of its status as affording some knowledge of reality.

According to these tenets, a certain vision of accountable research practice can be drawn out. Scientists can be *called to account within the scientific community* for their manner of proceeding (in order to warrant the label of "science" being attached to their activities). Their accountability is seen as linked to the commitment to use logic (inductive and/or deductive) to develop statements about reality grounded in the observation of facts. This commitment is threatened by biases being introduced into the practice of science. The scientific community in turn is *accountable to the wider society* for ensuring that "science" is indeed being practiced within the community. The wider society expects of this community that it can develop some knowledge of reality.

This manner of understanding the meaning of science, has been questioned from various angles. Chapter 2 discusses a number of positions that have developed in response to positivist argumentation. I do not attempt extrapolations of how positivism might be used in turn to react to the various challenges leveled against it. I am utilizing positivist argumentation in this book largely as a way of setting a starting position for entering into debates about researcher accountability. I do not direct attention to considering how positivism arose in reaction to what was taken to be "unscientific" ways of knowing, nor to considering how this might affect its positioning in relation to other arguments. However, in Chapters 4-7, when discussing some research strategies, I do show how positivism as a position can be used (along with other positions) to offer an understanding of the examples utilized. I explore in these chapters positivist as well as other assessments of the research processes in question and of the status of the findings produced.

The first response to positivist argumentation that is presented in Chapter 2, is what is called the critical rationalist position — so named by Popper. Popper devised this label in order to distinguish his approach from Comtian as well as other forms of positivism. He indicates that in developing his position he has drawn on some of the arguments developed within positivism, while criticizing others (1978). As far as the tenet of *scientism* is concerned, Popper suggests that the *same kind of logic* — that of deduction — is the logic properly characterizing all scientific activity, in the natural as well as in the social sciences. (He does not agree with what he sees as the positivist understanding of the way that induction can be used within scientific inquiry.) Popper supports the idea of the unity of the sciences in the sense in which he sees the (proper) operation of science (1959). In clarifying his own view of the unity of the sciences, he also points out that the different subject matter with which the social sciences deal, namely, regularities in social life, can be affected by self-fulfilling dynamics in ways not present in nature. That is, outcomes in the social world may be influenced by ideas that people have about possible outcomes that can occur. This implies that the tenet of *scientism* is somewhat qualified in Popper's argument (see also Delanty, 1997, p. 32).

The tenets of *phenomenalism* and *empiricism* are reviewed in critical rationalism as follows. It is argued that the evidence of experience (or rather, observation statements reporting hereon) can never be used to verify proposed hypotheses concerning the relationship between conditions and outcomes in reality. But hypotheses can become tentatively corroborated if they have withstood active attempts to falsify them. Deductive logic can be used for the task of subjecting hypotheses to testing — although it cannot be used to ever arrive at certainty regarding their veracity. Nevertheless, according to Popper, the knowledge of reality afforded by science (when properly practiced) still can be regarded as offering the highest level of certainty available to us.

The critical rationalist position advanced by Popper can be seen as bearing strong resemblance to the non-foundationalism espoused by, for example, Hammersley and Gomm (1997a). Hammersley and Gomm concentrate on showing implications of the(ir) non-foundationalist view of science for researcher accountability. Their argument is that as researchers proceed in their scientific inquiries, they have a responsibility to work under the constraint of trying not to produce accounts that are at odds with the evidence available about the relevant phenomena. They have a responsibility to, as Hammersley and Gomm put it, "maximize the chances of discovering the truth about the matter concerned" (1997a, paragraph 4.14). When researchers introduce biases (creating errors that could have been avoided), they may be accused of operating in an unaccountable way. In terms of Hammersley and Gomm's argument, collegial accountability — that is, accountability to colleagues in the scientific community — becomes the route to maximizing chances of truth-discovery.

The scientific realist position is the next one to be presented. As far as the tenet of the unity of science (*scientism*) is concerned, scientific realism advocates the methodological unity of the natural and social sciences (Keat & Urry, 1982, pp. 142-143). It is argued that the logic used in the study of nature applies also in the study of society. This does not entail the suggestion that people's attribution of meaning to events in social life is irrelevant for the study of society. However, in terms of this position, as explained by Keat and Urry, the meanings themselves can be seen as "causes" that generate outcomes; and furthermore, we can attempt to explain these meanings themselves in causal terms by considering why they have been adopted by agents (1982, p. 90). Keat and Urry argue that the method that Marx supported — especially in his *Das Kapital* (published in the second half of the 19th century) — provides a good example of how science can be practiced in terms of this understanding of the aim of science (Keat & Urry, 1982, p. 96).

Turning to the tenets of *phenomenalism* and *empiricism,* the scientific realist argument suggests that the focus of both positivism and (critical rationalist) non-foundationalism on "the evidence" as the court of appeal to judge the credibility of propositions, directs attention away from the real object of science. What needs to be investigated, according to scientific realism, are the structures and mechanisms that are hidden from observation. It is regarded as a crucial part of the process of scientific investigation to build theories about these realities. Sayer (1983) suggests that the logic of retroduction can be used to help further such investigation. In using this logic, however, statements about posited structures and mechanisms can neither be verified nor falsified. This lends more uncertainty to knowledge-production than is acknowledged by either positivism or non-foundationalism. Conceptions of accountability within a scientific realist view of science are nevertheless similar to views developed within non-foundationalism. They revolve around possibilities for scientists to advance knowledge in the midst of the (increased) uncertainty characterizing the endeavor to know.

As indicated above, positivism, non-foundationalism, and scientific realism, all in some way share the idea of the unity of the sciences (labeled above as *scientism*). This tenet has been strongly opposed from a perspective sometimes called interpretive social science, following Weber's call — in the beginning of the 20th century — for a sociology of history and culture, in which the understanding of meaning in social life is seen as a distinct enterprise (cf. Delanty, 1997, p. 49). In terms of the interpretivist argument, the study of society must involve a distinct way of doing science, due to the different subject matter of social science. Explanatory models developed by social scientists have to be rooted in an understanding of some "complex of meaning" that is ascribed to the actors involved. For Weber, "a correct causal interpretation of typical action means that the process which is claimed to be typical is shown to be both adequately grasped on the level of meaning and at the same time the interpretation is to some degree causally adequate" (1973, p. 133). The explanation of sequences of events must be rooted in the understanding (or interpretation) of meaning.

The accountability of social scientists is here linked to their efforts to develop plausible accounts of the motivating meanings that constitute social existence. According to this view, they should try to adopt an attitude of ethical neutrality in the process of developing their understanding, and ought not to adjudicate on what kinds of values should be furthered in social life. If they act otherwise, they are overstepping their role as scientists.

Positivism, non-foundationalism, scientific realism, and (Weberian) interpretivism can be seen as sharing the tenets of *value freedom* and of *instrumental knowledge*. They all suggest that knowledge of reality can be sought through adopting a scientific commitment toward apprehending it; and they all suggest that the development of such knowledge is useful for those who wish to act in terms of more informed understanding. These tenets have been criticized from various perspectives. The perspective of what is called "critical theory" (as, for example, expressed by Habermas, 1974) is based on a critique of the distinction between facts and values as upheld through the tenet of value freedom. According to this argument, "facts" are not to be treated as things to be apprehended in a more or less unbiased way through the application of scientific procedure. Habermas's argument is that unless people (in the "scientific community" and in public life) are able to bring into consciousness the interests/values that might be guiding their inquiries, they are unable to subject them to rational discursive examination (1974, p. 210). The accountabilities of those involved in critical theoretical inquiry is linked up with their commitment to generate a discussion around the values that they bring to bear when they proceed in "doing science". Without this discussion critical theorists argue that science can all too easily be used in the service of an administrative or technical approach to the solution of social problems.

Another way in which the tenets of *value freedom* and of *instrumental knowledge* can be criticized is through the feminist critique of the subject/object distinction. Problematizing this distinction means that it is considered as impossible to separate out an "object" of scientific inquiry that can be looked at "in a value free and neutral way" (Oakley, 1998, p. 710). The claim that it is possible to use the scientific process to get to grips with some object of inquiry about which information can be provided, is seen as leading to a hierarchical relationship in which, as Oakley puts it, "the knower is the expert, and the known are the objects of someone else's knowledge" (1998, p. 710).

Seen in this light, feminism is concerned that (what is called) "male-stream" social science is based on authorizing the supposed rationality of those defining themselves as scientists. Lather argues in this respect that the "master code of positivism" infiltrates the legitimation practices of those engaging in scientific practice, even when the hold of the code is supposedly loosened (1993, p. 674). Scientific practices aimed at getting closer to "the truth" (as representation of reality), operate to exclude other ways of defining "knowing" (Lather, 1991, p. 60).

Lather proposes that feminism can be associated with an anti-foundationalist position insofar as it does not posit some object of inquiry of which (representational) knowledge is sought. The "validity" of knowing rather inheres in the way in which provision is made in the research process for multiple visions to be developed. New ways of defining accountabilities within what is taken to be the scientific community, and between it and other communities, that encourage the development of heterogeneity as a way of organizing human relations, are called for within the anti-foundationalist feminist project.

Certain authors (some of whom align themselves with a form of anti-foundationalism) have concentrated on criticizing the tenets of positivism by focusing on the discursive practices of those involved in scientific inquiry. These authors highlight the way in which disputes in the scientific community beset the process of assessing the work of colleagues. The focus on this aspect of scientific inquiry can be seen as having roots in Kuhn's account of the way in which, as Delanty puts it, "science as a cognitive system is ultimately shaped by the institution of science" (Delanty, 1997, p. 35). Delanty notes that Kuhn attaches special importance to the role of the scientific community in the construction of science (1997, p. 35). However, Kuhn conceives this community differently from the manner in which it is seen by, for example, Popper. For Popper, the scientific community is the ultimate harbinger of the prevalence of reason within the practice of science. It acts as a check on unreason, through the mechanism of what Popper calls "mutual criticism". But for Kuhn, people's criticism of others' work does not simply amount to using reason to establish whether logic is being properly applied in the process of developing and testing theories.

According to Kuhn, scientists who are attached to a paradigm (and communities of scientists who are thus attached) can be blind to the implications of what others might regard as anomalous information (or disconfirming evidence). Discarding certain theories and endorsing or corroborating other ones that have apparently withstood the test of criticism, is not necessarily rooted in what "the evidence" seems to indicate. What "the evidence" indicates, can be subject to irresolvable dispute, especially if different paradigms of thought are brought into play in considering how the evidence should be read. In such cases, Kuhn avers, "a decision between alternate ways of practicing science is called for" (1970, p. 157). The thesis of incommensurability that Kuhn advances, suggests that there are no absolute standards by which scientists can compare ways of addressing the issues. There is no one set of standards that can be drawn upon for deciding between the merits of different paradigms (Kuhn, 1970, p. 148).

However, in the process of doing "normal" science, the scientific community works on solving problems or puzzles as set by a particular paradigm. As Kuhn notes, "for normal-scientific work, for puzzle-solving within the tradition that the textbooks define, the scientist is almost perfectly equipped" (1970, p. 166). What Kuhn calls revolutionary shifts of thought (entertaining new paradigmatic commitments) might become excluded at certain stages in the history of the scientific enterprise. The research community does, nevertheless, make revolutionary switches at other times.

Delanty suggests that while Kuhn's investigations into the operation of the scientific community can possibly be applied in the natural sciences, they cannot easily be applied in the social sciences. This is because, as Kuhn sees it, social science is characterized by multi-paradigm status — due to different paradigms co-existing alongside one another at any point in time. According to Kuhn, a practicing social scientist is faced with a diversity of paradigms and "has constantly before him [*sic*] a number of competing and incommensurable solutions ... that he must ultimately evaluate for himself" (Kuhn, 1970, p. 165). As these incommensurable options exist side by side, at any point in time social researchers have to make (personal) choices concerning how to define and approach issues of concern. These choices cannot be settled by recourse to "logic" — because each position brings alternative criteria to bear in defining ways of addressing what it regards as problematic (1970, p. 157).

Kuhn's discussion of the way in which normal science operates (in the natural sciences) has sometimes been seen as providing the basis for a critique of the way in which involvement in the scientific community occludes alternative possibilities for ways of knowing — both within the scientific community and in relation to discourse in the wider society (cf. Feyerabend, 1993). Feyerabend avers that the elevation of the status of science above other discourses in society is based on a legitimation of its processes that is belied by a Kuhnian vision thereof. He sees the Kuhnian position as opening the way for reducing the status of science in society — so that it is regarded as one belief system amongst others. This, for Feyerabend, in turn creates possibilities for the democratization of knowledge-production in social life (although Feyerabend admits that Kuhn does not carry his own argument quite in this direction).

Feyerabend's approach can be supplemented by Edwards' views concerning the operation of discourse in social life — including in "scientific communities". Edwards points out that while discourse is often seen (by those studying and by those enacting it) in terms of a metaphor of communication, this metaphor is unhelpful in enabling people to handle certain difficulties that arise within the metaphor. The metaphor is based on the assumption that the world can be more or less accurately pictured and that these pictures in turn need to be communicated through our discourse (1997, p. 17). But the problem of how we can ensure that the world is accurately pictured and the problem of how we can ensure proper message transmission between minds, then arise as difficulties that have to be addressed. Yet they can be addressed only by trying to establish some recourse to an appreciation of extra-linguistic reality. The procedures by which people undertake efforts to justify themselves with reference to such an appreciation, may, however, be in vain — for we do not have the recourse to reality implied by the metaphor of communication. Edwards therefore suggests a conception of discourse "as an activity" (1997, p. 17). Discourse is considered by Edwards as being a social activity in which people engage as they "become participants in event construction, offer ... versions of things, [and] choose amongst accounts" (1997, p 16).

In this book I take up Edwards' view of discourse in order to develop an approach that focuses on possibilities for enhancing what I call *discursive accountability.* My proposal in this regard involves conceptualizing the quality of knowing and living as resting on the development of a climate of inquiry whereby accountability is constituted through people's trust earning and trust awarding activities. I label this position a "trusting constructivism" (and I explore it in detail in Chapter 3).

Table 1: The development of positions in response to positivism.

Positivism: *Scientism; Phenomenalism; Empiricism; Value freedom; Instrumental knowledge.* **Scientists must develop statements about reality that are grounded in the observation of facts.**	
Critical Rationalism *Scientism* qualified somewhat in view of the possibility of self-fulfillment of propositions advanced. *Phenomenalism* and *Empiricism* reviewed: verification is impossible. But tentative corroboration (via the logic of deduction) is possible. And falsification is also possible. **Accountability of scientists: assessed in terms of their manner of using the scientific process to avoid errors in reasoning that could be avoided.**	Scientific Realism *Scientism* qualified in light of the recognition that social structures are perpetuated via, inter alia, human meanings. *Phenomenalism* and *Empiricism* reviewed: neither verification nor falsification of propositions advanced (about posited structures) is possible. **Accountability of scientists: assessed in terms of their manner of using the scientific process to theorize about (hidden) structures and mechanisms.**
Interpretivism *Scientism* opposed: focus on way in which meanings create possibilities for action. *Phenomenalism* and *Empiricism* re-interpreted: the "facts" of social life are meaningfully constituted. **Accountability of *social* scientists: assessed in terms of their manner of using the process of social scientific investigation to develop explanatory models grounded in social meaning-making.**	Critical Theory *Scientism* criticized: focus on contribution that critical theory can make in society. *Phenomenalism* and *Empiricism* reconsidered: facts and logic are constituted in terms of interests/commitments. *Value freedom* and *Instrumental knowledge* undercut: distinction between facts and values questioned. **Accountability of social inquirers: assessed in terms of their preparedness to embrace the value of communicative reason.**
Anti-foundationalist Feminism *Scientism* challenged as unduly privileging some vision of "science". *Phenomenalism* and *Empiricism* reconsidered: focus on the way in which facts and logic are constituted in male-stream social science. *Value freedom* and *Instrumental knowledge* undercut through reconsidering the relationship between knowers and known. **Accountability of social inquirers: assessed in terms of their preparedness to explore possibilities for contributing to a politics of difference through inquiry practice.**	Discursively-oriented Constructivism *Scientism* revisited: focus on the ways in which criteria for the "proper" practice of science are contended. *Empiricism* and *Phenomenalism* reconsidered: contention between alternative ways of seeing and using evidence highlighted. *Value freedom* and *Instrumental knowledge* undercut: focus on the value of discursive exchange between ways of constructing and working with information (as humanly mediated). **Accountability of social inquirers: assessed in discourse as social activity.**

Table 1 offers a rendition of the various arguments that I have mentioned thus far, with implications for the assessment of researcher accountability drawn out. The table shows how the views regarding (social) science can be situated in relation to the tenets of positivism.

This provides my starting point for the discussion in Chapter 2, which offers more detail on my placement of positions in relation to (ongoing) debates about the character of social inquiry, and reviews some of the points of contention in the debate. In the chapter, I set up a discussion between different arguments by considering their relation to the tenets of positivism and by showing how the various views on accountability that are discussed can be positioned too in relation to one another. Chapter 2 involves a process of surfacing differing views on accountability and their development in relation to alternatives, with special attention given to: critical rationalism/non-foundationalism; scientific realism; interpretivism; critical theory; and anti-foundationalist feminism. I consider as well how the different arguments can be positioned in relation to what I call "discursively-oriented constructivism" (which focuses on people's discursive practices). When comparing various positions in relation to constructivist thinking, I show how they address "constructivism" as seen in this way.

In order to spell out my views on "trusting constructivism", I create a separate chapter — Chapter 3. In that chapter, I draw upon arguments that I invoke to defend a form of constructivism that I believe is workable as a way of knowing and living. I regard this position — as I define it — as my preferred option in relation to the positions discussed in the book. Chapter 3 begins by offering an account of some of the roots of this option by exploring in more detail certain arguments mentioned in Chapter 2. Following on from and extending these arguments, I henceforward apply the term "trusting constructivism" to the position as developed through the course of the chapter.

Having set out the positions (and attendant views on accountability) that I discuss in Chapters 1-3, I proceed in Chapters 4-7 to discuss four actual examples of social research in the light of these various positions. Of course, I could not cover all possible positions that might be used to consider the issue of research(er) accountability; nor could I cover the assessment of all possible ways of organizing social inquiry. However, I hope at least to have drawn attention to some concerns that might be raised around the issue of researcher accountability, and to possible ways of addressing them.

Chapter 4 is aimed at considering from various angles the accountability of those involved in experimentation as a research strategy in the social sciences. The discussion is conducted by giving an example of the organization of an experiment (conceived as such by the researchers involved, namely Dovidio et al., 1997). I give a brief account of why the label of experimentation might be applied to the strategy (research process) adopted in this case. Here I draw on Gill and Johnson's (1991) account of how experimentation may be classified in relation to alternative approaches. I then offer what I regard as sufficient detail of the particular strategy used in this case (and also of techniques of observation utilized), in order to be able to undertake the ensuing comparison of various views concerning the researchers' accountabilities. In my outline of the example I have admittedly been selective, concentrating on aspects that I regard as relevant to the discussion subsequently organized. Using the example, I indicate in the course of the discussion how experimentation as a research strategy might be defended from various points of view (depending of course on how it is practiced). I

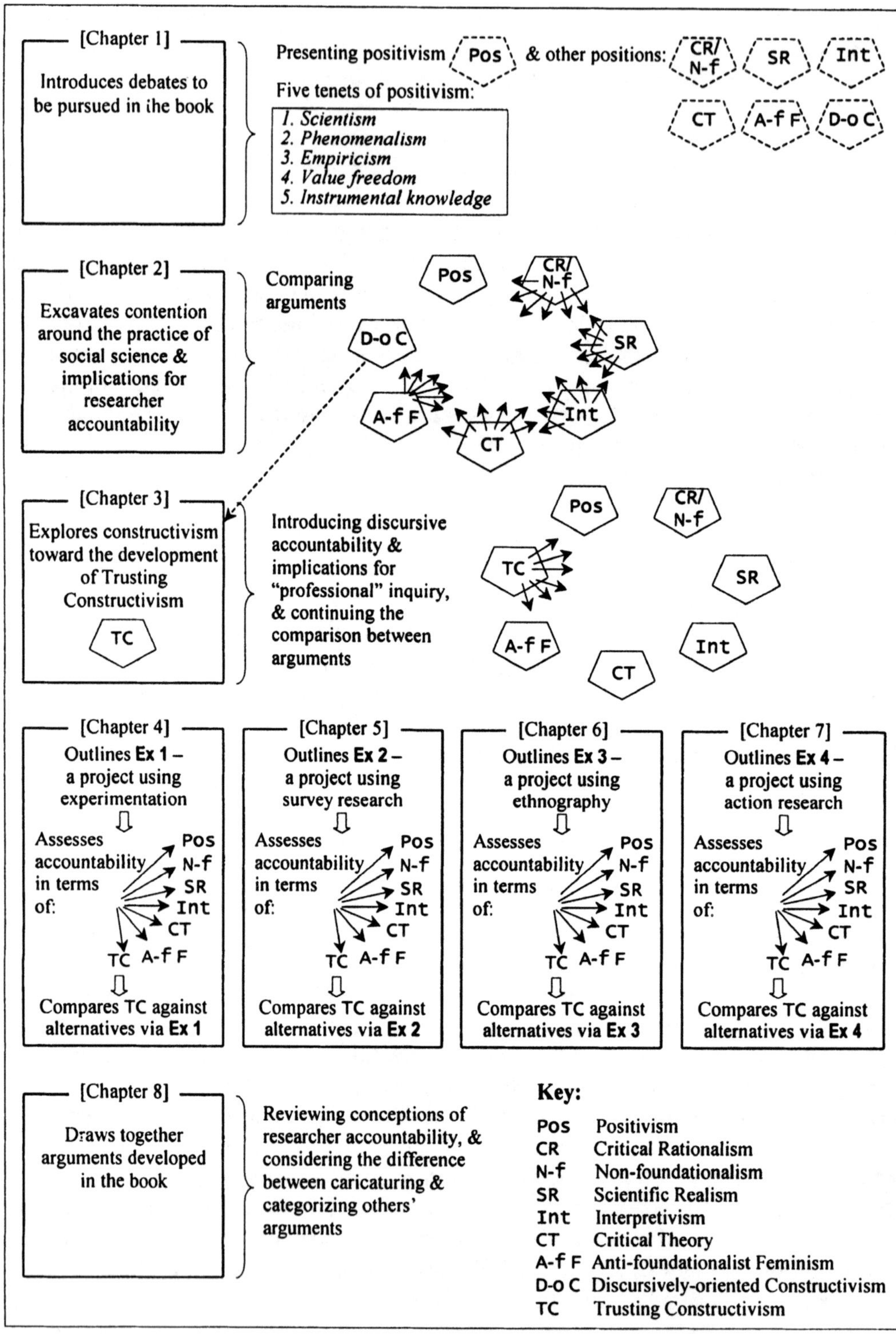

Figure 1: Layout of chapters and summary of their contents.

also build up the trusting constructivist view of experimentation by positioning it in relation to other positions, using the chosen example as a basis around which to organize the discussion.

In Chapter 5, I undertake an exploration of how survey research can be conceived and assessed from various angles. The discussion is conducted by giving an example of what I call survey research (using a broad definition thereof drawn from De Vaus, 1996). The conduct of the survey was part of a research evaluation in which I was involved (as principal researcher). The report by Romm et al. (1998) is used to direct the discussion regarding the particular style of survey research undertaken. The study is briefly outlined — again (as in Chapter 4) focusing on issues that I see as relevant to the ensuing discussion concerning different views of the accountabilities of the researchers in this case. By implication, different views of the accountabilities of those involved in survey research as a strategy in other cases too, are drawn out in the discussion. I provide commentary by referring to (my account of) various positions, including that of trusting constructivism (positioned in relation to alternatives).

In Chapter 6, following a similar format to the discussions undertaken in Chapters 4 and 5, I offer an exploration of how ethnographic research can be conceived and assessed from various angles. The discussion is conducted by exploring an example of what I see as ethnographic research (using definitions thereof drawn from the work of various authors). The research project — reported upon by Aphane et al. (1993) — is briefly outlined by focusing primarily on the researchers' manner of defining their role(s) as they interacted with participants. The ensuing discussion revolves around locating arguments concerning options for practicing ethnographic research in an accountable way. An account of the trusting constructivist position (positioned in relation to alternatives) completes the discussion.

In Chapter 7, I undertake an exploration of how action research can be conceived and assessed from various angles. The discussion is conducted by giving an example (from Weil, 1998). I indicate how the label of action research can help to situate the project in relation to alternative (research) approaches. I draw on various authors' arguments in regard to the meaning of action research (including the views of Weil). The particular style of action research undertaken is then briefly outlined — with a view to providing a basis for discussing different conceptions of the accountabilities of the co-researchers in this case, as well as in other cases of what might be considered as action research. Again, commentary is provided from (my discussion of) various positions, including that of trusting constructivism (positioned in relation to alternatives).

In Chapter 8, I conclude the book by iterating my way of organizing the debate in it. I indicate how (in Chapters 1-3) I have organized the positioning of arguments in terms of various categories of argumentation, and how "trusting constructivism" can be positioned in relation to certain alternatives in processes of argument. And I show, by way of the four examples spelled out in Chapters 4-7, how I have worked with my categorization of various positions (including the trusting constructivist one) to conduct a discourse around the examples.

I would like to note at this point that I have for the most part written in the "third person" in order to set out the various arguments discussed in the book. I do not hereby wish to deny my part in interpreting arguments as I express or report on them. My reports "about" arguments are admittedly my readings of the arguments, mediated by my own sense-making. I have tried in my readings not to denude the rationality of different arguments as I position them in relation to alternatives. Nevertheless, charges of "caricature" might still be leveled at some of my ways of reporting on positions. At the end of Chapter 8, I offer some commentary on my way of treating such charges (from within a constructivist perspective).

The structuring of the book is diagrammatically expressed in Figure 1. But as noted in the Preface, readers may decide to approach the book in a different order to the one that is laid out. It is quite possible, for instance, that a reading of one or more of the examples as discussed in Chapters 4-7 can precede an examination of the detail of the positions as explored in previous chapters.

2

The Practice of Social Science: Implications for Researcher Accountability

2.1 INTRODUCTION

In this chapter I present more detail on the views outlined in the previous chapter concerning what is properly involved in the practice of scientific (including social scientific) research. I also explore in more depth implications for the meaning of accountability in social research. I organize a debate between alternative positions as part of the process of this exploration. By locating the positions, I hope to foreground certain points of contention concerning the manner in which the accountability of social researchers might be considered (and assessed). The location of positions in this chapter is meant to serve this purpose.

The discussion begins by my considering some features associated with what is called positivism, and extrapolating on what accountability might mean in terms of this position. In order to elucidate the position, some of the statements made by selected authors are utilized and expressed in terms of their relationship to the tenets of positivism located by Delanty, namely, *scientism, phenomenalism, empiricism, value freedom, and instrumental knowledge* (1997, pp. 12-13). Naturally, some selectivity in choosing particular authors' arguments is inevitable. The focus in this case is primarily (though not exclusively) on the way Comte presents his positivist position, through his positivist self-understanding of the sociological enterprise. Swingewood avers in this regard that Comte "laid the foundations of a sociological positivism which was to remain the dominant paradigm during the course of the nineteenth century" (1984, p. 50).[2]

The elucidation of the tenets of positivism forms the context for opening up a discussion on alternative positions, and accompanying views on accountability. These positions are all outlined in terms of their relationship to the tenets of positivism, as a backdrop for exploring their relationship to one another in regard to researcher accountability.

---

[2]Swingewood offers an account of how the rise of positivism can be treated in its historical context by linking it with the process of industrialization. He discusses various authors' positivist arguments in these terms (1984, pp. 29-58).

The first alternative position to positivism that is discussed is that of critical rationalism, as Popper defines and labels it (1959, pp. 15-19, 1976, pp. 98-99, 1994, pp. 190-191). Popper's position is discussed in relation to the tenets of positivism. His argument has been followed up (and extended) by Hammersley (1995), and Hammersley and Gomm (1997a), where it is defined as a *non-foundationalist* stance. Hammersley and Gomm (1997a) offer a clear indication of implications of this argument for researcher accountability. Their argument, supplemented by material from Hammersley and Atkinson (1995), is utilized to elucidate the critical rationalist/non-foundationalist view on accountability.

The next position that is discussed is that of scientific realism. Keat and Urry's (1982) views are treated as one exemplar of this argument. Sayer's (1979, 1983) account of retroductive logic in turn explains why scientific realism postulates a different logic of scientific inquiry from the logic(s) considered within positivist and non-foundationalist positions. Layder's (1993), Davidson and Layder's (1994), and Pawson and Tilley's (1997) accounts of realism offer further details of the argument. Layder's (1993) and Davidson and Layder's (1994) arguments are particularly helpful in considering the way in which scientific realism can be positioned in debates concerning the issue of accountability.

To elucidate the position of interpretivism, Weber's work is drawn upon. He specifically indicates how a radical break with positivist *scientism* can be made via an "interpretive" social science (1949). The interpretive stance on what it may mean to render social science an accountable enterprise is outlined with reference to an argument developed by Henwood and Pidgeon (1993) concerning the quality of interpretive-oriented inquiries.

Habermas's response to what he (1978) sees as positivistically-oriented arguments, and to the interpretive tradition in sociology, provides an expression of the argument of "critical theory". Habermas (1972, 1982, 1996) also shows why critical theory differs to some extent from a structurally-oriented (realist) Marxism — which he regards as insufficiently communicative in its orientation to knowing. His views are relied upon to explore features of critical theory (including implications for accountability).

The anti-foundationalist arguments of certain feminists are focused upon next (rather than the arguments of, say, more realist-oriented feminists). This is done in order to highlight the way in which feminism can contribute to the development of an alternative epistemological argument (and attendant conception of accountability) from that provided by forms of foundationalism or non-foundationalism. Attention is given in particular to the (self-named) anti-foundationalist approach espoused by Lather (1991, 1993), and also to the argument of Maynard (1994).

I present the positions discussed in this chapter by positioning them all in relationship to the tenets of positivism and by considering their differing conceptions of accountability positioned in relationship to one another. The discussion in this chapter does not include detailed reference to what I have called a discursively-oriented constructivist position. However, in positioning various arguments in relationship to one another as I explore (and extrapolate) their views on researcher accountability, I also show how they can be positioned in relationship to discursively-oriented constructivism. (In Chapter 3, I explore in detail discursively-oriented constructivist arguments and develop these in the direction of what I call a trusting constructivist position.)

2.2 POSITIVISM AS A WAY OF DEFINING THE PRACTICE OF SCIENCE

Delanty's (1997) identification of the positivist tenets of *scientism, phenomenalism, empiricism, value freedom, and instrumental knowledge,* is used to organize this section. Delanty notes that not all of these tenets are embodied in any one positivist approach (1997, p. 13). But they provide a way of identifying certain ideas about science that might be adopted and that when drawn upon could warrant our characterizing the argument presented as "positivist" in orientation. Or, at least, they provide a way of opening a discussion on the relevance of the label as applied to an argument.

2.2.1 Scientism

The identification of this tenet draws attention to the positivist suggestion that the natural sciences can be considered as "the model for all the sciences" (Delanty, 1997, p. 12). This implies that the way of organizing scientific inquiry in the natural sciences, can be taken as more or less applicable in the social sciences too. Comte provides an expression of this tenet when he suggests that "the formation of social physics [to study phenomena in society] at last completes the system of natural sciences" (1970, p. 14). He expresses the hope that "this last branch of our knowledge [will become endowed with] the same positive character that already marks all the other branches" (1970, p. 13).

In response to the claim that social life might be more complex than can be understood via the approach adopted in the natural sciences, positivist argumentation suggests that there is no qualitative difference involved in the different subject matter of the various sciences. Thus Mill, for instance, suggests that even in certain physical sciences, such as meteorology and tidology (the science of tides), our understanding of lawful regularities in reality, does not allow for precise prediction. The existence of a lawful regularity implies that the phenomena under study are connected in a way that cannot be broken. Nevertheless, prediction in concrete situations requires knowledge of all the possible antecedent elements that might occur in that context, and this is rarely possible whether one is dealing with natural or social reality. Mill therefore concludes that social science is capable of achieving a comparable degree of prediction to that normally achievable via the physical sciences (1961).[3]

---

[3]Mill (1961) disagrees with Comte on exactly what is implied in developing an empirically based logic of science; but he agrees that a similar logic holds in natural and social scientific discovery. See Swingewood (1984, pp. 51-53) for an account hereof.

Donaldson, a current theorist of organizational development explicitly supporting a positivist position, likewise argues that the theories that a science of (human) organization is aimed at developing, should have explanatory power in the same sense as natural scientific theories have. The task of a science of human organization, as all science, is to create theories that have explanatory power in terms of their ability to show up "large correlations and ... clear causal sequences" (1997, p. 85). When phenomena are stated to be connected through a relationship of correlation, this means that a change or variation of one is accompanied by a variation in the other, although the direction of causality cannot be determined. A statement of causality between variables (phenomena that can vary), in addition specifies the direction in which causality is argued to be operating between the specific variables in question. Donaldson argues that although scientifically developed models depicting relationships of correlation or causality between variables isolated for attention may be abstractions or simplifications of what occurs in the real world, they can still advance our understanding thereof.

2.2.2 Phenomenalism

This tenet refers to the positivist suggestion that the reality to be studied by science can be reduced to "observable units", otherwise considered as "phenomena". Phenomenalism as a stance implies an "attitude to nature by which nature is regarded as existing outside science and can be neutrally observed" (Delanty, 1997, p. 12). Comte suggests, nevertheless, that we must be careful of seeing observation as a passive process. He argues that no observation is possible "except in as far as it is first directed, and finally interpreted, by some theory" (regarding the connection between facts) (1896, p. 243). But this active direction on the part of the knower, does not necessarily render the process of observation prejudicial. Comte suggests that as long as theoretical speculation and observation are continually combined (as they are within what he calls positive thinking), our theories are used to guide our observations and at the same time our factual observations become utilized to prevent our theorizing from slipping into speculations that have no basis in the realm of experience.

Donaldson suggests in this regard that it is possible to isolate variables such as, say, that of "organizational structure" or "organizational performance" and then use the scientific process to ascertain — through use of observable evidence — whether we can argue that there is a connection between them. The structure of organization (in this case) is treated as a variable that can be observed to vary in its type; and performance can be treated as a variable that can be observed to vary from low to high. Donaldson claims that scientific studies of organization have already advanced our understanding of the way in which, for example, "organizations move from misfit into fit under the guidance of their managers, i.e. away from low toward high performing structures" (1997, p. 85). This understanding is based not on speculation, but on the evidence as provided by reality.

2.2.3 Empiricism

This tenet is linked to the previous one in that it suggests that scientific inquiry requires that experience (that which can be observed by the senses) be used to adjudicate the veracity of scientific hypotheses (concerning facts and their connections in reality). Some mixture of inductivism and deductivism is normally seen within positivist argumentation as the route toward developing and testing such hypotheses. Via inductive and/or deductive processes, statements aimed at explaining certain observed outcomes (by relating them to antecedent conditions) can be developed and validated. Comte asserts that this does not amount to a "sterile empiricism" that denies the function of human reason in the organization of knowledge (1974, p. 731). The practice of science does not consist in the "accumulation of desultory facts", but in the development of explanations based on a conception of how the facts may be lawfully connected (1974, p. 799).

Donaldson's studies of organizational development, for instance, are conducted with the intention of examining the possible connection between certain variables that have been isolated for attention. He suggests that what is important in the science of human organization, as with all sciences, is that proposed connections are subjected to testing by using the scientific procedure — which consists of "quite simple approaches to examine phenomena and to test between competing theories" (1997, p. 85). According to Donaldson, the existing results of scientific endeavors show that science is able to illuminate "a great deal" when it proceeds according to the protocols of scientific procedure in order to develop models of connections between chosen variables (1997, p. 85).

2.2.4 Value Freedom

In terms of this tenet, the advance of science requires that scientific inquiries are not tarnished by the values or beliefs of those involved in the inquiry process. The principle of value freedom as developed within positivist argumentation refers to the idea that researchers (and the scientific community of which they are a part) should direct their inquiries toward striving to find out about the world as it exists outside of themselves. The world existing independently of our attempts to come to grips with it is the object of inquiry, of which knowledge is sought. As expressed by Delanty:

> Positivism involves a commitment to the pursuit of scientific truth, which is arrived at independently of ethical self-reflection or personal subjective elements since truth is a ... statement about an objectively existing reality. (1997, p. 12)

The suggestion is that scientists should themselves aim to be objective in their stance — in the sense that their particular values and beliefs should not influence their findings.

Comte argues that while it is possible that inaccurate theoretical conceptions can arise from the infiltration of a researcher's prejudices or biases into his or her scientific work, these are likely to become eliminated through further research (on his or her own part, or within the scientific community). He remarks that although there always exists the possibility that researchers may "pervert facts, by erroneously supposing them to verify some ill-grounded speculative prejudices of [their] own", there is nevertheless some "guard" against this in the further extension of the science (1974, p. 476). The status of scientific accounts, in any case, rests on scientists being able to demonstrate that their own predilections have not been brought to bear on the way they produced their findings.

2.2.5 Instrumental Knowledge

The pursuit of knowledge through science is presumed (within positivist thought) to lead to advancements in our knowledge of connections that exist between phenomena. Propositions that are forwarded about such connections, enable people in society to recognize under what conditions certain outcomes are likely to occur. This does not mean to say that this information will be utilized immediately. But the idea is that even theories formed for "purely scientific purposes" (as Comte puts it) can eventually be used for practical purposes (1970, p. 39). Knowledge of connections between conditions and outcomes is clearly useful for those wishing to consider how to avoid outcomes that are not desired, and, conversely, how to attempt to institute those that are desired. This is what is meant by saying that knowledge can be used as an instrument for goal seeking (see also Delanty, 1997, p. 23). The positivist suggestion is that it is only insofar as science is aimed at developing knowledge that is itself unmarked by specific political agendas, that it can be usefully applied. If particular political interests tarnish it, then it cannot offer the knowledge about reality that is necessary for actors to be able to make informed decisions about what goals can possibly be achieved and what the consequences might be of attempting to achieve them.

2.2.6 Considering Accountabilities in Terms of a Positivist Understanding of Science

The tenets associated with positivist argumentation are accompanied by a way of considering the accountabilities of those involved in the practice of scientific research.

In line with the *scientistic* tenet of positivism, scientists are seen as having to account for their endeavors as scientists by showing that they are, indeed, practicing science. Efforts at accounting for their work are to be aimed primarily at colleagues in the scientific community — who are seen as having the capacity to judge inquiries in terms of their relation to what is understood to be an accepted way of doing science. Accounting efforts can also be aimed at others in society who expect that those engaged in scientific practice are indeed "doing science" in terms of an accepted model.

In regard to the tenet of *phenomenalism*, scientists are required to show how they are generating observations that not only they, but also anyone else in the scientific community (and indeed anyone using similar measurements), would be able to generate. Attempts need to be made to "operationalize" the terms employed in scientific discourse, so that anybody performing the same operations is able to make the relevant observations (cf. Bridgman, 1952, pp. 7-8). The evidence of observation, and the way in which it is seen as referring to phenomena in reality, should be open to public scrutiny.[4]

The positivist tenet of *empiricism* suggests that scientists need to ensure that whatever conclusions they reach, follow from use of the appropriate evidence. Conclusions take the form of indicating why certain statements can be regarded as unlikely to represent truth, and/or of showing why other statements are regarded as offering explanations closer to the truth. The way in which these conclusions are reached is by recourse to the evidence of empirical experience. Scientists have to account for the way in which they build up their views about reality by showing their reasoning (through inductive and/or deductive logic) from the evidence.

The tenet of *value freedom* suggests that scientists need to display their commitment to the principle of objectivity. Their way of proceeding should not be organized around trying to endorse certain values or beliefs that they hold dear. While values and beliefs might provide a starting point for them to choose to investigate certain, rather than other, phenomena, from that point on these values and beliefs should ideally be held in abeyance so that the scientific process can proceed in accordance with scientific logic. Scientists should judge one another's work in terms of its adherence to the principle of practicing value freedom. The accountability of scientists to society at large is also a function of their expressing their commitment to value freedom in their scientific work.

The tenet of *instrumental knowledge* is linked in positivism to the conception of science as aimed at finding connections between phenomena in reality. Knowledge hereof may at some stage be useful to citizens wishing to make use of information provided concerning what outcomes can be expected to occur under specified conditions. Scientists should not, however, as scientists, take part in a discourse about whether certain outcomes are desirable or not.[5] Their remit consists in providing the necessary factual basis for others to create informed decisions regarding the pursuit of goals (or indeed for themselves to do so as citizens).

---

[4]Bridgman makes it clear that although he supports the identification of concepts in terms of instrumental operations that are repeatable, this does not imply that these operations constitute a sufficient condition of a concept's definition (1952, p. 8). Concepts may also have a "non-instrumental component" that they acquire as they enter into "some theoretical edifice" (1952, p. 8). Bryman indicates that the way in which the doctrine of operationalism has become carried into (positivist-oriented) social research is by proponents suggesting that, in any case, concepts have to be "rendered observable, i.e. measured" in some way (1992, p. 22).

[5]Swingewood notes that while Comte at one point created a (speculative) view of the progression of history, he later increasingly abandoned this conception (1984, p. 50). A philosophy of history was seen as having no place in a view of science as developing knowledge of specific conditions and their effects. Popper likewise points out that the making of unconditional prophesies is anathema to the concept of prediction, which is based on specifying the conditions under which outcomes can be expected to occur (1961, pp. 116-117).

In short, the tenets of positivism can be seen as linked up with certain views about how scientists might render themselves accountable in their scientific endeavors (to colleagues in the scientific community as well as to others in society).

- Scientists are required to practice science — whether as natural or social scientists — in proper fashion, using empirical evidence toward advancing knowledge of connections existing in reality.
- Proper practice can be maintained within the scientific community as scientists proceed to examine others' work in terms of its degree of adherence to scientific protocol.
- The scientific community taken as a whole can then be assumed to create advancements in knowledge about connections between phenomena as they exist in reality. This knowledge can rightfully be accorded some credibility by citizens in society.

The positivist understanding of science as expressed here (and the attendant conception of the accountabilities of those who wish to be seen as part of the scientific community) has been criticized from various angles. One kind of critique, which has strong currency in debates about the character of natural and social science, derives from the Popperian argument, labeled by him as a critical rationalist position.

2.3 CRITICAL RATIONALISM AS A WAY OF DEFINING THE PRACTICE OF SCIENCE

Popper criticizes certain aspects of positivism (as featured in, for example, Comte's views and in the views of the so-called Vienna Circle of positivist thought[6]), but draws on other aspects to develop what he calls a critical rationalist position. Popper's critical rationalist position can be outlined in relation to the tenets of positivism, as follows.

In regard to the tenet of *scientism*, Popper accepts the positivist idea of the unity of the sciences. He argues that all theoretical sciences — whether the natural sciences (which deal with non-human phenomena) or the social sciences (which deal with social phenomena) — have as their aim the gathering of knowledge about the regularities operating in their field of inquiry (1961, p. 130). However, he draws attention to the fact that the subject matter of the social sciences differs to some extent from that of the natural sciences. Whereas natural regularities cannot be affected through ideas influencing the action of the objects being investigated, in the social world it *is* possible for predictions to be self-fulfilling. This is because people in society can decide to act in the light of information presented about what effects might ensue from certain actions, and thereby contribute to the creation of these effects. As Delanty expresses it, "it is the aim of the natural sciences to make prediction possible, while in the social sciences

[6]See Romm (1991, pp. 33-36) for a discussion of Popper's relationship with the Comtian expression of positivism, as well as his relation to the so-called Vienna Circle (of logical positivists), whose arguments were developed in the early part of the 20th century. (See also Crotty, 1998, pp. 23-33.)

predictions can be self-fulfilling prophesies" (1997, p. 32). Nevertheless, Popper believes that the language of causality can still be properly applied in the natural as well as social sciences, with this proviso understood (1966, p. 362).

Popper's argument in relation to the tenet of *phenomenalism* is that observation statements (about units of observation) do not have the status of ever being provable beyond all doubt. Even such "basic statements" (as he calls them) about particular occurrences, themselves require a decision as to whether they should be accepted (1959, p. 105). Therefore, it is not possible to refer to observation as constituting a *firm foundation* on which science can rest.[7] Popper's contention is that none of the statements of science — ranging from the highest level of generality to "basic statements" about particular occurrences — should ever be considered as indubitable. Nevertheless, science can proceed because decisions can be made at certain points about whether or not to accept the statements (as tentatively corroborated). We may decide to accept them "as satisfactory, and as sufficiently tested" in the sense that we may "desist from justifying them by further arguments (or by further tests). But ... should the need arise, these [basic] statements [as others] can ... become tested further" (1959, p. 105).

Popper indicates sympathy with the positivist tenet (of *empiricism*) that "the foundation of science is observation" (as expressed by Delanty, 1997, p. 12). But he does not agree with what he sees as the positivist conception of the logic of scientific discovery. According to this conception, it is with the help of both inductive and deductive logic that statements aimed at explaining certain observed outcomes can be developed. Popper argues that inductive logic is not the logic by which science proceeds. He argues indeed that "*there is no induction:* we never argue from facts to theories, unless by way of refutation or falsification" (1978, p. 86). Although what is called induction may be useful as a way of generating hypotheses (by considering a number of observations and speculating about whether more general statements might help to explain them), it has no other scientific importance.

The practice of science proper, according to Popper, requires the testing of hypotheses via the process of deduction. Deductive logic is used to ascertain what observed outcomes could reasonably be taken as refuting a given hypothesis. If we are repeatedly unable to obtain such observations, we may consider our hypothesis as having withstood our repeated attempts to falsify it — and we may then take it as being "corroborated" (1969, p. 256). However, to say that it has been corroborated because it has withstood repeated attempts to falsify it, does not imply that it has been conclusively verified, for subsequent observations may still at some point lead to its being refuted.

Deductive logic — the logic of science — cannot be used to generate certainty of knowledge about reality. As Delanty (referring to the Popperian argument) puts it:

---

[7]Sapsford expresses a similar argument when he suggests, in considering the validation of processes used to observe evidence, that "no evidence will be valid absolutely, but we aim for as much proof of validity as we can obtain" (1999, p. 139). In offering advice to social researchers, Sapsford outlines a number of ways in which they may try to better validate their observations.

"Scientific knowledge is uncertain knowledge" (1997, p. 32). Nevertheless, as Delanty notes, Popper argues that it is the most certain kind of knowledge that human beings can aspire to. Popper suggests in this respect that the theory of scientific knowledge is "constantly faced with a near paradox" (1994, p. 100). On the one hand, it is true to say that "our knowledge is vast and impressive" (1994, p. 100). On the other hand, it is also true to say that "our ignorance is boundless and overwhelming" (1994, p. 100). Popper suggests that "both of these theses are true, and their clash characterizes our knowledge-situation" (1994, p. 100). For Popper, it is this clash that "inspires the advance of knowledge, and … determines its ever-moving frontiers" (1994, p. 100).

Concerning the accomplishment of the task of advancing knowledge, Popper accepts the positivist suggestion that *value freedom* needs to be striven for in the process of scientific inquiry. He emphasizes that the ideal of objectivity or "purity" of knowledge is, however, not fully attainable. Rather, "the purity of pure science is an ideal ... for which we constantly fight — and should fight — by means of criticism" (1976, p. 97). He indicates that in terms of his view of science, the "fight" for objectivity is not one that depends in particular on the dispositions of individual scientists. It depends more on the operation of the institution of "pertinent mutual criticism" within the community of scientists (1976, p. 96). "Objectivity" derives from the way in which decisions about whether to accept statements (of whatever level of generality) are made within the scientific community. Decisions should be made on the basis of a "well-conducted critical discussion" (1994, p. 160).

Popper rejects the claims of those who propose that because of the personal interests of social scientists in the results of their work, social science can never attain any objective information about reality. Objectivity in social — as in natural — science is dependent on the provision of mutual criticism within the community of scientists. He elucidates his position:

> It is a mistake to assume that the objectivity of a science depends upon the objectivity of the scientist. And it is a mistake to believe that the attitude of the natural scientist is more objective than that of the social scientist. ... . To put it another way, the objectivity of science [is a matter of] ... the friendly-hostile division of labor among scientists, of their co-operation and also of their competition. (1976, p. 95)

Merton, referring to (earlier) statements by Popper in this regard (with which he indicates his agreement), suggests that sociology, like all scientific enterprises, embodies the norm stating that scientists are "ultimately accountable ... to their compeers" (1964, p. 559). This norm implies that "scientific research is under the exacting scrutiny of fellow-experts" (1964, p. 559). The operation of this norm within the scientific community is what imparts a measure of objectivity to scientific findings.

Considering the utility of the knowledge that is produced through the scientific process, Popper concurs with the positivist suggestion regarding the practical use of the "prevision" afforded by scientific statements. He points out that this "prevision" should be understood in the sense of "conditional prediction" (Popper, 1961, p. 43). He suggests that scientific knowledge can be of *instrumental use* to people in that it is aimed at offering information concerning conditions and their associated effects. But he indicates that he does not agree with an instrumental conception of the value of scientific statements if this is taken as implying that the truth of statements is a function of their practical use. That is, he does not agree with (pragmatic) attempts to link truth to usefulness (1969, p. 99). Nevertheless, knowledge that is advanced both in the natural and social sciences clearly can have practical benefits. In the realm of the social sciences, for example, science can be used to aid us to ascertain "the effects of prison reform or universal health insurance, or the stabilization of prices by means of tribunals, or the introduction of new import duties, etc." (1961, p. 59). People are thus able to act in a more informed way in the light of the information that science can provide concerning the likely consequences of trying to accomplish a certain aim (such as reforming the prisons, offering universal health insurance, and so on).

In setting out his view of the utility of social scientific knowledge, Popper indicates that he prefers what he calls piecemeal engineering to large scale restructuring of society (1961, p. 67). In the latter case, according to him, it is impossible to monitor the effects of introducing the various changes. It is impossible to "disentangle causes and effects, and to know what [one] is really doing" (1961, p. 69). It is for this reason that Popper argues that "the holistic approach [as offered, for example, in certain Marxist approaches to social change] is incompatible with a truly scientific attitude" (1961, p. 70). Because of his claims in this regard, his position on the utility of science is associated (by some) with a reformist approach to addressing social issues.

2.3.1 Considering Accountabilities in Terms of a Critical Rationalist/Non-foundationalist Position

This section is organized around Hammersley and Gomm's (1997a) view of what accountability means within what they call a non-foundationalist view of science. Before elucidating their argument, it is necessary to show briefly the way in which their non-foundationalist position can be compared with Popper's critical rationalism. Hammersley, in outlining his view of non-foundationalism, expresses many Popperian arguments. Like Popper, Hammersley believes that there is scope for reason to prevail in scientific discourse, as long as certain norms are operative. He draws on Popper's suggestion that scientific rationality can be furthered primarily through the role of the research community.

Hammersley emphasizes (as does Popper) that "the institutional requirements for detachment" (or objectivity) within the scientific process should not be underestimated (1995, p. 15). Detachment is not so much a matter of "the psychological attitude of the researcher" as a matter of how the socialization process in the scientific community operates (1995, p. 115). As he puts it:

> Important as such asceticism [necessary for detachment] undoubtedly is, one should not underestimate the role of the research community in socializing researchers into it and maintaining their commitment to it, as well as in correcting at least some of the biases which it does not prevent. (1995, p. 115)

Hammersley here explicitly refers to Popper's argument in order to emphasize the importance of the research community in calling its members to account and in correcting errors that otherwise might have been left unchecked. Legitimate criticism of others' work (and self-critique of one's own) is part of the institution of science (Hammersley, 1997, paragraph 1.4).

Hammersley and Gomm (1997a) offer a detailed argument concerning what is meant by the term "collegial accountability" within the practice of research. Their argument is offered as a critique of two other types of epistemological positions: the first being "foundationalism" (paragraph 2.1), the second being "radical epistemologies" (in the form of standpoint epistemology and relativism — paragraph 2.10). The discussion below is aimed at elucidating their position in terms of its relationship to the positions isolated in Chapter 1. This is done by considering their account of how non-foundationalism differs from foundationalism, standpoint epistemology, and relativism, and by drawing on further material from Hammersley (1995, 1997), Hammersley and Atkinson (1995), and Hammersley and Gomm (1997b).

2.3.1.1 Positioning Non-foundationalism in Relation to Positivist Argumentation

To start off with, it should be noted that Hammersley and Gomm are themselves wary of employing the term "positivism" because they believe that it has become too elastic in current usages. It is not clear, according to them, what arguments are being referred to when one invokes the term (Hammersley & Gomm, 1997a, paragraph 2.2; Hammersley, 1995, pp. 10-12). "Foundationalism", for them, is a clearer concept: it refers to any argument that hopes to find some firm foundation on which knowledge-production rests. They summarize the foundationalist argument (in its extreme form) as the belief that "research, when it is properly executed ... produces conclusions whose validity follows automatically from the 'givenness' of the data on which they are based" (paragraph 2.3). Defined in this way, "foundationalism" can be seen to incorporate what Delanty (1997) labels as "positivism" (as discussed earlier). In terms of positivist tenets, the observation of data is not seen as a passive process, but is nevertheless argued to be capable of being a neutral one, unless tarnished by prejudice. It is when prejudice intrudes on the research process that the link between conclusions and evidence is regarded as severed.

Hammersley and Gomm point out that what they call "foundationalism" for a long time penetrated much of the thinking behind the understanding of research in social science (1997a, paragraphs 2.7 and 2.8). It was assumed that as long as researchers' attentiveness to "the evidence" is not hampered (through bias), true conclusions about the phenomena being studied in social reality (as in natural reality) could be attained.

Such foundationalist thinking according to Hammersley and Gomm does not take account of the continuing uncertainties that beset the scientific process (1997a, paragraph 2.6). They therefore propose what they call a non-foundationalist position. According to them, it is important to recognize that the scientific endeavor of creating conclusions will always be marked by uncertainty. Therefore, a more "moderate line" (than foundationalism) should be followed, by admitting the uncertainty of all attempts to arrive at conclusions (1997a, paragraph 2.6).

However, following Popper, Hammersley and Gomm suggest that this admission of uncertainty *should not commit communities of inquirers to take the position that there is no point in trying to avoid error.* On the contrary, it is still crucial (indeed, more so than ever) to develop a concern to attempt to "avoid the danger of accepting as true what is in fact false" (1997a, paragraph 4.2). Researchers engaged in practicing science, are required to "do their utmost to find and keep to the path which leads toward knowledge rather than error" (1997a, paragraph 4.3).

Hammersley and Gomm's view of accountability is based on the contention that the immediate goal of scientific inquiry must be the goal of producing knowledge — defined as statements that are in line with external realities. This implies that researchers should show some commitment to the principle of value freedom or objectivity. Hammersley emphasizes that this does not imply that researchers necessarily will succeed in achieving such objectivity. But they must at least "try to be objective" (1997, paragraph 1.10). The research community in turn should be geared to judging its members in terms of their demonstration of the commitment to the principle of objectivity. This is what is expected by "funders [of research] and lay audiences" in the wider society (1997, paragraph 1.12). As Hammersley argues, "if funders and lay audiences believed that research were primarily concerned with serving researchers' (or someone else's) political goals, it would not survive" (1997, paragraph 1.12). For it to survive, scientific activity must be seen to be directed toward the goal of producing knowledge.

Hammersley and Gomm make the point that *the accountability system in the scientific community does not operate in a fail-safe manner to exclude error.* They suggest that "where previously [in terms of a foundationalist position] procedural error was a matter of logic, it now becomes deviance from communal judgments about what is and is not reasonable behavior in pursuit of knowledge" (1997a, paragraph 4.9). Unlike in foundationalism, where it is suggested that adherence to scientific protocol implies following certain methodological canons, it is suggested in a non-foundationalist view that "protocol" is defined in terms of what is judged within the community to constitute reasonable behavior. These judgments too are fallible: they are "open to dispute and to subsequent revision" (1997a, paragraph 4.9). The possibility of fallibility of judgment nevertheless does not detract from the fact that people can still judge one another's behavior (though never with complete confidence) in terms of the way in which its reasonableness is presented.

The following bullet points provide a rendition of the non-foundationalist argument positioned in relation to positivism.

- It is not possible, as positivist foundationalism presupposes, to draw conclusions from the evidence in a foolproof way through the scientific process.
- Scientists are required to direct their activities toward following the path that is likely to lead to knowledge rather than error as they proceed with their inquiries.
- Insofar as they are socialized into the norms of the research community, researchers will be on the lookout for their own and others' errors that can be avoided.

- While in foundationalism accountability implies eliminating bias through adherence to scientific protocol, in non-foundationalism *accountability implies arguing for the reasonableness of behavior within the scientific community.*

2.3.1.2 Positioning Non-foundationalism in Relation to Scientific Realist Argumentation

Hammersley and Gomm do not explicitly refer to the scientific realist position as an alternative to foundationalism and non-foundationalism; and Hammersley refers only briefly to the position (1995, pp. 16-17, and p. 43). But from their statements about "standpoint epistemology", it is possible to glean something about their ideas on this position.

Hammersley and Gomm indicate that there exists what they call a "standpoint theory" that can be considered as being a "radical epistemology" (1997a, paragraph 2.10). A "standpoint" argument suggests that as long as a certain stand (for instance, that of protecting worker interests) is adopted, there is more chance of achieving true knowledge than when other stands (interests) are adopted. Hammersley and Gomm interpret some Marxist arguments as incorporating a standpoint epistemology. Traditional ways of practicing science are seen as having become infiltrated by a political interest to mystify the real structures of social reality: these ways of practicing science are not geared to aiding our appreciation of the historical character of the (observed) patterns and regularities in society that appear to be universal. A Marxist standpoint position rests on the idea that a different way of using logic and of practicing science may be able to afford such an appreciation (Hammersley & Gomm, 1997a, paragraph 3.8). Hammersley and Gomm note that in current expressions of the standpoint argument, the proposed way of practicing science is not seen as *necessarily* providing access to valid knowledge. Rather, it is seen as providing a way of searching for it: "it ... offers a *potential* for such knowledge" (1997a, paragraph 3.8).

Hammersley and Gomm compare what they call strong and weak versions of standpoint epistemology (1997a). In strong versions, the identification with the "right" standpoint in society is seen to offer access to knowledge of social reality. But the weaker version that is currently offered (and that would seem to apply to scientific realism as described in Chapter 1) provides no means for deciding when indeed knowledge is being advanced. Occupation of the "right" standpoint does not guarantee results in terms of knowledge-production.

In discussing certain Marxist claims about historically developed (and developing) structures, Hammersley argues that the source of validity of these claims is unclear. He asks: "What sort of knowledge is being claimed? We need to know this in order to decide what kinds of evidence would be sufficient to support it" (1995, p. 31). According to Hammersley, following Popper, in proposing any scientific hypothesis, one should at the same time provide an indication of what kinds of evidence will be taken as corroborating, and conversely what kinds of evidence will be taken as falsifying, it. The suggestion given within the scientific realist argument that we can develop theories that have only an oblique relationship with observed "evidence" is not helpful in this regard.

Hammersley and Gomm nevertheless concur with the scientific realist suggestion that the goal of science is to strive to produce accounts of (independently existing) realities. (Hammersley notes that the espoused goal of scientific realism is to "find a true description of unobservable processes that explain observable ones" — 1995, p. 15.) But extrapolating from Hammersley and Gomm's argument about the way the accountability system should properly operate in the scientific community, it seems that they would caution against trying to practice science in the way scientific realism proposes. They would caution that judgments about what is and what is not reasonable behavior in pursuit of knowledge of the unobservable entities that are posited to exist, might become made in an arbitrary manner. The appeal to a "logic" such as retroductive logic, seems to provide insufficient means to guide scientists in their assessments of whether unreasonable leaps of logic are being made.

The following bullet points provide a rendition of the non-foundationalist argument positioned in relation to scientific realism.

- While in scientific realism it is considered advisable to use the scientific process to theorize about underlying structures, non-foundationalism raises concerns about the way in which the logic of retroduction is used toward this end. The scientific realist appeal to this logic is not sufficient to guide the process of self- and mutual criticism in the scientific community.
- Non-foundationalism can appreciate the scientific realist argument that processes of advancing knowledge can never be said to guarantee results. But the way in which scientists can be rendered accountable (and assessed for their accountability) in the light hereof has not been given sufficient attention.
- According to scientific realism, it would seem that accountability of researchers implies their pursuing the possibility of developing an informed account of underlying structures constituting reality. But according to non-foundationalism, *the appeal to rationality via "retroduction" makes it almost impossible to assess the accountability of scientists in any coherent fashion.*

2.3.1.3 Positioning Non-foundationalism in Relation to Interpretivist Argumentation

Hammersley indicates that the non-foundationalist epistemology has many similarities with the Weberian interpretive argument in its view of the way in which values are to be treated in the research process. Hammersley refers to various ways in which values might legitimately enter the research process, in terms of Weber's account. Firstly, researchers can and do rely on certain values in selecting the topics they wish to investigate: their selection of topics is influenced by their values. Secondly, values might enter the research process by researchers deciding to evaluate policies in terms of "their consistency with the ultimate values espoused by those promoting them" (Hammersley, 1995, p. 114). That is, policies can be evaluated by considering the degree to which those promoting them are indeed acting in accordance with certain espoused "ultimate values". Thirdly, values can enter the research process by researchers "formulating policies conditionally, as the most effective and efficient means of achieving particular goals" (Hammersley, 1995, p. 114). The researcher's task here is to offer an account of the possible means that can be used (in the form of policies) to achieve goals that have been chosen in society.

Hammersley argues that none of these ways of conceiving the relevance of values to the research process need affect the fact that the aim of social research is to produce knowledge (1995, p. 115). He states that in his view, "Weber is correct to insist that the primary goal of research should be the production of knowledge, and that it should not involve the attempt to realize other values than truth" (1995, p. 115). Hammersley argues that Weber rightly recognizes the "need for detachment from political commitments", that is, the need for "objectivity" (1995, p. 115). What he does not concentrate sufficiently on, according to Hammersley, is the "institutional requirements [in the social scientific community] for this detachment" (1995, p. 115). Hammersley suggests that Weber "underestimates" somewhat these requirements. In other words, he does not draw sufficient attention to the importance of the operation of collegial accountability in the scientific community.

Regarding the way in which observation and theory construction operate in the social sciences, Hammersley and Atkinson argue that the fact that there is no "absolutely secure foundation" for making knowledge claims (here, as in the study of nature) does not render the process of science unworkable (1995, p. 17). Research into the patterning of social life is, admittedly,

> an active process, in which accounts of the world are produced through selective observation and theoretical interpretation of what is seen, through asking particular questions and interpreting what is said in reply, through writing fieldnotes and transcribing audio- and video-recordings, as well as through writing research reports. (1995, p. 18)

Despite the fact that the social world does not present itself to us without some activity on our part, we can still, as researchers, make advancements in our knowledge thereof, through our efforts to engage in "systematic inquiry" (1995, p. 17). This involves a process of developing and testing theories, a process that is not unlike the way in which other sciences properly proceed.[8]

In considering the effect that the systematic inquiry into the social world might have on this world, Hammersley and Atkinson take a similar position to Popper. They agree with him that "production of knowledge by researchers has consequences" (Hammersley & Atkinson, 1995, p. 17). They indicate that "at the very least, the publication of research findings can shape the climate in which political and practical decisions are made, and it may even directly stimulate particular sorts of action" (1995, p. 17). These actions may occur as the result of the fact that the research becomes published — they could be a consequence thereof. But (along with Popper)

---

[8]Hammersley argues in this regard that just because the positivist view of there being a secure foundation for science has "collapsed" does not imply that "natural science has no relevance to our work as social researchers". He indicates that in his view "it [natural science] should remain the primary model, even though we can also learn much from the humanities" (1995, p. 18).

Hammersley and Atkinson do not draw the conclusion from this that research is, or should be, a political process in itself. According to them, "the primary goal of research is, and must remain, the production of knowledge" (1995, p. 17). Research activity should be directed to this goal, rather than being directed by a concern with how it may itself impact on society. As they contend, when we are "more concerned with the practical effects of our research activities" ... this may lead us to be 'economical' with the truth" (Hammersley & Atkinson, 1995, p. 20).

Hammersley and Atkinson emphasize that they do not wish their argument to be interpreted as a statement that "researchers should be unconcerned about the effects of their work on the world" (1995, p. 21). What is important is that researchers' activity is not directed by a consideration of these effects, but rather by a consideration of trying "to minimize any distortion of their findings by their political convictions or practical interests" (1995, p. 21). This, for them, is the import of Weber's call to uphold the principle of value freedom.

Hammersley and Atkinson deal with another argument regarding the possible effect of social research on subjects in society. This argument concerns the reactivity effect within social research — that is, the effect(s) produced through the fact that "people respond to the presence of the researcher" in the research process (1995, p. 18). Hammersley and Atkinson make certain proposals in regard to ways of handling the reactivity effect. For a start, they point out that not all research procedures have the same level or direction of effect. They suggest that "in order to understand the effects of the research and of research procedures, we need to compare data in which the level and direction of reactivity vary" (1995, p. 19). In this way full use can be made of the reactivity effect in helping us to establish patterns of human behavior in the research situation. As they propose:

> Data should not be taken at face value, but treated as a field of inferences in which hypothetical patterns can be identified and their validity tested. Different research strategies can be explored and their effects compared with a view to drawing theoretical conclusions. (1995, p. 19)

They indicate (1995, p. 19) that in making these suggestions about the way the reactivity effect can be treated as a research topic, they are re-affirming their commitment to realism (understood in a non-foundationalist sense and not in the sense given in "scientific realism" as explained in the previous section). Theoretical statements about the effects of researcher presence within different research strategies can be made. These statements, like all others proposed, are to be tested through a scientific process in which scientists regard themselves as accountable to colleagues for the conclusions that they reach about the topic under investigation (in this case, the reactivity effect).

The following bullet points provide a rendition of the non-foundationalist argument positioned in relation to (Weberian) interpretivism.

- In the quest to get to grips with social life, non-foundationalism emphasizes the importance of researchers adhering to the principle of value freedom that is indeed espoused within Weberian interpretivism.
- Non-foundationalism sees the search for knowledge of social reality as the search for patterns: the aim of social scientific inquiry is to develop an appreciation of social patterns (which admittedly in this case may be influenced by factors not present in the case of the study of nature).
- While in interpretivism the accountability of researchers is linked to their efforts to create plausible accounts of social life, in non-foundationalism *accountability requires drawing on a logic that can be judged for its reasonableness within the scientific community in terms of the way propositions are developed and tested.*

2.3.1.4 Positioning Non-foundationalism in Relation to Critical Theoretical Argumentation

Hammersley refers explicitly to what he calls "critical research" as an alternative to "traditional research" (1995, p. 30). He refers to Connerton's (1976) indication that one of the claims of critical researchers is that, unlike traditional researchers, they "incorporate in their work an analysis of their own social function" (Hammersley, 1995, p. 30). This involves accounting for inquiry processes in terms of this function, and directing the inquiry consciously toward furthering "emancipatory goals" (1995, p. 30). Hammersley argues that the problem with such "critical" approaches is that it is difficult to justify the claim that they are able to "gain genuine knowledge of social reality rather than being deceived by appearances like everyone else" (1995, p. 30).

Hammersley suggests that the Frankfurt School of critical theorists (including Habermas) have tried to offer a way of addressing this issue, that breaks somewhat with Marx's argument. According to the Frankfurt School, the source of human alienation is no longer seen as "simply capitalism". This means that it is recognized that "the overthrow of capitalism will not in itself overcome alienation" (Hammersley, 1995, p. 32). Hammersley indicates that this argument on the part of the Frankfurt School "has the effect of making the social transformation [toward a posited better society] even more radical than in the case of Marx" (1995, p. 32). At the same time "the ground for seeing the proletariat as the epistemologically privileged and emancipatory agent" becomes "undercut" (1995, p. 32).

Hammersley claims that Habermas recognizes that one cannot simply posit an agent of both knowledge and social emancipation. Habermas therefore provides a "systematic attempt" to offer an alternative position (1995, p. 32). He does this by setting out to

> found critique on the logical implications of communicative acts ... . He argues that communications involve implicit norms which require people to try to ensure that their communications are truthful, comprehensible, sincere and justifiable. (Hammersley, 1995, p. 32)

But Hammersley avers that Habermas's justification for suggesting that in the ideal speech situation certain normative rules apply, entails a circular argument — in that he presupposes that these norms must apply in genuine communicative interaction. Hammersley does not find convincing Habermas's attempts to unite the process of knowledge-production with the possible realization of a form of social life in which people are able to take on board the standards of "ideal" communication to which he refers.

Hammersley and Gomm argue that the problem with this kind of "radical" epistemological argument is that it still provides no basis for "judging the validity of statements about the source of a knowledge claim" (1997a, paragraph 3.8). The question of how to determine quality of knowledge-production cannot be resolved in terms of this epistemology — except by claiming that access to a particular ethical and political orientation offers a route to (potential) knowledge-production. The critical theorizing that presents itself as transcending both commonsense knowledge and conventional ways of doing science, is unable to ground its statements about social reality — other than by suggesting that they are rooted in a supposedly superior ethical and political stance (Hammersley, 1995, p. 30).

The following bullet points provide a rendition of the non-foundationalist argument positioned in relation to critical theory.

- Critical theory explicitly ties social theorizing to the goal of generating certain standards of social interaction within society. The task is to ensure that research processes are not divorced from the quest to contribute to the development of a better society.
- In tying its theorizing to this goal, however, it fails to provide a coherent view of how researchers might render themselves accountable in their scientific work.
- Non-foundationalism provides the suggestion that *the accountability of researchers requires their reasonable behavior as scientists — guided by the quest to generate knowledge of realities under investigation.*

2.3.1.5 Positioning Non-foundationalism in Relation to Anti-foundationalist Feminist Argumentation

In considering different forms of feminism, Hammersley and Gomm indicate that feminists by and large wish to focus our attention on the fact that gender positioning (as well as other social positioning) can affect the experiences and information to which those considering themselves as researchers have access. Hammersley and Gomm's complaint against this kind of focus as proffered within feminism, is that it directs our attention away from the requirements for scientific research to be successful. Following a Popperian position, Hammersley and Gomm indicate that the personal characteristics (including social position) of researchers is not what is at stake when considering the conduct of scientific inquiry. What is at stake is that mechanisms of collegial accountability exist within the scientific community to create conditions for an increase in the objectivity of findings produced.

Hammersley and Gomm aver that feminists (as others adopting some kind of "radical" epistemology) often vacillate between on the one hand propounding a *standpoint* epistemology and on the other propounding a *relativist* position. In adopting a *standpoint argument,* feminists propose that a particular standpoint, such as gender positioning, offers a potential source of knowledge-production. Thus, for instance, "women are treated as the oppressed group occupying a standpoint that provides epistemologically privileged knowledge" (Hammersley & Gomm, 1997a, paragraph 3.5). Hammersley and Gomm's critique of the feminist idea that the adoption of a particular standpoint (some social position or some value position) affords the best chance of generating knowledge, is that it becomes difficult for others to contest the proposed claims to knowledge.

Hammersley and Gomm note that feminists sometimes move toward adopting a *relativist* position. The relativism of feminism springs from the suggestion that, as Hammersley and Gomm put it,

> all accounts of the world reflect the social, ethnic, gendered, etc. position of the people who produced them. They are constructed on the basis of particular assumptions and purposes, and their truth or falsity can only be judged in terms of standards that are themselves social constructions, and therefore relative. (1997a, paragraph 3.1)

As Hammersley and Gomm see it, such claims made on the part of feminists seem to commit them to the relativist claim that there is no "truth". As Hammersley and Gomm argue, truth now can mean only "that which is taken to be true within some community whose members share a particular perspective" (1997a, paragraph 3.2). But they argue that "it simply makes no sense, from a relativist point of view, for a member of one epistemic community to accuse members of another of being biased because their views deviate from what he or she takes to be true, rational, etc." (1997a, paragraph 3.2). Criticism of others' views (and reasons for holding these) becomes occluded within a relativist position, according to Hammersley and Gomm.

Hammersley and Gomm see feminists who adopt a relativist epistemology as oscillating between two contradictory positions. On the one hand, they "recognize multiple perspectives as each true in its own terms" (1997a, paragraph 3.3). On the other hand, they become dogmatic when they shift into assuming that somehow the truths that they put forward are to be regarded as offering insight into social reality. This is the point at which they might turn to standpoint epistemology to justify their assumption of what Hammersley and Gomm call "epistemologically privileged knowledge" (paragraph 3.5).

Hammersley and Gomm argue that the epistemologies provided within feminism (relativism and standpoint epistemology in different mixtures) serve as a hindrance to the mechanism of collegial accountability. Neither of these epistemologies indicates how the development of statements about reality can be subjected to mutual criticism in the scientific community.

Against the feminist suggestion that reasonable behavior within the scientific community has been traditionally defined by certain people (mainly men!) who set the parameters of what is regarded as rational, Hammersley insists that it is still important that would-be scientists are geared to offering arguments about how they are deriving conclusions based on relevant (cogent) evidence. Colleagues should be open to listen to and address these arguments as part of the process of debate within the community. There is no guarantee that criticisms of one's work that are made will all be "legitimate". And even the nature of what is or is not "legitimate criticism" is admittedly open to dispute (1997, paragraph 1.4). There is, as Hammersley puts it, "room for reasonable (as well as unreasonable) disagreement about this" (1997, paragraph 1.4). Hammersley proposes that at least members of the community should try to be reasonable when they organize their criticisms and when they listen to that of others. He suggests that:

> In the absence of a foundation of empirical givens whose validity is absolutely certain [that is, as is supposed by foundationalism], it is [of course] possible continually to raise doubts about any claim to knowledge: ... there is always further scope for criticism. However, it is not desirable to pursue criticism when, on any reasonable judgment, an account is valid. (1995, p. 43)

Scientists need to exercise reasonable judgment when deciding whether it is appropriate to operate the process of continuing criticism (of knowledge claims).

In considering the feminist suggestion that Hammersley's (and others') conception of the scientific community serves to perpetuate cognitive and political hierarchies in society, Hammersley indicates that he has no qualms with the scientific claim to "intellectual authority" (1995, p. 57). The authority is based on the premise that

> the findings of research are, on average, less likely to be in error than information from other sources. And this stems from the operation of the research community in subjecting research findings to scrutiny and thereby detecting and correcting errors. (1995, p. 57)

While everyone in society is entitled to opinions on issues, not all of these are "equally likely to be true" (1995, p. 57). It is wiser according to Hammersley to assign more weight to research findings than to the "opinions of people who have no access to the relevant information" (1995, p. 157). Hammersley argues that the right of all people to "define reality for themselves" that some feminists seem to postulate as an ethical principle, is not necessarily a better ethical position to adopt than a more realist-oriented one (realism here understood in terms of non-foundationalism — see Hammersley, 1995, p. 107). According to Hammersley, scientists have a responsibility to direct themselves toward understanding realities existing in the external world, and to "ensur[ing] that, as far as possible, the information provided [about reality] is valid" (1995, p. 59). This responsibility, as far as Hammersley is concerned, cannot and should not, "be shifted on to the people studied, or on to anyone else" (1995, p. 59).

Hammersley contends that it is a shirking of responsibility on the part of scientists if they do not direct themselves toward trying to find out about the phenomena in reality under study. To deny that scientists have some expertise in processes of knowledge-production, is to deprive lay people of benefiting from the value of the informed accounts created within the scientific community. The status of these accounts should not be downplayed by referring to some ill-defined "commitment to equality" in the process of generating knowledge of social reality (1995, p. 58).

In considering the question of researchers' responsibilities, Hammersley and Atkinson add the point that researchers might of course, in addition to acting as researchers in certain situations, decide to be helpful to people in other ways (that is, other than trying to develop knowledge). If such situations arise for them, and their research leads them into additional activities, this is because at this point they are deciding as citizens to take on this role. However, such activities should not be regarded as part of their remit as researchers (Hammersley & Atkinson, 1995, p. 286). The research remit is to serve the goal of advancing knowledge.

The following bullet points provide a rendition of the non-foundationalist argument positioned in relation to anti-foundationalist feminism.

- Contrary to qualms raised within feminism, there is no reason to be concerned when researchers see it as their task to advance knowledge of social reality.
- As long as mechanisms for collegial accountability are operative in the scientific community, findings generated are likely to be able to afford better insight than the uninformed opinions of lay people.
- Within anti-foundationalist feminism, accountability seems to be linked to adopting an ethical principle of opening spaces for exploring multiplicity of visions. In a non-foundationalist position, accountability of researchers rather implies *embracing a responsibility to work in terms of the norms of mutual criticism within the scientific community as a way of increasing the chances of generating knowledge about reality that can be of value to social actors.*

2.3.1.6 Positioning Non-foundationalism in Relation to Discursively-oriented Constructivist Argumentation

As indicated in Chapter 1, Kuhn's argument is sometimes considered as one of the sources of a discursive/constructivist approach to the process of knowledge-production. Hammersley notes that Kuhn has been one of the influential figures in the "drift toward relativism" (1995, p. 13). But he also notes that "there has been much debate about what Kuhn meant" by referring to the paradigmatic character of research (1995, p. 12). He points out that social scientists have sometimes drawn the conclusion from Kuhn that "all knowledge is founded on assumptions which are arbitrary from a rational point of view, and that ultimately it is a matter of taste or politics which paradigm one adopts" (1995, p. 13). Hammersley sees this relativist argument as linked to the view that is sometimes called "constructivism". This view suggests that "researchers must be regarded as actually constructing the phenomena they describe, on the basis of cultural [linguistic] resources available to them". Hammersley expands on what is meant by this, namely that "their descriptions and explanations reflect their own perspectives rather than, or as much as, the nature of the phenomena they claim to be describing" (1995, p. 16). Hammersley points out that such a constructivist position amounts to arguing that scientific inquiry cannot justify any claims to describe or explain realities existing outside of the knowing process. Scientific inquiry is then seen as a process of constructing, rather than representing, the realities that it describes and explains.

As discussed in Section 2.3.1.5 above, Hammersley and Gomm (1997a) offer a critique of the relativist/constructivist epistemology as developed within feminism. In this section, the details of their position against constructivism are explored further. Hammersley argues that the main problem with constructivism is that it provides no understanding of the difference between naive realism (phenomenalism) and the kind of realism that is espoused by non-foundationalism. The tenet of phenomenalism (as adopted in foundationalist thinking) suggests that people "respond to external stimuli in a mechanical fashion" (1995, p. 16). This implies that the world of phenomena can be mirrored in consciousness. There is thus a secure empirical foundation on which science can rest. Hammersley argues that this naive realism "should not be confused with all forms of realism" (1995, p. 107). As Hammersley and Gomm also indicate, just because we have to play an active part in making sense of the phenomena presenting themselves to us, does not mean that the distinction between "accounts and the phenomena they purport to represent" should be dismantled (1997a, paragraph 4.2).

Hammersley and Gomm emphasize that the research process is (must be) directed toward the pursuit of knowledge, defined in realist terms. As they argue: "The pursuit of knowledge, defined in realist terms, is unavoidable. We are not free to interpret reality just however we like, that is part of the meaning of the word 'reality'" (1997b, paragraph 1.9). They argue that the pursuit of knowledge (directed toward trying to find out about reality) "is one among many activities that every one of us routinely engages in, and is an essential feature of human life" (1997b, paragraph 1.9). People could not live by the decision to redefine reality as something that is simply a construction of social discourse, with no independent existence. People live in terms of the idea that they require some knowledge of (outside) reality in order to orient themselves in the world. And researchers can engage in systematic attempts to find out more about reality than is likely to be found in other arenas of social life.

Hammersley and Gomm aver that when constructivists appeal to the idea that we live in a world of multiple realities (expressed as different constructions), they develop an unsustainable position. As soon as they say that the world "is" a world of multiple realities, they are making a claim about what the world really is like (outside of discourse). They are making a statement "about how the world is independently of their decisions about it" (1997b, paragraph 1.9). This, according to Hammersley and Gomm, does not tally with the constructivist insistence that the world is created through the way people speak about it.

Hammersley and Gomm note that sometimes constructivists attempt to justify their epistemological position on moral grounds. It is then suggested that it is preferable to treat the world as a realm of constructions so as not to unfairly authorize particular ways of seeing at the expense of others (1997b, paragraph 1.4). But Hammersley and Gomm argue that by attempting to justify the position on moral grounds, constructivists simply assert the superiority for their position, judged in terms of their own standards of ethical conduct. They comment that when undertaking a comparison between positions, Romm, for example (1997a), "judges *our* approach in terms of *her* principle, and, not surprisingly, finds it wanting" (1997b, paragraph 1.5). They propose that instead of discarding non-foundationalism for its supposed moral inferiority, it would be better for people to engage with the arguments that they have advanced for it. One of their arguments is that an evaluation of the factual adequacy of claims is essential as a way of helping people to consider which views should rightfully be accorded more credibility (1997b, paragraph 1.8). Hammersley and Gomm argue that constructivism does not, and cannot, deal with the question of how different knowledge claims are to be assessed. It has no way of calling people to account for the constructions that they might put forward, other than in terms of the requirement to allow others too to express (differing) viewpoints. This still does not explain how researchers might be called to account for the quality of the constructions that they generate in the course of their inquiries.

Thus, the constructivist idea that all accounts are created in processes of discourse whose relationship with external reality cannot be subject to assessment, takes us off the route toward increasing the possibility of knowledge-production, according to a non-foundationalist position. Unless researchers direct themselves to criticizing one another in terms of some reference of their constructions to outside reality, there is no point in talking about "research" as an activity. The word "research" (as with the word "reality") loses its meaning.

The following bullet points provide a rendition of the non-foundationalist argument positioned in relation to discursively-oriented constructivism.

- According to non-foundationalism, constructivist arguments suggesting that scientific inquiry is a process of constructing realities, gloss over the question of how different constructions are to be assessed in terms of their credibility.
- Within a discursively-oriented constructivist approach it is proposed that the accountability of researchers can be developed without any reference on their part to extra-linguistic realities (realities outside of discourse) to justify accounts created.
- In terms of a non-foundationalist position, *accountability of researchers must make provision for creating "constructions" in the scientific community that are likely to be of more value in terms of their factual adequacy than discourses developed in other arenas of social life.*

2.4 SCIENTIFIC REALISM AS A WAY OF DEFINING THE PRACTICE OF SCIENCE

Self-named scientific realists (for example, Bhaskar, 1975; Keat & Urry 1982; Sayer, 1983) argue that one of the main contributions of their view of science lies in the appreciation of the nature of the causality that science seeks (or at least should seek) to discover. According to a scientific realist approach, to offer causal explanations does not merely involve indicating that regular relationships exist between phenomena (variables) that have been isolated. It also involves answering questions about the underlying structures responsible for the production of these relationships (and of other effects).

In order to elucidate how this position can be used to examine scientific activity, including social scientific activity, Keat and Urry's argument is drawn upon in detail in this section. At points, other arguments are invoked to further explain the position. Keat and Urry's argument can be outlined in relation to the tenets of positivism, as follows.

Keat and Urry accord to some extent with the positivist claim (expressed through the tenet of *scientism*) that social life can be studied using the same approach as is used in the natural sciences. But they disagree with the positivist conception of what is involved in the doing of science (as is explained later in discussing their relationship to empiricism). Their argument concerning the unity of the sciences is set in relation to the claim of interpretive social scientists that the human world requires a distinct mode of investigation. While they agree with interpretivism that part of the study of social life requires that one makes an attempt to understand the "agent's viewpoint" regarding his or her actions (1982, p. 90), they believe that the scientific realist conception of causality is equipped to address this aspect of social life. They argue that one of the reasons why it is necessary to understand an agent's viewpoint (set of meanings) is that this might act as a cause influencing the agent to act in the way s/he does. Once we understand meanings as causes, then they can readily be accommodated within a realist approach to the study of society. A realist approach makes provision for theorizing about any "social object" that has "causal efficacy" (Keat & Urry, 1982, p. 167).

But Keat and Urry add that we must also be aware that the existence of meanings themselves may require further explanation. According to them, explaining why particular meanings have become developed in a society, requires proposing an account of mechanisms in the society that could be seen as having the causal power to trigger these meanings. It is possible that the agents themselves are not aware of these mechanisms (which are indeed hidden). Keat and Urry refer to an example from Marx's work to elucidate this idea. They note that Marx is able to indicate that while it may *appear* to members of a (capitalist) society that the exchange between capitalists and workers is a "free and equal exchange relationship", this appearance is misleading. Marx also offers an explanation of why this misleading (false) appearance has been "systematically generated" through the internal structure of capitalist society: it serves the interests of capital to conceal the exploitative relationship between capital and labor by disseminating an alternative conception (Keat & Urry, 1982, p. 195).

Keat and Urry suggest that Marx developed a particular procedure in terms of which "correct, undistorted accounts of different social formations" could be sought (1982, p. 206). But use of the procedure implies a different view of empirical evidence from that which is argued for within a positivist position. And it implies a different view of the way evidence relates to the development of theoretical accounts. These differences can be expressed by discussing the scientific realist position in relation to the tenets of *phenomenalism* and *empiricism*.

Keat and Urry review the positivist *phenomenalist* tenet that observation can be regarded as constituting a firm foundation on which science rests. They argue that because observation is theory-laden from the start, it is more difficult (than is supposed by positivism) to use it to test between competing theories. Nevertheless, they argue that in the process of scientific inquiry it should be possible for theorists offering alternative theoretical arguments to arrive at certain "agreed descriptions" (1982, p. 52). Even though their prior theoretical visions might indeed influence their perceptions, certain observations that are "free of the particular theoretical beliefs at issue" in the controversy can be arrived at (1982, p. 52). It is this fact — the fact of being able potentially to arrive at acceptable observations — that allows for the possibility of "rationally assessing" competing theories (1982, p. 56).

Keat and Urry contend that the phenomena that are observed should in any case not be related to theorizing in the way suggested by positivism. In terms of a positivist conception, observation of phenomena is used to ascertain whether or not some regular connection between them can be argued to exist. Causality refers to the fact that phenomenon Y is seen to follow regularly in the presence of phenomenon X. Scientific realism asserts, in contrast, that "causality" refers to an element of necessity that arises because of the nature and constitution of certain underlying "generative mechanisms". Scientific explanation involves knowledge of the underlying structures responsible for triggering the outcomes that become produced.

Layder (acknowledging his indebtedness to authors such as Keat and Urry for the development of the realist position) points out, for instance, that the configurations of power operating in the social world "are generally at a low level of visibility" (1993, p. 159). In terms of this argument, Layder suggests that researchers should be sensitive to the problem of visibility; otherwise they risk "underestimating the influence of structural components" (1993, p. 159). Offering the example of power configurations, he proposes that:

> The researcher should be aware of forms of power and control relations that operate "behind the scenes" of the observable interactions of everyday life. Some forms of power and control are built into the settings and contexts of activity, like work or occupational organization. These power relations influence behavior in a subtle manner and may not be fully revealed by the more overt exercises of power that can be readily observed in behavior. (1993, p. 170)

Layder here illustrates the importance of assigning some analytic weight within scientific theorizing to structural components of reality. He suggests that the positivist way of relating "theory" to "empirical observation" needs to be complemented by other conceptions of this relationship.

Keat and Urry tackle the question of the relation between theorizing and empirical observation by considering (and criticizing) two types of positivist position: that held by so-called "confirmationists", and that held by so-called "falsificationists". They point out that positivist "confirmationists" believe that it is in principle possible — through empirical inquiry — to establish the truth of particular theoretical statements (1982, p. 18). For confirmationists, use of the logic of science allows scientists to make inferences from statements about observation to theoretical statements about the lawful regularities operating in reality. (Inductive logic is here seen as important in the process of scientific inquiry, supported by deductive logic.) Positivist-oriented falsificationists by contrast (under which Keat and Urry label the Popperian position[9]) believe that it is logically impossible to establish the truth of particular statements; all that is logically possible (through use of deductive logic) is to establish when a particular statement is false — when it is disconfirmed by the evidence.

While Keat and Urry agree with positivism that empirical observation certainly plays a role in the process of scientific inquiry, they do not accept that it can function either to definitely confirm or to definitely falsify particular theoretical statements. They revise the positivist tenet of *empiricism* in order to take account hereof. They indicate that they do not wish to identify themselves with the positivist suggestion (including the Popperian argument) that

> there is any specifiable set of logical relationships between theories and perceptual [observation] statements such that the truth, probable truth, or falsity of the former can be definitely determined by their relationship to the latter. (1982, p. 233)

How then do they envisage ways in which statements (about theoretically postulated entities) may be developed? They suggest in this regard that a statement of the form "x is [characterized by] Q" may be proposed as referring to the internal constitution (or structure) of an entity whose existence would help explain the occurrence of observed phenomena (1982, p. 38). The logic used in supporting this kind of reasoning has sometimes been called "retroductive" logic (cf. Sayer, 1979, 1983). Sayer notes that retroductive logic proceeds by: "postulat[ing] mechanisms which should they exist would explain how the phenomena under investigation come to assume the forms in which they are experienced" (1979, p. 40). In terms of retroductive logic, the logical link between "H" (a hypothesis about an unobservable causal mechanism) and "P" (perceived events) consists in the fact that "H" posits a mechanism which, if it existed, would offer an explanation for P_1,P_2,P_3, and so on (1983, p. 116).

Pawson and Tilley summarize the way in which scientific realism argues that scientific theorizing can progress using case material as evidence. They note that what is at stake is a kind of theoretical "generalization" built upon the possibility of abstraction. (They draw on various authors' views on the meaning of theorizing in order to make this point.) They point out that

> cumulating knowledge is a process in which we move from one specific empirical case to a general theory and back to another case and so on. What are transferable between cases are not lumps of data but sets of ideas. The process works through the development of a body of theory which provides an *organizing framework* which "abstracts" from a program [that is, a way of triggering observable events] a set of essential conditions which make sense of one case after another. (1997, p. 120)

---

[9]Keat and Urry (along with other authors) regard Popper's position as positivist in orientation (despite Popper's reservations in regard to the appropriateness of this label). However, Keat and Urry do distinguish his orientation from what they call a "confirmationist" one (1982, p. 15).

In this way, they suggest, empirically informed theoretical accounts can be developed over time.

As far as the tenet of *value freedom* is concerned, scientific realism suggests that the aim of scientific study is to increase the accuracy of theoretical accounts (regarding the structures operative in reality). Keat and Urry remark that scientific realism shares with positivism the belief that science strives to attain objective knowledge of reality. As they put it: "Both [positivism and scientific realism] share a general conception of science as an objective, rational inquiry" (1982, p. 44). They agree with positivism that when scientists attempt to construct their theories about reality, these must be "objectively assessed with reference to the empirical evidence" (1982, p. 44). They also concur with the positivist belief that rational assessment between competing theories is possible (via recourse to empirical evidence) (1982, p. 97). This does not mean that they wish to underestimate "the degree of difficulty, and uncertainty, involved" in the enterprise of doing science (1982, p. 43).

Turning to the question of the utility of the scientific enterprise, Keat (1981) offers an indication of the way in which social theories being developed can be regarded as having practical relevance in society. He discusses this in terms of three points.

Firstly, he points out that in the process of moral decision-making, one often relies on certain factual information. For instance, the moral protestation against the fact that a given class in society is exploiting another, is dependent on "the character of the economic relationships between the two, such as the distribution and control over the means of production" (1981, p. 44). The viability of the moral judgment depends on the credibility of the relevant factual knowledge.

A second reason why scientific knowledge can become relevant in the domain of moral decision-making is by virtue of its being able to proffer "conditional predictions about the consequences of various courses of action" (1981, p. 44). Keat cites the case of a social theory aimed at explaining the distribution of social goods in capitalist society: "Such a theory might show that it [the distribution of goods] results from the class-distribution of the means of production, and that it cannot be radically altered ... without a transformation of the capitalist mode of production" (1981, p. 64). The demonstration that this entity ("capitalist mode of production") is accompanied by (and seems to be responsible for generating) these consequences, suggests that any desired removal of these consequences requires an attendant social transformation of the entity.

A third reason why knowledge of social reality may be morally relevant, springs from the relationship between the "morally desirable" and the "possible" (1981, p. 145). Any moral statement proposing that a certain state of affairs ought to be considered as a social goal, can be set aside if we can show the absolute impossibility of ever implementing such a state of affairs. But Keat argues that we need to treat "with considerable caution" statements that people may offer concerning what is indeed possible or impossible to achieve in reality. He points out that statements of apparently factual "impossibility" often conceal certain normative judgments concerning what sorts of actions are worthwhile (or not) to pursue in society.

In commenting on the practical relevance of the scientific enterprise in society, Keat indicates disagreement with those (such as critical theorists) who claim that upholding the fact/value distinction leads to the encouragement of a "reformist" approach to addressing social problems. He specifically attempts to show that a social science directed toward uncovering realities existing independently of our knowledge thereof, might lead actors toward considering options for radical social transformation. The *instrumental use* of science to address issues of concern to actors is thus not necessarily associated with a reformism — especially if science is undertaken in accordance with a scientific realist self-understanding of its purpose (namely, to uncover hidden structures and generative mechanisms in reality). The knowing endeavors of scientists can become instrumental in offering intellectual support for the engagement in radical transformative activity. This can be done by isolating possible arenas for action, taking into account the constraints that may inhere in existing social structures.

2.4.1 Considering Accountabilities in Terms of a Scientific Realist Position

This section is organized primarily around the work of Layder (1993), including his work with Davidson (1994). This work is used to provide an indication of how a scientific realist position might consider researcher accountabilities in response to alternative arguments. Layder's (1993) and Davidson and Layder's (1994) arguments in relation to positivism, non-foundationalism, interpretivism, critical theory, anti-foundationalist feminism, and a discursively-oriented constructivist approach are discussed. Although Layder (1993) and Davidson and Layder (1994) do not discuss their relationship to different arguments using the terminology of researcher accountability, it can be suggested that a position concerning accountability is implied in their argument. This is explored in the following sections.

2.4.1.1 Positioning Scientific Realism in Relation to Positivist Argumentation

As indicated in Section 2.2.6, the positivist conception of scientific practice implies that the accountability of researchers consists in their observing phenomena without prejudice, and deriving conclusions that are grounded in observable evidence. Layder indicates that he appreciates that theories need to be validated in relation to appropriate evidence and that scientists can be called to account in terms of this requirement (1993, p. 52). However, in line with scientific realist argumentation, he argues that what is properly involved in the process of scientific inquiry is different from what is supposed within a positivist view of science. He points out that

> a key aspect of the [scientific] realist project is a concern with causality and the identification of causal mechanisms ... in a manner quite unlike the traditional positivist search for causal generalizations [stating a relationship of causality between variables]. (1993, p. 16)

When causal generalizations in the positivist sense of the term are sought, then, as Layder notes, scientific theorizing is reduced to examining specific connections between variables that have been isolated. Layder argues that despite the lip service that "empirical researchers" pay to the "importance of theory", the way in which they define this reduces its province to "forms of theory ... that narrowly specify the relationships between precisely measurable variables" (1993, p. 15). Layder suggests that where a research project is based almost exclusively on this "narrow sense of theory", there should be room for complementing this with other kinds of theorizing.

His suggestion is that general theory can be built up by combining different ways of theorizing so as to contribute to the cumulative development of theory (1993, p. 31). Besides testing specific hypotheses about relationships between variables, space should be made for theory-constructing approaches that go beyond looking for such relationships. Theorizing should also not, however, be confined to the emergent theorizing suggested by certain interpretive-oriented ethnographic researchers, because the view of structure implied here still does not allow us to assign analytic weight to structures beyond the "immediate environment of actors" (1993, p. 56). Layder suggests that a more complete picture of reality can be built up through the influence of different forms of theorizing:

> It needs to be recognized that theorizing may take a number of forms, and may be related to the empirical world of research in a number of ways. A truly cumulative approach to theory would seek to integrate these seemingly diverse efforts. (1993, p. 31)

One of the reasons why these "diverse efforts" are not integrated, is because of the "radical academic division of labor" between researchers taking different views of the way in which "theorizing" should be practiced (1993, p. 150). Layder believes that a "genuine dialogue" between the positions should be encouraged (1993, p. 15). This can be combined with a multi-strategy approach to research, including different methods and methodological traditions that bring with them "different perspectives which allow the researcher using them to see empirical reality in slightly different ways" (1993, p. 122). He suggests that by cutting into the data from different angles in terms of "different research strategies (methodological or analytic)", a "variety of 'slices' of the research site" can become revealed (1993, p. 123).

Layder's complaint against a positivist-oriented research strategy (defined as one using a narrow view of theorizing to seek relationships between variables) is that it becomes insulated from other forms of theorizing. It is thereby unable to offer anything other than a partial view of reality — from the perspective of its own analysis. Layder therefore calls upon researchers to account for their practices not only by showing how their research strategies allow them to seek connections between variables.

Davidson and Layder follow up this argument by suggesting that researchers do not need to be "*blind* to any picture of reality other than that which mirrors their own preconceived vision" (1994, p. 51). As human beings they should be capable of recognizing "that they themselves view the world through a particular lens" (1994, p. 51). Once this is recognized by them, they should be capable of using their self-consciousness in a way that allows them to contribute to producing a picture of reality that does not simply reflect their initial theoretical preconceptions or preoccupations.

Davidson and Layder also suggest that researchers should make concerted efforts to

> reflect upon the conditions under which they are producing knowledge, thinking, for example, about the ways in which their desire to win praise from the academic community or to comply with funding bodies is shaping the picture of reality that they are painting through the research. An awareness of these issues is central not just to producing better research, but also to critically evaluating the research of others. (1994, p. 53)

Davidson and Layder concur with the positivist suggestion that it is important for researchers to commit themselves to trying to build up a picture of reality that is free of the influence of their particular extra-scientific values (not domain to the goal of producing knowledge). This implies for Davidson and Layder, however, that they are alert to considerations of whether a set of colleagues' criteria for assessing their work may be too narrow. It also implies that they are able to adopt, if necessary, a critical attitude toward funding bodies — with a view to recognizing when following their criteria for acceptable research may reduce the possibility of doing better research. Davidson and Layder's caution here is set in the context of their concern that both in the scientific community and in the wider society there may be factors at play preventing a more accurate picture of social reality to emerge.

The following bullet points provide a rendition of the scientific realist argument positioned in relation to positivism.

- Scientific realism concurs with positivism that scientific inquiry involves striving to develop theories untarnished by particular scientists' partiality. But scientists must also be careful not to accept narrow definitions of theorizing (as offered within or outside of the scientific community).
- Because of divisions and disagreements that exist concerning what is involved in doing science, the accountabilities of researchers cannot be measured in terms of their adherence to what is taken to be "scientific protocol" in (some section of) the scientific community.
- Researchers need to find a way of organizing a dialogue across different divisions within the research community as part of the development of their role. *Accountability implies being self-conscious about the way in which research may become biased, so as to pre-empt some of the biases. This includes a process of becoming open to engage with different perceptions of evidence and different forms of theorizing as part of an involvement in dialogue within the scientific community.*

2.4.1.2 Positioning Scientific Realism in Relation to Non-foundationalist Argumentation

Layder offers an indication of how a scientific realist position relates to the non-foundationalist position as espoused by, for example, Hammersley. He notes that according to Hammersley, what marks *scientific* theorizing off from other forms of discourse is the way in which it is used in the context of testing theoretical knowledge. For example, Layder refers to Hammersley's suggestion that one could develop a theory, based on ethnographic studies of schools, concerning the polarization of pupils in certain kinds of school setting.

> The theory [proposed by Hammersley] claims that if pupils are allotted to different classes according to their general intellectual abilities, or abilities in particular subjects, their attitudes toward these rankings will become polarized. Those given the highest rankings will accept them and the values they embody, while ... those given the lowest rankings will reject them and as a consequence this will lead to various kinds of "disruptive" behavior. (Layder, 1993, p. 30)

Layder indicates that Hammersley suggests that a research program could be developed to test this theory. This could be organized by considering what kinds of further evidence would function to support, or conversely, to disconfirm (or lead to the qualification and modification of) the theory.

Layder compares Hammersley's argument concerning theory-testing as the hallmark of science with that of Merton. Merton has argued for what he calls "middle range theory" as a way of organizing research so that theories can be put to the test via empirical studies. Layder notes that Merton's idea is based on the suggestion that "a testable hypothesis or proposition [can be] logically deduced from an existing set of assumptions. The empirical data that is then collected either confirms or disconfirms the original hypothesis or proposition" (1993, p. 26).

But Layder argues that the practice of converting what he calls "larger theories" into what can be regarded as testable in this way, "may lead to an inadequate empirical application of the theory" (1993, p. 28). He refers to the example of Marx's theoretical ideas on class inequalities in society. As Layder sees it, the Marxist-oriented focus on social structures and their historical development cannot be translated into observable portions in the way that Merton's conception of middle range theory implies (1993, p. 27). Against the complaint that the frameworks provided by larger theories become untestable (unless broken into smaller portions), Layder contends that their analytic utility can still be explored and challenged in the process. Working with such theoretical frameworks need not amount to proceeding simply "to 'explain' all empirical evidence within the terms of the framework" (1993, p. 53). According to Layder, this way of using evidence does indeed reinforce the boundaries between different frameworks and is not conducive to "cross-fertilization of ideas from different frameworks" (1993, p. 53). The problem is that in social theorizing thus far the promise of dialogue across frameworks has been "largely unfulfilled" (1993, p. 53).

Layder indicates that "if all fieldwork data is interpreted in terms of a prior framework favored by the researcher then it will lead to a blinkered outlook. In this kind of situation knowledge cannot progress" (1993, p. 52). But he argues that the theory-testing approach supported by Merton (following Popper) can "produce the same effect" (1993, p. 52). Knowledge cannot progress to its full potential in this manner. Only those layers of reality that the analytic tools (of the theory-testing approach) are equipped to handle, will be shown up. Other layers that do not submit to being understood by being broken up in the way suggested by Merton, will remain outside the domain of social theorizing.

Layder states that there may be situations where Merton's middle range theorizing can "provide useful information. For example, such an approach would be useful in testing hypotheses about the relation between ... variables such as 'income level' and 'rates of participation in community activities'" (1993, p. 29). But he contends that it would be beneficial also to widen the scope of social research in order to try to develop, for example, "some understanding of the broader structures of power and domination, the nature of labor markets and so on" (1993, p. 29). To do this might require working with a different understanding of the way in which theorizing can properly relate to empirical research efforts.

In considering Hammersley's argument for the importance of theory-testing, Layder remarks that in certain respects it is "strikingly similar to Merton's" (1993, p. 31). Although Hammersley has indicated that he is concerned with the social factors that give shape to observable events, he has still argued for a "tightly specified theory and the testing of specific hypotheses derived from it" (Layder, 1993, p. 31). He has also insisted that in order for theory-testing to work, a narrow research focus is needed. Layder cites Hammersley in this regard as stating that "if the development and testing of theory is to be pursued effectively, the research focus has to be narrow" (Hammersley, 1985, as quoted in Layder, 1993, p. 31). Layder's position is that:

> In the same manner as MRT [middle range theory], Hammersley's approach can be regarded as one among a number of possible approaches to research which would provide useful information and contribute to the cumulative development of theory. However, this cumulative development has to be understood in rather wider terms than is presently implied in Hammersley's model. (1993, p. 31)

In discussing his relationship with Hammersley's argument regarding communal assessment of research practice in the scientific community, Layder cautions that what is considered reasonable behavior by scientists should not be judged in terms of a narrowly defined conception of theorizing. With this proviso, he argues that the scientific community does provide a forum for developing "a truly cumulative approach to theory" (1993, p. 31).

Davidson and Layder (1994) comment that they are in accord with Hammersley that although the scientific community can never propose to guarantee the validity of findings that are generated within it, it still should be regarded as having some claim to authority in society. They state that they agree with Hammersley that

> to deny the "intellectual authority" of the researcher on the grounds that everyone's opinion is as important as everyone else's ignores the whole point of research, which is to generate opinions informed by evidence gathered in ways which are open to the scrutiny of (any) other researchers. (Davidson & Layder, 1994, p. 183)

Davidson and Layder emphasize that for the research community as a whole to build up an integrated picture of reality, scientists need to strengthen their capacities for scrutinizing their own work and that of others, with a view to contributing to the cumulative development of theory. It is in the strengthening of these capacities that their accountability lies.

The following bullet points provide a rendition of the scientific realist argument positioned in relation to non-foundationalism.

- Scientific realism supports the non-foundationalist suggestion that scientists should examine one another's (and their own) work critically so as to reduce instances of empirical evidence being used simply to buttress prior theoretical conceptions.
- According to scientific realism, the non-foundationalist support for middle range theorizing itself may reduce the possibility of introducing a wider analytic focus into the research enterprise — which can become stunted from the start.
- While in non-foundationalism accountability implies arguing for the reasonableness of behavior within the scientific community, in scientific realism *accountability implies taking part in a process of engaging with different perceptions of evidence and different forms of theorizing as part of a communal effort to contribute to the cumulative development of theory.*

2.4.1.3 Positioning Scientific Realism in Relation to Interpretivist Argumentation

In discussing the scientific realist position in relationship to interpretivism, Layder notes that in what he calls humanist (interpretive-oriented) arguments, "there is a tendency to reject concepts (like that of structure) which seem to imply that there is a social realm which is independent of people" (1993, p. 57). Layder states that he can concur with humanists insofar as they are concerned with the problem of reified thinking, in which social forces are treated as "things" in the same way as natural objects are. These things are then seen as facts that seem to have an existence outside of the human activities that make them up. Layder appreciates this concern of humanists. But he comments that the humanist concern with reified ways of thinking about social structures, involves some confusion over the use of the term "structure". As he puts it: "However, often there is a confusion involved here, to the effect that the very use of terms like 'structure' *automatically* involves a reified mode of analysis. This is not the case" (1993, p. 57).

Without reifying structures, Layder suggests that it is still important to recognize that the properties of structures in social life do transcend "everyday encounters" of social interaction. Layder indicates that insofar as a humanist (interpretive-oriented) argument focuses its attention on "everyday encounters" of social interaction, and shies away from paying attention to "structural dimensions" and their properties, a scientific realist position would want to complement its analysis with other ones (1993, p. 59).

Davidson and Layder observe that Weber's method of *verstehen* involves an attempt to interpret people's actions "in the light of the meanings, motives and intentions behind their action" (1994, p. 31). The focus of *verstehen* is on the "inner, subject world of human beings" (1994, p. 31). Interpretivism suggests that this focus is necessary in order to understand "why people act as they do" (1994, p. 31). Davidson and Layder point out that the method of *verstehen*

> produces empirical data of a very different type to that generated by positivist research techniques [aimed at operationalizing variables in order to examine their relationships]. Talking to people, observing their interactions, analyzing the language they use and the documents they produce in order to achieve empathetic identification and interpretative understanding yields qualitative data and evidence. (1994, p. 32)

However, they note that Weber does not stop at the point of offering a plausible account of people's meanings. He also believes that a check on the meanings as discerned by the researcher can be made through the observation of statistical regularities. Insofar as the statistically derived regularities "correspond to an intelligible meaning", the account of meaning as it operates in social life can be regarded as (more or less) "verified" (1994, p. 32). Davidson and Layder's argument in relation to Weber's position is that he still does not accord the necessary attention to causal mechanisms written into particular social structures. He does not offer the analytic tools to appreciate the manner in which interactions come to be shaped in everyday encounters, or to appreciate the way in which regularities in social life become generated. The picture of reality that is revealed through his suggested mode of inquiry is thus too one-sided.

The following bullet points provide a rendition of the scientific realist argument positioned in relation to interpretivism.

- Interpretivism does not pay sufficient attention to a consideration of the way in which the properties of social structures might exert causality in social reality.
- In interpretivism, accountability involves researchers' arguing for the adequacy of social scientific interpretations of action on the level of causality (through statements of statistical probability) and on the level of meaning (through offering accounts of motivating meanings).
- In scientific realism, *accountability implies researchers' subjecting to argument their accounts of social life, including their ways of theorizing about, inter alia, the structural dimension of society and its properties.*

2.4.1.4 Positioning Scientific Realism in Relation to Critical Theoretical Argumentation

Davidson and Layder's argument in relation to the critical theoretical suggestion that facts and values are inextricably linked in the knowing enterprise, can be examined by considering their view of how researchers should deal with their own values. In line with a scientific realist position, Davidson and Layder reject the critical theoretical proposal to include moral considerations at the moment of knowledge-creation as part of the criteria for defining the "validity" of research. Research for them cannot and should not be assessed in any other terms than its ability to forward insight about independently existing realities.

Layder (1993), and Davidson and Layder (1994), emphasize that the point of scientific theorizing is to be able to build a picture of reality that is, as far as possible, free of the values and preconceptions of those involved in the inquiry. Being critical in the research process does not for them mean setting out to create theories whose aim is to change the social world in the name of certain values. Being critical, can, however, involve unmasking moral and political values that are hidden behind a "veil of 'neutral' pseudo-scientific terminology" (Davidson & Layder, 1994, p. 27). Insofar as those supposedly practicing "science" are engaged in a value-laden pursuit hiding behind a veil of value freedom, their way of doing science needs to be subjected to critical scrutiny. Indeed, all scientific endeavors need to be subjected to such scrutiny in order that the picture of reality that comes to be created in the scientific community can be argued to bear some relation to realities existing outside of the community.

Davidson and Layder appreciate that

> social researchers draw on their everyday knowledge and on their political and moral values in the process of research; they use them to set the research agenda and to design classification systems; they use their social, as well as professional, skills to obtain information; they employ their knowledge as members of society and their political values to analyze and interpret their findings. (1994, p. 28)

Despite all these ways in which the personal position and value-orientation of researchers might affect their research practice, Davidson and Layder argue that the principle of organizing research through the commitment to generate knowledge of reality, should not be forgotten. It is still possible to commit to "research that is rigorous and reflexive" (reflecting back on the ways in which its biases can possibly be pre-empted) (1994, p. 28). This way of conducting research can be argued to "produce knowledge that is more objective than research which is sloppy and uncritical" (1994, p. 28).

They summarize their position by stating that, "in short, there are better and worse ways of conducting research" (1994, p. 28). Researchers should regard themselves as accountable in the sense of committing to "better" ways of conducting research. Judgments about the quality of research can be made through a rational dialogue within the scientific community — which should function in terms of a healthy communication across different ways of theorizing and different ways of anchoring this in the empirical world. Participation by researchers in such a dialogue means that the research community as a whole is able to develop more integrated knowledge about reality.

Davidson and Layder note that besides having to consider their role as part of a research community, researchers are also faced with what they call "the question of ethics" as they conduct their research (1994, p. 55). One crucial question that arises is "how to ensure that academic research is independent" (1994, p. 56). This question is particularly relevant in a context in which "sponsors ... are able to influence ... the research agenda and methodology by making awards on the basis of certain criteria" (1994, p. 56). For example, they may favor research that is less theoretically "abstract". Researchers might need to find a path for proceeding with integrity in the light of these felt pressures.

Furthermore, Davidson and Layder note that sponsors are also able to "limit the freedom of the researcher by imposing controls on what is published and when" (1994, p. 56). Davidson and Layder regard it as "ethically unsound" for researchers to "allow a funding body to push them to produce research which contributes to the more effective exploitation of already vulnerable groups, or which merely strengthens the resource base of a powerful vested interest group" (1994, p. 56). When "pushed" in this way, researchers need to consider the social context in which they are doing the research — they cannot immunize themselves from such considerations. Research is, after all, set in a context of unequal power relations in society, and researchers need to take this into account in the research that they do (1994, p. 56).

Researchers need to accept that at times they will face difficult ethical decisions in their research practice. For instance, if they wish to gain access to study relatively powerful groups in society, they might have to adopt "covert" methods of inquiry. The question then arises whether

> the social researcher's commitment to the pursuit of truth and the advancement of knowledge about such powerful groups [is] a more important ethical principle than that which condemns the use of deception and covert methods. (1994, p. 58)

Davidson and Layder indicate that one way of approaching this question is by considering whether powerful people or public figures need to publicly account for the consequences of their own actions. This being seen as the case, it can be argued that their "right to privacy" can be overridden by different considerations, namely those relating to the requirement for public accountability for their actions (1994, p. 58).

Davidson and Layder suggest that in considering ethical questions such as these, there are no clear-cut answers. They point out that "the specific circumstances of a research project and the moral and political values of the researcher will inevitably have a powerful effect on the ethical stance which is taken" (1994, p. 58). They indicate that as far as they are concerned it is important for researchers to remember, in any case, that research can, potentially, "contribute to struggles against inequality" (1994, p. 60). *Nevertheless, to recognize that this is the case is not to tie up the criteria for valid research, with the wish to contribute to the reduction of inequality in society.* Davidson and Layder here echo Keat's suggestion concerning the importance of defining truth as logically independent of values (1981, p. 44). Keat has indicated in this regard that he cannot agree with Habermas's suggestion to "test" the validity of statements with reference to their practical success in securing social goals (1981, p. 158). The critical force of scientific statements rests, for Davidson and Layder, as for Keat, on their reference to (real) features of society, rather than on any supposed tie between knowledge and values at the moment of knowledge-creation.

The following bullet points provide a rendition of the scientific realist argument positioned in relation to critical theory.

- Researchers need to face ethical decisions in regard to the topics they choose to investigate and the way in which they choose to organize their research. These decisions are rendered difficult because research is set in a context of unequal power relations. Nevertheless, whatever decisions are made, social scientific research is to be judged for its validity in terms of its contribution to advancing knowledge of realities.

- In terms of (Habermasian) critical theory, accountability implies a commitment to exploring ways of communicating that allow people (including scientists and others in society) to excavate the values implicated in statements being made about "reality".
- In scientific realism, *accountability implies that (social) scientists organize a dialogue across different ways of theorizing in relation to the empirical world with the aim of generating a progressively more complete picture of external reality: this can potentially help actors in their struggle against inequality.*

2.4.1.5 Positioning Scientific Realism in Relation to Anti-foundationalist Feminist Argumentation

Davidson and Layder note that one of the aims of feminists is to organize research in such a way that the possibility of setting up non-hierarchical relationships between researcher and researched can be explored (1994, p. 124). They point out, for example, that insofar as interviewing of subjects is concerned, feminists "assert that the orthodox methodologist's emphasis on control, hierarchy and the impersonal nature of scientific research reflects a masculine view of the world and of human relationships more generally" (1994, p. 125). In trying to explore non-hierarchical relationships during the research process, feminists consciously endorse a less "masculine" approach, and aim to exemplify less controlled relationships in the way social research is practiced (1994, p. 124).

Davidson and Layder's response to this feminist argument is that the issue of "control" is a very ambiguous issue (1994, p. 124). For instance, they indicate that in the process of interviewing respondents, the researcher needs to try to gain information from respondents and therefore might need to become in some sense "submissive" (1994, p. 124). To say that in traditional research researchers alone exercise the control is not to understand the complexity of the relationships involved (1994, p. 125). These relationships are made the more complex by the "power relations external to the interview situation" — these can "further complicate the issue of control". For example, "a female researcher interviewing males may be in control in the sense that she is initiating the interview and using it to her own ends, but may simultaneously feel vulnerable because of her status as a woman" (1994, p. 124).

In view of the ambiguity of the issue of control, Davidson and Layder suggest that the interviewing process (or use of any other research technique) should not be guided by the practical quest to establish non-hierarchical relationships. Rather, it should be guided by the effort to "obtain ... information" (1994, p. 125). According to Davidson and Layder, the rules and skills that might be appropriate to interviewing (as to any other technique of inquiry) depend on the requirement to obtain information. Researchers should be aware, for instance, that "they will not glean much information ... if they use the interview as an opportunity to hold forth on their own life experiences or political beliefs" (1994, p. 127). If they do this, the interviewees will be unlikely to find the chance to offer their own viewpoints — so the point of the interview will be lost. Alternatively, interviewees, when they do speak, may decide to participate in the interview by expressing that which they (now) know the researcher wishes to hear. In either case, they will be prevented from "stating that which he or she wishes to state" (that is, his or her views) (1994, p. 147).

Davidson and Layder emphasize that what is important when establishing a relationship with respondents in the social world is not to "manipulate respondents into saying what the interviewer wishes or expects to hear" (1994, p. 128). The aim of organizing interviews with respondents is to "collect data". No matter what kind of interview is taking place, it should be remembered that it is never a "chance encounter or an aimless chat. It is initiated by the researcher and takes place with a particular goal in mind, namely data collection" (1994, p. 128).

Davidson and Layder are also critical of the feminist suggestion that the research process should be used as an opportunity to offer expression to groups in society (such as women) whose understandings are not normally (in social discourse) accorded much credibility. They reject the feminist argument that research should be treated as a process of "giving voice" to research subjects (1994, p. 184). Research for Davidson and Layder is defined by specific requirements that need to be met if it is to warrant the label of research. In order for research to meet certain "standards of integrity and validity" it must pass through what they call a "methodological domain" (1994, p. 35). This domain consists of three elements: the first is certain techniques (such as techniques of interviewing or of questionnaire-construction); the second is "the research community" (which consists of researchers in dialogue with one another and with information gleaned through the research process); the third is methodological rules (such as rules relating to picking of samples, formulating of questions, and coding and analysis of data). Davidson and Layder argue that if this methodological domain is functioning well, it "provides markers which allow us to make judgments about which research is better or more valid given the present state of knowledge on a particular topic or area" (1994, p. 35). Provided that the methodological domain of research is functioning well, the research community is likely to produce better accounts of reality than the expressions of other voices within society. Although Davidson and Layder admit that the existing methodological domain is "a far from perfect filter", they suggest that researchers should unite in the struggle to improve the quality of the filter (1994, p. 59).

The following bullet points provide a rendition of the scientific realist argument positioned in relation to anti-foundationalist feminism.

- Contrary to the claims of (anti-foundationalist) feminism that the research process should be treated as a way of contributing to the development of less controlling relationships in society, scientific realism treats it as an opportunity to contribute to the advancement of knowledge about social reality.
- Researchers must pay prime attention to the purpose of developing their theorizing about social reality.
- In anti-foundationalist feminism, accountability of researchers is tied to their commitment to develop ethically defensible human relationships in the research process as part of the practice of social inquiry. In scientific realism, *accountability implies that (social) scientists recognize the importance of allowing their research to pass through the filter of the methodological domain (of techniques,*

methodological rules and argument within the scientific community concerning the use of evidence).

2.4.1.6 Positioning Scientific Realism in Relation to Discursively-oriented Constructivist Argumentation

Davidson and Layder refer to Kuhn's argument concerning the fact that in the social sciences there is a "lack of unanimous consensus over the appropriate methods for acquiring knowledge" (1994, pp. 32-33). They note that Kuhn uses the word "paradigm" to refer to a "kind of general agreement over methods and procedures" (1994, p. 33). Kuhn suggests that within the natural sciences there is normally (in periods of "normal science") some paradigmatic consensus. However, the social sciences differ from the natural ones in that they do not operate in terms of a "clearly defined research paradigm" (1994, p. 33). Davidson and Layder agree with Kuhn insofar as he points to the variety of methodological and analytic strategies that seems to characterize the social sciences. But they do not thereby relinquish their (realist) belief that science — including social science — can be geared to producing "a more reliable [and valid] form of knowledge than, say, that produced by journalists or lay people" (1994, p. 33). They leave in abeyance the question of whether Kuhn would concur with this. In any case, they do not believe that Kuhn's argument should be taken to the "relativist" conclusion that social scientists are "permanently locked into a narrow vision of the world, determined completely by their prior commonsense or theoretical assumptions" (1994, p. 33).

Considering scientists' involvement in constructing the realities that they perceive and theorize about, Davidson and Layder suggest that Scholte's (1972) argument for the practice of reflexivity in anthropology "is pertinent to other social scientific disciplines" (1994, p. 52). Drawing also on the work of Callaway (1992), they argue that anthropologists and other social scientists should "reflect on his or her own part in constructing a view of reality through the interpersonal relations of fieldwork and also reflect more generally on the ... conditions and modes of producing knowledge" (1994, p. 52). Researchers should reflect upon the way "in which their own social identity and values affect the data they are gathering and the picture of reality they are producing" (1994, p. 52). They should also reflect on the social conditions (such as pressures from funding bodies) that might affect their way of producing knowledge. But Davidson and Layder do not consider that an acknowledgment of researchers' possible involvement in constructing "data" and "pictures of reality", should lead them (or their audiences) into a relativist epistemological position. As they indicate:

> It may seem that there is a danger of reflexivity [reflecting on one's own part in the production of knowledge] leading to a form of relativism, that we are asking researchers to be so hypersensitive to their own role in constructing the data that they would lack all confidence in their findings. How can even the most reflexive white male researcher, for example, ever be sure that the picture of social reality produced by his research is generally valid and not a biased or partial portrait? (1994, p. 53)

Davidson and Layder's answer to this question lies in their belief in triangulation of research strategies. They refer in this respect to Denzin's (1970) conceptualization of triangulation as a form of cross-checking of findings. They point out that the idea behind triangulation is that "the more sources and types of data we can gather and compare, the surer we can be of the validity of our overall findings and interpretations" (1994, p. 53). (In referring to Denzin's views on this, they remark that they do not wish to associate themselves with his later writing — where he seems to suggest that the "local knowledge" of research subjects should be given priority in the research process — 1994, p. 55.)

Davidson and Layder indicate that triangulation can be implemented in various ways. For instance, it can "include the use of different investigators on the same research project and the use of different theories as a means of understanding and interpreting the data" (1994, p. 54). They are concerned that in more conventional ways of doing research, triangulation "usually means trying out different and fairly narrowly defined explanations of the same data" (1994, p. 54). But they suggest that it can (and sometimes does) "involve the application of a number of more general theories or approaches" (1994, p. 54).

Triangulation for Davidson and Layder is a way of avoiding the danger of a constructivist approach that leads to a relativist approach. They argue that there is no need to take constructivism to the point of suggesting that the only "reality" that can be known is as seen from some subjective viewpoint (whether that of researchers or the research subjects). The fact that there are "different versions of the truth, and that social researchers do not and cannot observe neutrally" is not sufficient ground for losing all hope of building a picture of reality that can be assumed to represent it in some way (1994, p. 55). Researchers can still unite in the project of trying to improve the quality of the filter that filters out "biased and defective research" (1994, p. 59).

The following bullet points provide a rendition of the scientific realist argument positioned in relation to discursively-oriented constructivism.

- Constructivism should not be taken to the relativist conclusion that reality can be known only through some subjective viewpoint. Such a constructivism is inimical to the search to produce a better understanding of (social) reality.
- A discursively-oriented constructivist approach assumes that the quality of people's accountability can be defined in and through social discourse, with no recourse to extra-linguistic reality.
- In scientific realism, *accountability implies that (social) scientists appreciate the importance of allowing their research to pass through the filter of the methodological domain of scientific inquiry so as to contribute to the advancement of human knowledge about (extra-linguistic) reality.*

2.5 INTERPRETIVISM AS A WAY OF DEFINING THE PRACTICE OF SCIENCE

Weber offers a definition of sociology (as the study of the social world) that opposes the positivist tenet of *scientism*. His definition is as follows:

> Sociology ... is a science which attempts the interpretive understanding of social action in order thereby to arrive at a causal explanation of its course and effects. In "action" is included all human behavior when and in so far as the acting individual attaches a subjective meaning to it. (1973, p. 128)

In arguing that this is how sociology should be defined, Weber rejects Comtian (and other positivist) views regarding the similarity of the social and natural sciences. He argues that the "uniformities" or "generalizations" of the kind that positivistically directed social science uncovers, can be understood as sociological ones "only when they can be regarded as manifestations of the understandable subjective meaning of a course of social action" (1973, p. 133). Knowledge of statistically worked out probabilities that "a given observable event ... will be followed or accompanied by another event" are to be treated as "an aid" for sociologists to develop their knowledge of meaningful action (1973, p. 133, and 1949, p. 80). The imputation of meaning to action has a greater chance of being correct (or more adequate) if it can be shown that courses of action held to be meaningful are likely to occur. This lends more plausibility to the interpretation of meaning (1973, p. 132).

In line with his focus on interpretive understanding of action, Weber is also concerned with what he regards as the reification, or assigning of impersonality, to social "collectivities" in scientific discourse about them. He argues that in order to give attention to the interpretation of action in sociological work, any social "collectivity" should be "treated as *solely* the resultants and modes of organizations of the particular acts of individual persons, since these alone can be treated as agents in a course of subjectively understandable action" (1973, p. 135). Collectivities should therefore properly be thought of in terms of the "development of actual or possible social actions of individual persons" (1973, p. 135).

Weber's argument in relation to the positivist tenet of *phenomenalism* is based on his suggestion that in studying the (social) world, the point is not only to "observe human conduct but ... also to understand it" (1949, p. 83). Studying the social world will always be guided by conceptualizations of what are matters of significance for investigators (and will also take place within some historical age). As soon as investigators distinguish "the important from the trivial" they already have a point of view from which they make this distinction. For Weber, "cultural science [always] ... involves 'subjective' presuppositions insofar as it concerns itself only with those components of reality which have some relationship, however indirect, to events to which we attach cultural *significance*" (1949, p. 82). Standpoints concerning what is significant cannot be "derived from the facts themselves" (1949, p. 82). To adopt a phenomenalist position that assumes that this is the case, is, for Weber, to engage in "naive self-deception" (1949, p. 82).

Weber also argues that it is preferable to admit that the point of view adopted by investigators is of "great importance in the construction of the conceptual scheme which will be used in the investigation" (1949, p. 84). Considering the way in which social scientists should go about constructing conceptual schemes, he suggests that the concepts constructed in the social sciences can be generated by the "analytic accentuation of certain elements of reality" (1949, p. 90). Weber calls concepts that are constructed in this way "ideal types". He proposes that:

> An ideal type is formed by the one-sided *accentuation* of one or more points of view and by the synthesis of a great many diffuse, discrete, more or less present and occasionally absent *concrete individual* phenomena, which are arranged ... into a unified *analytical* construct. (1949, p. 90)

Ideal typical concepts are not to be treated as descriptions of reality. They are devices that aid us to examine features of empirical reality by considering the extent to which they match the ideal typical exposition. "It [the ideal type] has the significance of a purely ideal *limiting* concept with which the real situation or action is *compared* and surveyed for the explication of certain of its significant components" (1949, p. 93). The construction of the ideal type requires some imagination: it "cannot be found anywhere empirically in reality" (1949, p. 90). But it must bear some relationship thereto, for it is an *accentuation of certain empirical features* that are more or less present in "concrete individual phenomena" (1949, p. 90). The quality of the ideal-typical constructions offered by social scientists can be judged in terms of "the adequacy of our imagination, oriented and disciplined by reality" (1949, p. 93). Weber suggests that the character of social science as an *empirical science* should be seen in this light — if we are to fairly apply judgments of validity to the conceptualizations generated by investigators in trying to understand social reality.

When Weber states that conceptualizations should be recognized to express the point of view of investigators and to involve their way of accentuating certain features of empirical reality, he does not take this to the point of denying the principle of ethical neutrality or *value freedom*. He concurs with the positivist suggestion that scientists should strive for value freedom in their work. He argues that social scientists should be dedicated to the task of "recognizing facts, even those which may be personally uncomfortable, and to distinguish them from [their] own evaluations" (1949, p. 5). He emphasizes that, as he sees it, whenever scientists introduce their personal judgments, "a full understanding of the facts *ceases*" (1970, p. 146). Scientists should proceed with the understanding that facts and values cannot be conflated. Once they operate with an acknowledgment of the distinction between facts and values, they are more likely to be able to recognize facts that may be "inconvenient" for their opinions (1970, p. 147).

Just because social scientists deal with a subject matter that contains certain values (those of the people being studied), does not mean to say that they themselves need to become involved in making value-judgments. Weber suggests indeed that

> the investigator and teacher should keep unconditionally separate the establishment of empirical facts (including the "value-oriented" conduct of the empirical

individual whom he [sic] is investigating) and *his* own practical evaluations … . (1949, p. 11)

The investigator may, however, become interested in identifying how certain convictions held by people have become "causally very significant" in terms of their influence on social action (1949, p. 13).

The aim of social science, thus, is to develop knowledge of people's "convictions" and of their consequences for social action. In considering the *instrumentality* of such knowledge, Weber sees this as deriving from its ability to examine consequences of adhering to certain values — consequences that might not be immediately apparent to actors in the social world (1949, p. 13). Just as scientists can, in terms of their vocation, recognize facts that may be inconvenient for their party opinions, so they can "accustom" others to the existence of such facts (1970, p. 147). By locating consequences that actors may not have intended or expected, social science is able to offer information that can help people in reviewing their actions in the world.

Weber avers that the utility of social science does not consist in trying to give scientific approval to "widely accepted value-judgments" (1949, p. 13). On the contrary, he suggests that the specific function of science is to "ask questions about these things which convention makes self-evident" (1949, p. 13). Weber thus disassociates his position from being seen as lending support to what are "conventionally" accepted values. The immediate obligation of those involved in social scientific activity is, as he puts it, "to keep a cool head in the face of the ideals prevailing at the time" (1949, p. 47).

2.5.1 Considering Accountabilities in Terms of an Interpretivist Position

This section explores the question of how accountability might be envisaged within an interpretive-oriented approach. The section is organized around an elucidation of Henwood and Pidgeon's view of what it might mean to conduct research in terms of what they call an interpretative self-understanding of the purpose of social science. They concentrate in their discussion on the distinction between interpretative approaches and "'experimental', 'hypothetico-deductive', or 'positivist' ... approaches" (Henwood & Pidgeon, 1993, p. 15). Following the organization of the discussions on accountability in previous sections, their argument (and attendant conception of researcher accountability) is presented below by positioning it in relationship to positivist, non-foundationalist, scientific realist, critical theoretical, anti-foundationalist feminist, and discursively-oriented constructivist argumentation.

2.5.1.1 Positioning Interpretivism in Relation to Positivist Argumentation

Henwood and Pidgeon criticize the (positivist) view that all scientific endeavors should follow the pattern set by the natural scientific approach, which "emphasizes universal laws of cause and effect based on an explanatory framework" (1993, p. 15). In this view of science, they note, reality — whether natural or social — is seen as consisting of phenomena that can be observed and precisely measured, so that relationships between them can then be explored. In practice, this means that "much of the work of natural scientists concerns the methodological minutiae of operationalization and measurement" (1993, p. 15). Concepts used in theoretical schemes or in hypotheses (hypothesizing relationships between phenomena) need to be "observable" and "manipulable" (1993, p. 15). The concepts need to be manipulable in the sense that we can consider how variation in the phenomena to which they refer, might lead to (observable) changes in other phenomena.

Henwood and Pidgeon suggest that this view of the way science is properly practiced, leaves little room for what is sometimes called "softer" ways of addressing the subject matter of the human sciences (1993, p. 24). They concentrate on showing how researchers employing "softer" practices (as opposed to "harder" ones), may be called to account for the quality of their research. They identify seven ways in which such research practice might be judged in terms of the interpretative tradition.

Firstly, in line with Weber's view that conceptualizations must be "disciplined by reality" (Weber, 1949, p. 93), Henwood and Pidgeon suggest that categories used in the process of theorizing must have "closeness" to the data. They should "fit the data well" (1993, p. 24). Both researchers themselves and their peers can "evaluate fit" (that is, whether the categorizations are seen as fitting the data).

Secondly, they suggest that theorists must attempt to make connections between the data and different levels of conceptual abstraction (1993, p. 24). They draw on Glaser and Strauss's work (1967) on "grounded theory" to make this point. Grounded theorizing involves researchers' building up (and accounting for) their conceptual schemes in relation to the data of human experience in a "rich, complex and dense" fashion by integrating "diverse levels of generality" (1993, p. 24).

Thirdly, Henwood and Pidgeon argue that when researchers operate in terms of an interpretative self-understanding of science, they recognize that their decisions about what is significant (to use Weber's terminology, 1949, p. 82), might affect the way they conduct their research and present results. Henwood and Pidgeon note that some authors, such as Lincoln and Guba (1985), have argued that in order to account for the way they have proceeded, researchers need to keep a "log of methodological decisions and accompanying rationales" (Henwood & Pidgeon, 1993, p. 25). This is a way of enabling both themselves and others to revisit the investigation by (re)considering the choices made during the research process.

Related to the point about researchers' keeping a log of decisions made during the research process, Henwood and Pidgeon refer to a requirement for "documentation" (which can be seen as a fourth requirement). They indicate that "keeping a reflexive journal is but one facet of the important process of building up documentation ... . This exercise provides an account of what is done, and why it is done, at all phases in the research process" (1993, p. 25). By leaving what Lincoln and Guba (1985) have called a "paper-trail" through such documentation, Henwood and Pidgeon suggest that researchers open themselves up to an "external audit ... by immediate colleagues and more distant peers" (1993, p. 25).

The fifth way in which researchers can fulfill requirements for accountability is by accounting for the way they draw conclusions from "samples" analyzed. Henwood and

Pidgeon note that according to the interpretative tradition, sampling is governed by the aim of generating emergent theorizing and extending and modifying it through further investigation. What is called "theoretical sampling" by Glaser and Strauss (1967) is organized around the need to consider whether further case material might constitute grounds for modifying or extending some "emergent model" that the investigator is developing. Further cases are thus "selected for their disconfirming potential" (Henwood & Pidgeon, 1993, p. 25). Henwood and Pidgeon emphasize that the aim of such sampling — in accordance with grounded theorizing — is to "systematically explore differences in the expanding data corpus" (1993, p. 26). Strauss and Corbin (acknowledging their indebtedness to Glaser) summarize the argument in this respect as follows:

> All of these instances [of unexpected data] qualify our original questions and statements of relationship [between phenomena]. **They don't necessarily negate** our questions or statements, or disprove them, rather **they add variation and depth of understanding.** (1990, pp. 108-109)

In line with this argument, Henwood and Pidgeon suggest that theorizing in the social sciences can be judged in terms of whether enough theoretical variation has been applied to an understanding of the phenomena.

Henwood and Pidgeon indicate that within the interpretative tradition, another (sixth) means for assessing the quality of research has sometimes been linked to the requirement for it to be "readily recognizable to participants" (1993, p. 26). The idea is that social scientific theories can be judged as adequate only insofar as respondents can consider the interpretations of reality offered within them as acceptable. Henwood and Pidgeon note that this requirement, although frequently cited, is "hotly contested" (1993, p. 26). They propose instead that it should be qualified by the possibility of "exploring the reasons why respondents' and researchers' interpretations may differ" (if this is the case) (1993, p. 26). Such an exploration of differences in interpretation serves as a "further source of data with which to elaborate the developing theory" (1993, p. 26). Seen in this light, researchers do not need to consider that they are accountable in the sense of organizing a match between their own and respondents' interpretations of social reality. They are accountable rather in the sense of accounting for possible differences in interpretation as part of the development of their theorizing. This of course does not preclude their keeping open a dialogue with research participants and other audiences around continuing differences in the interpretation of social realities.

Lastly — as a seventh issue that has to be considered in interpretive-oriented research — Henwood and Pidgeon refer to the extent to which researchers need to account for the "general significance" of their findings. They note that working within the interpretative tradition means that researchers cannot presume to create generalizations about human behavior that transcend specific contexts of social action. However, it is possible to speak of "transferability of findings" — once this is understood as referring to the idea that findings might be applicable "in contexts similar to the context in which they were first derived" (1993, p. 27). Once one understands transferability in this way, then it is incumbent upon researchers to "fully report on the contextual features of a study" (1993, p. 27). This can thus be seen as a seventh requirement for accountable research practice (within the social sciences) as isolated by Henwood and Pidgeon.

The following bullet points provide a rendition of the interpretive argument positioned in relation to positivism.

- In terms of a positivist position, the quality of theorizing derives from the attention paid by inquirers to uncover facts and the relationship between them. Interpretivism suggests that the quality of social theorizing is to be considered in a different way.
- It seems that in positivism, the accountability of researchers is linked to their commitment to observe and precisely measure (natural and social) phenomena, and to find ways of (logically) testing relationships between them.
- In interpretivism, *accountability of social researchers implies exploring "softer" approaches to the study of society — toward the development of conceptually rich, dense theory that is grounded in human experience.*

2.5.1.2 Positioning Interpretivism in Relation to Non-foundationalist Argumentation

As indicated earlier, Henwood and Pidgeon do not specifically separate out the critical rationalist or non-foundationalist argument from positivism in organizing their account of the distinctiveness of interpretive-oriented approaches to research. They align what they call "'experimental', 'hypothetico-deductive' or 'positivist' ... approaches" (1993, p. 15). However, in this section, their reference to the "Popperian strategy" of using falsification, and its difference from "negative case analysis", is highlighted in order to consider in particular their relationship to (Popperian) hypothetico-deductivism as a non-foundationalist position.

Henwood and Pidgeon refer to the fact that in developing "contextually grounded theory", cases regarded as instances of some phenomenon are chosen with a specific purpose (1993, p. 25). The purpose is to seek cases that can be used to "challenge initial assumptions and categories" that are being put forward in the theory (1993, p. 25). Cases that have the potential to act as "negations" of existing theoretical elaborations, are chosen so as to be able to generate "conceptually dense" theory (1993, p. 26).

Henwood and Pidgeon compare the strategy of using negative case analysis with the "Popperian strategy of ingeniously seeking wherever possible to falsify working hypotheses" (1993, p. 25). In terms of the Popperian strategy, hypotheses about the relationships between phenomena can supposedly be tested by deducing what events or cases of empirical occurrence should *not* be observed if the hypothesis is true. If such events or empirical cases can be found in reality, then they serve to disconfirm the initial hypothesis. However, if repeated attempts at falsifying the hypothesis fail, then it is taken as a tentatively corroborated statement about regular connections existing between phenomena.

Henwood and Pidgeon note that negative case analysis functions in a different way in the social sciences. It is not meant to serve the purpose of corroborating some existing theory by indicating "failure to falsify" it (1993, p. 26). Rather, it serves to make a contribution to the variegated character of the conceptualizations of social reality. It helps theorists to build up a sense of the world of social phenomena as created differently in different contexts, depending on the meanings being developed in various contexts of interaction.

Henwood and Pidgeon concur with the Popperian view that logic — whether used in the natural or social sciences — can never be used as a means to create indubitable knowledge. However, they argue that the fallibility of human knowledge as isolated by Popper can be taken further. This is by recognizing that any instance that is regarded as a "falsifying instance, will always be ambiguous in its implications" (1993, p. 29). We can never be sure that what is argued to be a "falsifying instance" in relation to some conjecture, really does disprove the conjecture. There is always the possibility that certain supporting assumptions could be made, which cast new light on the way the empirical evidence is treated.

Henwood and Pidgeon appreciate the Popperian suggestion that one can use "the data" (if chosen carefully) to re-examine claims being made in some emerging theory. But for them, the possibility of social theorists' revising the way conclusions from "the evidence" are developed, has a somewhat different purpose than in the natural sciences: such theorizing should be properly aimed at developing a "dense" understanding of the constitution of social life.

The following bullet points provide a rendition of the interpretive argument positioned in relation to non-foundationalism.

- While non-foundationalism endorses a deductive approach as a way of testing (and possibly falsifying) particular statements about patterns that are hypothesized to exist in social reality, interpretivism treats what is called "negative case analysis" with a different intention.
- In non-foundationalism, the accountability of social (as natural) researchers requires their arguing for the reasonableness of behavior within the social scientific community in terms of the purpose of advancing knowledge of social regularities/patterns.
- In interpretivism, *accountability of social researchers implies their using case material as a means of creating an argument for the development of conceptually rich and dense theory that is specifically sensitive to contexts of social interaction.*

2.5.1.3 Positioning Interpretivism in Relation to Scientific Realist Argumentation

While Layder (1993) — following a scientific realist position — has argued that the grounded theorizing approach does not provide sufficient weight to theoretical analyses at the level of social structure, Henwood and Pidgeon do not concur with this argument. Their position, following Weber, is that analyses undertaken at the level of social structure imply that "structures" can operate above the heads (and meaningful experience) of actors involved in contexts of interaction. They argue that analyses at the level of structure, which purport to explore the properties of posited social entities, tend to "overwrite ... internally structured subjectivities by externally imposed 'objective' systems of meaning" (1993, p. 16). So although scientific realism might give credence in its analysis to systems of meaning that become developed in different social orders,

Henwood and Pidgeon would regard its treatment of meaning as somewhat inappropriate. Henwood and Pidgeon would see scientific realism as not accounting sufficiently for analyses of "the meaning of experience and behavior in context in its full complexity" (1993, p. 16).

In terms of Layder's realist argument (1993), analyses at the level of structure could still be integrated with the contributions made through a more grounded theorizing approach, through accumulating the knowledge gained in each different style of analysis. But Henwood and Pidgeon's view of grounded theorizing suggests that the kind of analyses endorsed by scientific realism, might already preclude proper recognition of the complexity of the meaningful character of social life. Interpretative approaches, as discussed by Henwood and Pidgeon, are based on the (ontological) view that structure has no meaning apart from the meanings given by actors. Structure cannot be separated out from the *experience of structure* by actors. This follows Weber's view (1973, p. 135) that it is only insofar as actors orient to collective entities *as if* they exist, that they come to exert causal force in society. If actors orient to structures as if they have properties apart from the assignation of meaning, then they might indeed take on the character suggested within a scientific realist approach. However, if they are seen differently (by actors) then the experience of structure will be different. What is crucial, in any case, is that we recognize that entities in society cannot be considered apart from the human experience thereof.

The following bullet points provide a rendition of the interpretive argument positioned in relation to scientific realism.

- The scientific realist view of (social) structures does not accord sufficient attention to the human experience of the world.
- In scientific realism, accountability of social researchers requires their reasonable (or argued-for reasonable) use of retroductive logic to build up informed theoretical accounts of the properties of social entities, including social structures.
- In interpretivism, *accountability of social researchers implies their developing an orientation toward theorizing that is specifically grounded in human meaning-making.*

2.5.1.4 Positioning Interpretivism in Relation to Critical Theoretical Argumentation

Henwood and Pidgeon do not refer to the critical theoretical argument in their discussion. However, their differences from this position can be extrapolated from their account of the purpose of social theorizing.

Henwood and Pidgeon defend their account of the requirements for "good practice" in (interpretive) theorizing as follows:

> Theory that is represented at diverse levels of abstraction [allowing for the specificity of concrete instances to add theoretical density], but which nevertheless fits the data well [and is grounded therein], should be challenging, stimulating, and yet highly plausible in the sense of clearly reflecting substantive aspects of the problem domain. (1993, p. 27)

Here Henwood and Pidgeon indicate that social theorizing should be challenging and stimulating: through the way it presents its interpretations, it can pose new questions for consideration in society. While being challenging and stimulating, though, it should not be removed from "substantive aspects of the problem domain": it needs to offer a plausible way of viewing the issues being studied. Henwood and Pidgeon refer in this respect to Marshall's suggestion that productive research should lead to "the 'aha' experience of discovery" (Marshall, 1985, as cited in Henwood & Pidgeon, 1993, p. 27). They also suggest that accounts can be rendered plausible through the fact that they "get close to the bone" (Rennie et al., 1988, as cited in Henwood & Pidgeon, 1993, p. 27).

In order to gain some understanding of the social realities being investigated, Henwood and Pidgeon suggest that researchers should attempt to construct within the research process "a joint reality with participants" (1993, p. 26). This is a way of validating interpretations made. In cases where different interpretations of social realities are proffered by researchers and actors, some "negotiation" around these differences should be attempted (1993, p. 26). But Henwood and Pidgeon remark that democratic ways of organizing the "negotiations" of definitions of reality are likely to be rendered difficult due to the "relationships of power which exist between and around the researcher and researched" (1993, p. 26). Having said this, though, they do not indicate whether, or how, moves toward more democratic relationships could be made. Whether or not they see it as important for researchers to seek ways of setting up more democratic relationships, the point of doing this, for them, would not be — as in critical theory — to *cultivate the value of democracy* in society as part of the research remit. Rather, the point would be to make *a better contribution to knowledge of reality* by rendering more plausible the interpretations that are being developed.

The following bullet points provide a rendition of the interpretive argument positioned in relation to critical theory.

- While for critical theory the aim of social inquiry is to make a contribution to the instantiation of democratic social relationships, this is not the prime purpose of social science, according to interpretivism.
- In critical theory, accountability of social inquirers implies a commitment to exploring ways of communicating that allow people to cultivate as a social value the democratic potential of human reason.
- According to interpretivism, *the remit of social researchers (for which they are held accountable) is to develop accounts of social reality that can be seen to be related to the human experience of the world.*

2.5.1.5 Positioning Interpretivism in Relation to Anti-foundationalist Feminist Argumentation

Henwood and Pidgeon note that they appreciate the feminist suggestion that social research should not pretend to be "objective" in the sense of being able to offer an impartial account of social realities (1993, pp. 24-25). They follow Weber's view that social research involves organizing research from a point of view and that this already means that it is somewhat "subjective". However, also following Weber, they do not take this argument to the point of denying the possibility of researchers' developing what are considered to be plausible accounts of social realities. Their interpretations are to be aimed at offering insights into the domains of society that are being investigated. Henwood and Pidgeon thus point to the need for researchers to be concerned about reflecting realities (including the realities of social meaning-making) that exist independently of the research process.

Their argument differs from anti-foundationalist feminism insofar as they do not regard it as a matter of ethical concern for researchers to reconsider the authority that might be accorded to the accounts of social life generated by them. They contend that the proper goal of research is to generate insights that "clearly reflect substantive aspects of the problem domain" (1993, p. 27).

Henwood and Pidgeon admit that within interpretivism there is "hot contestation" around the way in which researchers' ability to reflect realities is accounted for (1993, p. 26). One way of grounding the validity of interpretations made, is to try to organize some "fit" with participants' interpretations of social reality. But respondents' accounts also should not be "taken at face value" (1993, p. 26). Researchers might need to be cautious of accounts offered by participants. For instance, "people may not always be fully aware of reasons for their actions, and accounts may be offered to perform a variety of ... functions (for example, allocating blame to others, [or] warranting particular claims to truth)" (1993, p. 26). Henwood and Pidgeon suggest that normally some effort to "negotiate" definitions of social reality with participants should be attempted, while keeping in mind that these negotiations themselves take place within "power relationships" (1993, p. 26).

However, what is important for Henwood and Pidgeon is that the research process is appraised in terms of its purpose to theorize around the social realities being investigated. The cogency of the theorizing is aided if researchers can organize some dialogue between their own and respondents' views concerning the social realities of which they both are part.

The following bullet points provide a rendition of the interpretive argument positioned in relation to anti-foundationalist feminism.

- According to interpretivism, the aim of social research is not first and foremost (as in anti-foundationalist feminism) to try to instantiate a type of politics via the research process.
- Difficulties in setting up cooperative relationships with participants in the research process are relevant to the research endeavor because they affect the quality of information that can help researchers in constructing their accounts of social reality.
- While in anti-foundationalist feminism accountability of researchers is linked to an ethical requirement to revisit the human relationships created in the research process, in interpretivism *accountability implies that (social) scientists recognize the requirement to try to validate accounts by generating some kind of dialogue around different interpretations of social reality.*

2.5.1.6 Positioning Interpretivism in Relation to Discursively-oriented Constructivist Argumentation

Henwood and Pidgeon argue that researchers working within the interpretative tradition "acknowledge the ways in which research *inevitably* shapes and constitutes the object of inquiry" (1993, p. 24). In stating this, Henwood and Pidgeon lend some support to the constructivist idea that the reality seen by researchers is a reality shaped through the way in which they decide to conduct their investigation. That is, they accept that researchers' manner of interacting with those being researched, and their decisions concerning how to proceed within the research process, have an effect on the results generated — and in this sense "constitute the object of inquiry" (1993, p. 24). But they do not believe that we need reach the conclusion from this that knowledge of social reality is rendered impossible. As long as researchers are aware that knowledge is always from a point of view, and as long as this is well recorded, researchers can proceed to develop their insights into "substantive aspects of the problem domain" (1993, p. 27).

Henwood and Pidgeon consider the relevance of Kuhn's argument concerning the way in which discoveries about reality come to be defined as such. They refer to Kuhn's idea concerning the way in which the scientific community can function to define acceptable scientific practice and attendant "discoveries" arising from such practice (1993, p. 19). Their reference to Kuhn's argument is set in the context of their recognition that what they consider to be requirements for good social scientific research, are "new" in relation to the positivist or hypothetico-deductivist self-understandings of the purpose of science. For "new" ideas about science (and scientific practice) to be constituted as legitimate, Henwood and Pidgeon suggest that a certain openness on the part of members of the scientific community to the "insertion" of new discourses is required (1993, p. 19). Whether or not the assignment of legitimacy implies a "full scale paradigm shift", it is important for people to be able to recognize when "existing theory is incomplete, inappropriate, or entirely absent" (to tackle the issues being studied) (1993, p. 20).

They agree with Kuhn that when "normal science" is operative, existing models of "good practice" are consistently endorsed. For this situation to alter, new discourses need to make their way into becoming legitimized. Applied to an understanding of the history of the social sciences, they argue that the "preset" views on theorizing offered within, say, positivist or hypothetico-deductivist positions, are ill-equipped to deal with the complexity of human meaning-making that characterizes social existence. The theorizing offered within these traditions is "incomplete" in relation to this complexity, and should be recognized as such (1993, p. 20).

Henwood and Pidgeon do not take Kuhn's views on paradigms to the point of suggesting that we have no means of assessing the relative adequacy of the different discourses. They believe that it can be shown indeed — through cogent argumentation — that interpretive-oriented thinking embraces ideas that help us to understand the meaningful character of social reality. And they believe that social scientific researchers should argue for the plausibility of their accounts of social reality in terms of their hoped-for capacity to reflect substantive aspects of the problem domain.

Henwood and Pidgeon also take into consideration Feyerabend's claim that it is impossible to assess theorizing in terms of its relationship to "the data" — because "legitimate data are necessarily defined through theory" (1993, p. 22). In response to Feyerabend's suggestion, they contend that because "grounded theory" is built up as an emergent process, there is a "constant interplay between data and conceptualization" (1993, p. 22). They argue that this mode of theorizing requires that researchers avoid "premature closure or fixing of theory whenever new insights might arise" (1993, p. 22). As the analysis proceeds, the researcher should be enabled to develop lenses (for ordering the data) that are "more sharply focused" than when the investigation begun (1993, p. 22).

The following bullet points provide a rendition of the interpretive argument positioned in relation to discursively-oriented constructivism.

- Theorizing about the social world has traditionally been dominated in terms of positivist or hypothetico-deductivist discourses as ways of understanding (and practicing) social science. It is necessary to render legitimate in the social scientific community interpretivist ways of thinking about science.
- Although accounts that are developed by (social) researchers can in some sense be seen as constructions based on a way of organizing investigations, they can and should still refer to aspects of the problem domain being investigated.
- In a discursively-oriented constructivist approach, accountability of researchers is linked to their finding a way of working with others' constructions of reality. In terms of an interpretivist position, *accountability of researchers implies that they submit to the requirements that are needed for the creation of compelling and plausible accounts of the realities of social life under investigation.*

2.6 CRITICAL THEORY AS A WAY OF DEFINING THE PRACTICE OF SCIENCE

"Critical theory" is associated with the way in which the so-called Frankfurt School of critical theorists argued for the development of human reason in society. Held (1980, p. 29) notes that the Frankfurt School was comprised of a group of authors working under the auspices of the Institute of Social Research, which was founded in Frankfurt (Germany) in 1923. He indicates that the thought of the Frankfurt School "has been a major source of stimulus to the man who has ... become the leading spokesman for a new generation of critical theorists — Jürgen Habermas" (Held, 1980, p. 249).

Habermas sympathizes with many of the concerns of other members of the School in exploring the operation of human reason in society. But he distances himself from the "Marxist philosophy of history", which he sees as underlying most of their work — especially the work undertaken before the 1940s (Habermas, 1982, pp. 231-232). In terms of this philosophy, the critical theoretical hope of instilling a critical consciousness in society is regarded as being "supported" by the historical process. Habermas points out that by 1941 some of the members of the School (and later other ones too) began to recognize that the motor of critical consciousness lies rather in the potentialities possessed by humans themselves. Habermas states that his own efforts (at developing critical theory) "can be understood in connection with the undertaking that critical theory broke off at the start of the 1940s" (1982, p. 232).

Habermas argues that the task of critical theorizing is to develop a theory guided by a specific intention. This intention is that of "recovering the potential for reason encapsulated in the very forms of social reproduction" (1982, p. 221). He indicates that while in his earlier writings he located this potential through a theory of human knowing, he now prefers to locate it through a theory of human language (1982, p. 233). The former way of locating was a "roundabout way" of approaching the issues. The location of the potential for reason through an analysis of the structure of human speech is a more direct route (1982, p. 233). Delanty expresses the relationship between Habermas's earlier and later work as follows:

> In *Knowledge and Human Interests* [1972] Habermas demonstrated the basis of social science in the cognitive interest in emancipation but did not demonstrate how this cognitive interest itself was constituted. ... . Habermas's mature social theory of communication demonstrates how emancipatory interests are constituted in the critique of "distorted communication". (1997, p. 87)

Habermas's argument — briefly expressed — is that the idea of communicative rationality is contained in the structure of human speech. In the case of genuine discourse (which he opposes to "distorted communication"), the interacting parties are oriented to developing a consensual understanding of the issues under consideration. In the process of developing such understanding, claims to (propositional) truth and (normative) rightness, as well as to sincerity and comprehensibility of statements made, are opened to validity checking (1982, p. 235). Habermas argues that a regulative principle of consensus is operative in discursive exchange — whether or not the people involved believe an actual consensus will be possible through the course of discussion: in proposing "validity claims" redeemable through discourse, one has always to "keep open the possibility of consensus" (Habermas, 1993, p. 94). Habermas's argument is that in cases of disagreement, the people engaged in the discussion can operate by upholding "the idea of reasonable disagreement". This means that they can "leave contentious validity claims undecided" — while still recognizing that the possibility of consensus is guiding the discussion (1993, p. 94).

His views on the way in which a critical social science might make a contribution in society are elucidated below with reference to his discussions on the philosophy of social science and his development of the theory of communicative action. His position in relation to the tenets of positivism is addressed first.

Habermas approaches the positivist tenet of *scientism* by focusing on the social consequences of adhering to it. His objection to the *scientistic* argument is that the positing of a similarity between natural and social scientific investigation, itself has certain effects in society. The suggestion that natural and social life can be studied in the same way, results in an engineering approach becoming endorsed for the treatment of social problems. Democracy in turn becomes reduced to the operation of developing an efficiently organized system aimed at correcting "dysfunctions" through the application of an administrative machinery (1979, p. 169).[10]

Habermas is concerned with these developments, for he believes that democratization is sought in society "not in order to increase the system's efficiency but in order to change the structures of power" (1979, p. 145). He suggests that the tension between systems efficiency and real democratization of goal making in society is one that needs to be highlighted. As it stands, the tension between the pressures for efficiency and the possibility of (discursive) democracy becomes obscured by a social science that models itself on the natural sciences. Due to the operation of such a social science in society, the capacities of the human species to solve *technical problems* is given precedence over, and colonizes, the communicative infrastructure of society. In order for social science to contribute to the revitalization of the communicative infrastructure of everyday life, a form of critical inquiry that does not model itself on the natural sciences is essential (1981, p. 7).

As far as the positivist tenet of *phenomenalism* is concerned, Habermas argues that the manner in which facts and their connections are treated within the sciences, is indeed a particular way of apprehending them. For instance, he argues that when scientists apparently discover in reality "facts structured in a law-like manner", they are apprehending this in the light of a certain human interest — the interest in prediction and control (1972, p. 305). It is this interest that makes "the phenomena" appear as they do.

In line with his revision of the tenet of phenomenalism, Habermas reconsiders the tenet of *empiricism*. He notes that "empiricists" (who support the tenet of empiricism) create concepts and theories that are meant to be "mimetic adaptations to reality as given" (1996, p. 9). Their aim is to ground their theorizing in what appears to be empirically "given". But by doing so, they already are defining reality in terms of certain (unacknowledged) standards. Habermas argues in this regard that facts and standards are dependent on each other for their acceptance. When an acceptance of "facts" is being considered, some standard is invoked in deciding the truth of the (factual) statement. Conversely, when an acceptance of a standard is being considered, the argument often includes appeals to some empirical material (in the form of "factual" propositions). When the truth of a proposition is under consideration, normally the choice of underpinning standards is not subjected to discussion. Likewise, when the choice of standards is under consideration, the supporting empirical material is not normally questioned (1978, p. 214). However, it is always possible in principle to re-open discussion around underpinning standards or underpinning empirical material in the course of argumentation.

Habermas's suggestion is that the problem with (conventional) social science as it currently operates in contemporary society, is that it *fails to raise for discussion the*

---

[10]Habermas's argument is posed in relation to societies where, in his view, the ideology of science and technology has come to dominate social thinking. To the extent that this ideology is considered as being prevalent in so-called less developed countries, the argument applies here too. (See Romm, 2001, for an account hereof.)

standards in terms of which it operates. The standards are obscured behind the empiricist presupposition that they have no part to play in the vision of reality that is being developed. An example will serve to illustrate Habermas's point. In outlining the political theory of Elster, Habermas expresses a concern that he still appears to "cling to empiricist premises" (1996, p. 338). Indicating the need to challenge these premises, Habermas argues that a political theory working in these terms already implies a certain standard for the operation of politics. It implies that the task of politics is to "eliminate inefficient and uneconomical regulations" (1996, p. 337). Habermas questions this view of politics, arguing that the concept of democracy implies public discussion around the development of social goals and the direction of social change. But the possibilities for such a discussion become occluded from vision unless the "realities" of political life are apprehended via a different, less technically-oriented, normative standard.

According to Habermas, an empiricist consciousness that screens out normative questions as it approaches the study of society is unable to cross the divide between facts and values (1996, pp. 3-4). He notes that when he uses the concept of "communicative reason" in connection with his reconstructive social theory, it is meant to serve the function of offering

> a guide for reconstructing the network of discourses that, aimed at forming opinions and preparing decisions, provides the matrix from which democratic authority emerges. From this perspective, the forms of communication that confer legitimacy on political will-formation, legislation, and the administration of justice appear as part of a more encompassing process in which the lifeworlds of modern societies are rationalized under the pressure of systemic imperatives. At the same time, such a reconstruction would provide a critical standard, against which actual practices — the opaque and perplexing reality of the constitutional state — could be evaluated. (1996, p. 5)

Habermas suggests that the pressure for systems efficiency poses a threat to the operation of communicative rationality — which his reconstructive theory posits as being the source through which democratic authority emerges (or should emerge). The theory highlights — from the perspective of what Habermas calls an "emancipatory interest" — the potential for communicative power within the social fabric. (See also Delanty, 1997, p. 87.) *The theory presents openly its interest (value commitment) to extend discursive democracy in society.* It does not strive to generate value free accounts of social realities.

Habermas's vision of the way in which critical social science might aid public discussion about the goal direction of the historical process, also entails a rejection of the (positivist) tenet of *instrumental knowledge*. The tenet of instrumental knowledge is based on the idea that science can offer knowledge of possible means to achieve goals that have been defined by citizens. But Habermas argues that working on the basis of this vision of scientific reason, already implies a submission to the ideology of science and technology (1981). Through the power of this ideology, a technical agenda comes to frame the parameters of any social discussion. Communicative reason becomes restricted accordingly.

Habermas's indication of how instrumental reason is endorsed within the positivist tenet of *instrumental knowledge*, is thus directed at showing how it can operate to restrict and hamper the realm of discursive communication in society.

2.6.1 Considering Accountabilities in Terms of a Critical Theoretical Position

Habermas's suggestions concerning ways in which critical theorists can be called to account for their theory construction are explored in detail in this section. In order to provide further commentary on the implications of his position in regard to the accountabilities of critical theorists, Delanty's (1997) reference to Habermas's argument is also mentioned. The section is organized by detailing Habermas's critical theoretical view in relation to the alternative stances of positivism, critical rationalism (non-foundationalism), scientific realism, interpretivism, anti-foundationalist feminism, and discursively-oriented constructivism.

2.6.1.1 Positioning Critical Theory in Relation to Positivist Argumentation

In his *Knowledge and Human Interests* (1972) Habermas undertakes a critique of the arguments of what he calls "early positivism" (using as examples Comte's texts as well as others). He notes that the positivist position rests on the claim that scientific propositions can be justified as long as they are reality-bound and not merely speculative (1972, p. 74). But he contends that positivism has no way of justifying its view of what reality consists of, except by resorting to the idea that the empirical sciences can allow us access to this reality. Reality is defined as constituted of that which empirical science allows us to discover. He states the point as follows:

> Everything that can become the object of rigorous science counts as a fact. Therefore the delimitation of the object domain of science leads back to the question of how science itself is to be defined. At the only level positivism allows, science can be defined only by the methodological rules according to which it proceeds. (1972, p. 74)

According to Habermas, the positivist view that science allows us to get to grips with reality, involves a circular argument — because reality is already pre-defined as that to which science offers access. As indicated in Section 2.6 above, Habermas is particularly concerned that the positivist presupposition that scientists can make "discoveries" about facts and their regular connections in society, affects social life in ways that positivism fails to take into consideration.

Habermas argues that in the case of the social sciences, critical inquiry expresses its difference from a positivist self-understanding of science by acknowledging that the "truth" of its statements is "linked in the last analysis to the intention of the good and true life" (1972, p. 317). By acknowledging the way in which its statements about "reality" are bound up with questions of "the good life", it introduces a different way of accounting for its theorizing about (and in) society. The quality of theorizing is derived partly from its contribution to extending options for conceiving "the good life" in society. This means that the enterprise of science cannot be accounted for separately from a discussion of its relationship to society. Delanty (1997) argues that Habermas has still not fully explored what might be implied in establishing an appropriate relationship. But he suggests that O'Neill (1995) provides a good starting point for developing our thinking about this. As Delanty notes:

> O'Neill argues that democratic legitimation is formed in the constitution of a communication community in which scientists and politicians are open to responses from the public. Science thus has a crucial role to play in the articulation of social goals. (1997, p. 143)

In order that science can play a role in the expression of social goals, Delanty (drawing on, while extending, Habermas) suggests that scientists need to consider how their manner of doing science relates to definitions of collective problems in society. As Delanty expresses it, "social science is shaped in the definition of problems" (1997, p. 140). According to Delanty, however, the positivist self-understanding of science allows scientists to "hide behind a protective veil of methodology" in the hope to advance theorizing free of a concern with social goals (1997, p. 139). But in doing so, they become unaccountable for the way in which their scientific practice already is implicated in the definition and addressal of what are taken to be "problems".

The following bullet points provide a rendition of the critical theoretical argument positioned in relation to positivism.

- The positivist belief that science is geared to uncovering facts and their connections in reality, leaves unaccounted for the manner in which science relates to the definition and addressal of social "problems" in society.
- Scientists need to develop a relationship with the public in which the articulation of social goals can become part of the remit of doing science.
- While in positivism accountability of scientists is defined through their adherence to a form of theorizing firmly rooted in empirical reality, for critical theory *accountability in critical social science implies that theorists consciously incorporate in their inquiries the normative intent to contribute to the revitalization of public discourse in social life.*

2.6.1.2 Positioning Critical Theory in Relation to Non-foundationalist Argumentation

Habermas (1976a,b) remarks that Popper occupies a peculiar position in relation to positivism. He indicates that Popper rightly points out that statements about facts — as all statements — are not indubitable and depend for their acceptance on decisions made within the scientific community (1976a, p. 151). We therefore cannot consider that there is any firm "basis" (or foundation) on which science can be said to rest. (This is why Hammersley and Gomm apply the label of non-foundationalism to characterize this type of argument.)

But Habermas avers that Popper's manner of treating the basis problem is out of step with his insistence that knowledge still consists of a correspondence between scientific propositions and realities argued to exist outside of the knowing process. Habermas comments that Popper maintains a "deep-seated positivistic prejudice" in this respect (1976b, p. 203). Habermas proposes that we should be able to incorporate the merits of Popper's critique of (positivist) empiricism without resorting to an untenable correspondence theory of truth (1976b, p. 205).

Habermas approaches Popper's argument by taking to a different conclusion his claim that the acceptance of statements within the scientific community depends on the community's decisions as to whether to regard them as "true". Habermas believes that Popper's concession that the statements of science are rooted in intersubjective processes within the scientific community, creates an opening for us to reconsider the purpose of "scientific" knowing endeavors (especially as appertaining in the social sciences). Habermas wishes to create this opening in the light of his concern that the very orientation that scientists adopt in their inquiries, can all too easily prefavor an engineering outlook to ways of treating what become defined as problems in society.

Habermas charges Popper with the same failure as positivism in his refusal to acknowledge the interest directing the operation of inquiries geared to uncovering regularities in the world (1976b, p. 199). When science is defined as aimed at discovering such regularities, the information that it advances is conducive to being used for instrumental purposes. Habermas proposes that it is these purposes that need to be subjected to critical reflection. He thus calls for new ways of organizing accountability within the practice of science, to make provision for appropriate discourse around the way in which scientific inquiries might tap into "the public identification and definition of collective problems" (Delanty, 1997, p. 140).

The following bullet points provide a rendition of the critical theoretical argument positioned in relation to non-foundationalism.

- The critical rationalist (non-foundationalist) belief that there is no firm basis on which science can rest, should be freed of positivist prejudices regarding the need for scientists to discover facts and their connections in reality.
- The standards governing social inquiries should be made explicit and subjected to discursive assessment so that they do not operate unwittingly to serve a technical agenda in the addressal of social problems.
- In non-foundationalism, accountability of scientists implies that they subject to mutual criticism within the scientific community all statements being developed, so science can advance progressively closer to the truth. According to critical theory, *accountability implies that scientists recognize that they are already involved (through the way they conduct their inquiries) in defining ways of treating social "problems" in society. This needs to be accounted for within a critical social science.*

2.6.1.3 Positioning Critical Theory in Relation to Scientific Realist Argumentation

Habermas's views on critical theorizing can be juxtaposed against certain interpretations of Marxism, including interpretations offered within scientific realism. Habermas argues that his own critical theory offers a specific way of seeing the value of the Marxist approach to the study of society. He opposes this to a Marxism that is interpreted as a science modeling itself in some way on the natural sciences. Such a Marxism, he believes, is unable to link its accomplishments to the extension of (discursive) democracy in society (1996, p. 46).

Habermas indicates that what he draws from Marx's analysis of society is the insight that "proletarian forms of life [involve] the distortion of a communicative form of life ... [they involve] an abuse of a universal interest reaching beyond the particular" (1982, p. 221). Habermas suggests that Marx's analysis of the domination of the proletariat in (capitalist) society points at the same time to the idea of a "universal" form of life reaching beyond particular interests. By expanding on Marx's insights in this regard, Habermas develops his own critical theory guided by a specific intention. The intention is that of "recovering a potential for [discursive] reason" in the communicative fabric of society (1982, p. 221). Habermas insists that when undertaking theorizing with this intention, the task is clearly not solely a theoretical one, for it incorporates a practical agenda.

As mentioned in Section 2.4.1.4, Keat (1981) argues — from a realist perspective — that Habermas's specification that knowing itself must be tied to practical "results" in the social arena, is not workable. According to Keat, scientific knowing can and must be judged for its validity without introducing such considerations. However, in opposition hereto, Habermas argues that the realist attitude to social theorizing, already poses as a threat to the communicative infrastructure of everyday life in society. Delanty indicates in this regard that the critical theoretical fear is that a Marxist realist approach becomes immune to considering the possible effects of its very manner of conducting and presenting its inquiries within society. The link between "science" and "society" is not communicatively accounted for through a discursive process. As he notes: "The problem of [realist] Marxism from the perspective of a conception of social science as discursive practice is that its cognitive system derives from a non-communicative understanding of knowledge" (1997, p. 142).

According to a scientific realist position, argument and the drawing of conclusions by scientists, should be informed by evidence whose import may be debated in the scientific community. But the critical theoretical concern is that the way in which "scientific" conclusions are developed, already implies a specific treatment of the relation between science and society. In terms of this relationship, people in society may still be encouraged to defer to scientifically developed visions of the operation of social (structural) forces. While scientific realism suggests that its acknowledgment of social entities does not amount to reifying them (as things that need to be reckoned with by actors in defining courses of action), critical theory is still wary of the problem of reification (Habermas, 1996, p. 46). Within the "realism of the Marxist model", actors' self-understanding of their lifeworld is necessarily given less precedence than scientifically authorized visions regarding the character of social reality (Habermas,

1996, p. 47). A more communicative relationship between science and society — where science is accounted for in terms of its manner of relating to everyday discourses — is thus called for.

Critical theory also presents an alternative view of "ideology" to that provided within realist-oriented scientific discourse. According to Keat and Urry, as noted in Section 2.4, Marx's scientific approach is aimed at developing "correct, undistorted accounts of different social formations" (1982, p. 206). These can be opposed to the ideological conceptions that pervade society. Scientists are charged with the task of developing more accurate accounts than the ideologies pervading everyday life. Habermas defines ideology differently. He defines ideological "falsehood" by drawing on the (critical theoretical) idea that it can be regarded as expressing the *suppression of discursive encounter in society*. Falsehood expresses the failure of reasoning to be validated through processes of discursive debate within society. (See also Romm, 1991, pp. 140-141, and 1996a, pp. 209-210.)

Habermas suggests that the fundamental uncertainty that characterizes the project of gathering knowledge, arises because of the unstable character of discursive encounter. The instability of scientific (as all) accounts of reality should be recognized as springing from the way in which reasons are used in discursive exchange (1996, p. 35). Reasons are constantly exposed to "being invalidated by better reasons and context-altering learning processes" (1996, p. 36). As he puts it elsewhere, "negotiated descriptions of situations, and agreements based on the intersubjective recognition of criticizable validity-claims, are diffuse, fleeting, occasional and fragile" (1982, p. 235). The grounding of assertions in (fragile) intersubjectivity is what lends them their "fundamentally hypothetical character" (1982, p. 223). A salutary byproduct of this (in terms of critical theorizing) is that "it shifts decisions and responsibilities unambiguously to the side of those who have to bear the risk of the consequences of their action" (1982, p. 223). By accounting for their knowing endeavors via a "fallibilistic consciousness", social inquirers can make (more) room for people (including themselves) to make choices about how to see, as well as how to act, at points in time.

The following bullet points provide a rendition of the critical theoretical argument positioned in relation to scientific realism.

- Because scientific realism defines the knowing enterprise as geared toward the representation of external realities, the potential of social scientific inquiry to contribute to the development of the communicative fabric of social life is hampered.
- The fact that the knowing enterprise is characterized by uncertainty, should not lead people to turn to a realist-oriented science to minimize this uncertainty. Instead, the uncertainty that springs from knowing being defined as discursive encounter, can be dealt with by people acknowledging that choices have to be made about ways of seeing and acting.
- While in scientific realism accountability of scientists can be assessed only by upholding a logical separation of "truth" from "practical results" in society, for

critical theory *accountability of social scientists is linked to their efforts to cultivate in practice communicative ways of "knowing".*

2.6.1.4 Positioning Critical Theory in Relation to Interpretivist Argumentation

Habermas (1972) considers the knowing endeavors of those undertaking cultural scientific inquiry into social meaning-making. He challenges, inter alia, Dilthey's quest to create "a virtual simultaneity of the interpreter with the object" (that is, the object of study) (1972, p. 189). He argues that this quest obscures the knowledge-constitutive interest underlying the inquiries. He suggests that the interest that guides such inquiries is indeed a "practical" one of achieving intersubjective communication (1972, p. 195).

Insofar as those involved in cultural scientific studies fail to take cognizance of this interest guiding their inquiries, Habermas suggests that their work might have conservative, rather than emancipatory, implications in social life. It can have the effect of "defending sterilized [cultural] knowledge against the reflected appropriation of active traditions and [of] locking up history in a museum" (1972, p. 316). Habermas argues that if the practical interest can be recognized and incorporated consciously in cultural scientific study, then it can be used critically to locate the extent to which the possibility for genuine intersubjectivity in society is currently suppressed by distorted communication. This involves a critical examination of society that includes a normative intention — the intention to restore the cultural impoverishment of the lifeworld.

In considering the Weberian interpretivist position, Habermas sides with Weber's analysis of the way in which the purposive-rational action of the modern state and economy has come to overpower meaning-making in society. Habermas argues that Weber rightfully recognizes that this represents both a "loss of meaning" and a "loss of freedom" in social existence (1984, p. 243). He believes that Weber's recognition of this, can become incorporated in a critical social theory concerning the tension between purposive-rational action (geared toward systems efficiency) and communicative rationality (where meaning and freedom can potentially be revitalized). Such a critical theory allows us to appreciate, from the standpoint of a normative standard, the growing gap between the lifeworld (as linguistified forms of action-co-ordination) and the expert spheres of the (increasingly) autonomous systems of the state and economy. Habermas's theory of communicative rationality is posed as a way of conceiving this gap that allows people to reflect on both the achievements and the failings of reason as it is currently operating in society.

Habermas argues that the advantage of engaging in some kind of interpretive understanding of social reality within a critical social science is that it draws attention to the possibility of communication serving the purpose of "mutual understanding". This can be distinguished from "mutual influencing" via strategic action — which lacks the intention to validate claims with reference to discursive exchange (1982, p. 234). Participants' experience of the importance of justification as being part of social life becomes evidence for Habermas of what he calls the "medium" of the "lifeworld shared by the participants" (1982, p. 234). Habermas suggests that "the everyday appeal" to justification of validity claims points to the possibility of a kind of communication that is discursively oriented. This is not to say that discourse (as he defines it) is a probable form of communication. He admits that it is indeed improbable. As he notes: "Discourses are islands in the sea of communication; the everyday appeal to validity-claims implicitly points, however, to their possibility" (1982, p. 235).

Habermas believes that insofar as interpretive social inquiry is undertaken with normative intent, it can offer a way of appreciating the potential for communicative reason that is ever-present in society. He indicates that critical social inquiry must therefore incorporate some openness toward "interpretive explication and conceptual analysis" (1996, p. 6). In addition, it should be open to admit "description and empirical explanation" (as long as this is not cut off from a concern with participants' experiences). It should furthermore include the perspective of different (social) roles: "judge, politician, legislator, client and citizen". In this way the investigations can "stretch over a wide field" (1996, p. 7); and they can be geared to generating a discursive relationship with society. A commitment to the development of a vibrant public sphere built on the possibility of revitalizing communicative forms of social life, can be expressed through the theory.

The following bullet points provide a rendition of the critical theoretical argument positioned in relation to interpretivism.

- According to critical theory, interpretive-oriented social inquiry should not be based on the hope to put aside normative intentions in the act of knowing.
- Interpretive inquiry can contribute to a critical social science through its investigations into the lifeworld (as linguistic forms of action co-ordination). It can highlight the potential for communicative reason written into the linguistic structures of everyday life.
- While in interpretivism accountability of social researchers implies their developing accounts of social life rooted in an understanding of people's meaning-making, in critical theory *accountability implies developing a critical social science with the normative intent of highlighting processes whereby the lifeworld of communicative action might be activated in the face of alternative pressures in society.*

2.6.1.5 Positioning Critical Theory in Relation to Anti-foundationalist Feminist Argumentation

Habermas indicates that he appreciates certain feminist ways of challenging the manner in which the "welfare capitalist society" treats social rights. He agrees with the (feminist) suggestion that in the context of considering the notion of equality, "rights" should not be seen as "collective goods that one consumes in common" (1996, p. 419). He cites favorably Young's feminist argument that "rights are not fruitfully conceived as possessions. Rights are relationships, not things ... . Rights refer to *doing* more than *having*" (Young, 1990a, as cited in Habermas, 1996, p. 419). When rights are defined in terms of social relationships, then justice in turn does not refer "only to distribution, but also to the institutional conditions necessary for the development and exercise of individual capacities and collective communication and cooperation" (Young, 1990a, as cited in Habermas, 1996, p. 419).

Feminist argumentation is geared to criticizing the reduction of both the concept and practice of "justice" within the welfare state — where the question of justice is reduced to the issue of how goods can be better distributed. In terms of this way of conceiving and practicing justice, "increasingly the activities of everyday work and life come under rationalized bureaucratic control, subjecting people to the discipline of authorities and experts in many areas of social life" (Young, 1990a, as cited in Habermas, 1996, p. 420). Habermas, citing again Young, suggests that the processes whereby people are inhibited or prevented "from participation in determining their actions or the conditions of their actions" need to be explored (Young, 1990a, as cited in Habermas, 1996, p. 420).

Aligning himself further with what he sees as feminist argumentation, Habermas is concerned about the stereotypical constructions of gender identity that are perpetuated in legislation supposedly promoting the equal status of women. Here he cites Rhode's complaint that:

> At the most basic level, traditional approaches have failed to generate coherent or convincing *definitions of difference*. All too often, modern equal-protection law has treated as inherent and essential differences that are cultural and contingent. (Rhode, 1989, as cited in Habermas, 1996, p. 423)

Habermas gives the example of the manner in which the question of whether women should serve in the military is treated. He notes that in considering this question, as with questions related to "civilian occupations", the "normal work relation" of the fully employed male conventionally serves as the standard for "deviations" that need to be offset (1996, p. 424). This standard can be reconsidered by (re)locating "the normative core of the system of rights in the autonomy both of the individual and of the associated citizens" (1996, p. 562). To develop fully such a theory of rights, for Habermas, implies recognizing the *intersubjective character* of rights as a social concept. Public discourse is necessary to define the meaning of the comparisons being made between people and their "characteristics" in different contexts. He summarizes his argument:

> Institutionally defined gender stereotypes must not be assumed without question. Today these social constructions can be formed only in a conscious, deliberate fashion; they require *the affected parties themselves* to conduct public discourses in which they articulate the standards of comparison and justify the relevant aspects. (1996, p. 425)

It is only through participating in articulating and justifying standards of comparison between "the sexes", that people can contribute to furthering what Habermas calls the "emancipatory meaning" of equal rights (1996, p. 420). The feminist highlighting of competing views about the identity of the sexes and their relation to each other, as well as the indication of the "fallible character" of gender classifications, point for Habermas to the need to consider different views via open "public discussion". The contest over definitions of difference "cannot be delegated to judges and officials, nor even to political legislators" (1996, p. 426). Habermas thus draws on what he sees as the concerns of feminism in order to reiterate his point regarding the need for a vibrant political public sphere.

Habermas is aware that not all feminist argumentation (including "radical" argumentation), would support his view of the "emancipatory meaning" of equal rights. He argues that some feminist critiques "miss their real target" when they discuss the way in which difference should be treated in society (1996, pp. 424-425). Some of them fail to see the discussion of difference as being either possible or desirable in the manner suggested through Habermas's conception of communicative rationality in the public sphere. Anti-foundationalist feminists, for instance, would wish to oppose Habermas's posited regulative principle of consensus as the basis for social discourse. For them, truth, rightness, sincerity, and comprehensibility, as social constructs, cannot be grounded through positing some consensus principle written into the structure of human speech. Habermas's rejoinder to this argument is that in any instance of discourse, those involved might decide to postpone for the time being the possibility of reaching consensus: the parties involved may indeed "have equally good grounds for their inability to reach consensus" (1993, p. 94). But this does not mean that they have discarded the *principle* of consensus.

The following bullet points provide a rendition of the critical theoretical argument positioned in relation to anti-foundationalist feminism.

- Habermas appreciates the feminist concern with the way in which the language of social rights has become reduced in welfare state capitalism to the question of how goods can be (re)distributed.
- Critical theory points to ways in which rights can be exercised via an activation of the political sphere of public discourse. This discourse should be regulated by the principle of consensus. The anti-foundationalist feminist concern that this implies a form of control, is unfounded — because participants are not forced to try to reach consensus in any particular discursive exchange.
- While in anti-foundationalist feminism accountability of social inquiry is linked to the intention of transfiguring "control" relationships (including control exercised through the principle of consensus), in critical theory *accountability in social inquiry is linked to the possibility of revitalizing public discussion via discursive relationships grounded in human (communicative) rationality.*

2.6.1.6 Positioning Critical Theory in Relation to Discursively-oriented Constructivist Argumentation

Habermas concurs with the constructivist suggestion that it is only through language that the worlds given to humans can be accessed. As Brand indicates, he rejects the idea that knowledge is created through the way in which subjects encounter some extra-linguistic "object world" (Brand, 1990, p. 16). Even the world that is objectified as we make propositions about empirical reality, is still a world created through the orientation of speakers toward it. The object world can be separated from other worlds (the social and subjective ones) through the way in which speakers organize their discussions around claims being made in the process of discourse. Claims to "truth" invoke discussion around the "objectification" of the world. Claims to "normative rightness" invoke discussion around the "social world" of interpersonal relationships — which embodies norms for legitimately ordered social relationships. Claims to "sincerity" or "authenticity" raise for consideration the person's own world of personal experience (Habermas, 1982, pp. 271-272).

Habermas suggests that in the course of everyday life, reasons are utilized in the process of making and testing different validity claims. Reasons, however, should not be conceived merely as "dispositions to have opinions". This view of reason fails to appreciate the way in which it functions to organize subject-subject — that is, *intersubjective* — relations in the knowing process. According to Habermas:

> Reasons ... are the currency used in a discursive exchange that redeems criticizable validity claims. Reasons owe their rationally motivating force to an internal relationship between the meaning and the validity of linguistic utterances. (1996, p. 35)

Habermas notes that the symbolically structured lifeworld of society — which is "mediated by interpretations and beliefs" — is rendered unstable due to the fact that "reasons are constantly exposed to the risk of being invalidated" (1996, p. 36). The lifeworld is "shot through with fallible suppositions of validity" — suppositions that have a very "precarious kind of stability" (1996, p. 36). Habermas holds on to the idea that the communicative infrastructure of the lifeworld — albeit somewhat unstable — is the mechanism for affording social integration in complex societies. According to his understanding of communicative action, "the important function of social integration devolves on the illocutionary binding energies of a use of language oriented to reaching understanding" (1996, p. 8). But because of the "burdens that the ideal character of claims to reason impose on finite minds" people cannot always bring to mind the best reasons for statements made; and continuing disagreements over competing worldviews can easily occur (1996, p. 60).

Habermas considers the question of how the pluralism of worldviews is to be treated in society. He indicates that one solution that has been offered in the literature on pluralism is to suggest that participants should simply learn to tolerate their differences, without trying to organize a consensus across their alternative competing perspectives. This, as Habermas sees it, is the solution of a constructivist argument that pleads for the toleration of the incommensurable (1993, p. 95). But he opposes this solution by suggesting that there is no need to assume that different languages and vocabularies for understanding are "utterly incommensurable" (1993, p. 95). People engaged in discourse can at least orient their efforts toward effecting translations across their competing views. Even when people accept the "co-existence of competing worldviews", this does not imply that they need "resign" themselves "to a mere *modus vivendi*" (1993, p. 94). They can rather uphold their own validity claims for the time being, while "postpon[ing] the possibility of consensus, kept open in principle, to an indefinite future" (1993, p. 94).

Habermas thus suggests that the constructivist idea that worlds are socially constructed in language, should not be taken to the conclusion that the quest for consensual understanding between competing worldviews is in vain. This solution might render actors less accountable than if they operate in terms of an idealizing assumption directed toward developing such understanding. In elucidating Habermas's argument in this respect, Rehg comments that if communities are to have a degree of stability then there is a need for participants to, at least, attempt to organize discourse around the quest for rationally motivated agreements (Rehg, 1996, p. xvi). According to Habermas, then, a constructivist position that highlights the de facto pluralism of worldviews without indicating how interacting parties can be oriented toward developing consensual understanding, might be unable to recognize the potential for reason encapsulated in forms of social reproduction.

The following bullet points provide a rendition of the critical theoretical argument positioned in relation to discursively-oriented constructivism.

- As noted within constructivism, the de facto competition between worldviews cannot be settled with reference to extra-linguistic realities. But the task of critical theory in the face of this competition is to explore the potential for reason written into the structures of human speech.
- A discursively-oriented constructivism defines the accountability of inquirers in terms of the way in which they are able to account — in instances of social discourse — for their constructions.
- According to critical theory, *accountable social inquiry does not shy from issuing suggestions concerning the way (communicative) rationality might be used as a standard to assess validity claims put forward for discursive redemption.*

2.7 ANTI-FOUNDATIONALIST FEMINISM AS A WAY OF DEFINING THE PRACTICE OF SCIENCE

Anti-foundationalist feminism explores what it might mean to engage in social inquiry without seeing this as a quest for "truth" — whether truth is defined in terms of representation of external realities or in terms of consensual understandings sought in the process of discourse. Anti-foundationalist feminism links up with both so-called postmodernist and poststructuralist argumentation in its critique of the presupposition that knowing involves the search for "the truth" (Lather, 1993, p. 673).[11] In developing an anti-foundationalist position as a way of conceiving research in society, Lather asks the question "what are the anti-foundational possibilities outside the limits of the

[11]In considering the relationship between postmodernism and poststructuralism, Denzin indicates that postmodernism proffers the suggestion that there can be no definitive criteria for judging knowing endeavors. As he states: "This [postmodernist] position doubts all criteria and privileges none, although those who work within it favor criteria such as those adopted by some poststructuralists" (1997, p. 8). Denzin suggests that what he calls "critical poststructuralism" can be characterized by its contention that "an entirely new set of criteria, divorced from the positivist and postpositivist traditions, need to be constructed" (1997, p. 9). Denzin treats Lather's (1993) proposals as an exemplar hereof (1997, p. 9).

normative framings of validity in the human sciences?" (1993, p. 674). She expresses concern that framings of validity currently grounding the self-understanding of science in society, are largely under the hold of the "master code of positivism" (1993, p. 674). This master code continues to shape even research that regards itself as postpositivist (1993, p. 674). Her own goal is to develop postmodernism/poststructuralism in a way that allows us to consider new conceptions of validity within the process of social inquiry. As she notes, "my goal is to reinscribe validity in a way that uses the anti-foundational problematic to loosen the master code of positivism" (1993, p. 674).

In referring to the anti-foundationalism of both postmodernism and poststructuralism, Lather notes that the distinction between the two can be made in various ways. One way of seeing the difference is to suggest that postmodernist discourse points to a way of life where "control" becomes destabilized, whereas poststructuralism is "used more often in relation to academic theorizing 'after structuralism'" (as an academic argument) (1993, p. 688). But she also notes that the terms are often used interchangeably and that she does not wish to overstate the differences. She suggests that what is important (for her, following certain other feminists) is to keep in mind a possible tension between on the one hand postmodernism/poststucturalism and on the other "feminist emancipatory discourse". The latter discourse specifically embraces commitments to envisaging possibilities for change and cultivating participation in "transformative praxis" (1993, p. 681). As far as Lather is concerned, anti-foundationalism should not be taken to the point of denying the search for what might be conceived as "better" ways of knowing and living.

Lather argues that what is specific to "feminist praxis in an anti-foundational time" is the way in which it cherishes a "politics of uncertainty" (1993, p. 683). Social research can contribute to such a politics by "foregrounding the availability of multiple discourses and [exploring] how they can be used to decenter the researcher as the master of truth and justice" (1993, p. 680). She cites the case of Woodbrooks' (1991) exploration of African-American identity as research practiced in a way that helped her to "unlearn her own privilege" (Woodbrooks, as cited in Lather, 1993, p. 680). She conceptualizes Woodbrooks' strategy as one that "refines our sensitivity to differences, introduces dissensus into consensus, and legitimates via fostering heterogeneity" (1993, p. 680). Using this, and other examples, Lather suggests that attempts made by people to open up spaces to explore "difference" through the research process can be seen as an expression of their investment in transfiguring social relationships. It is in this sense that research is an expression of a political stance.

Anti-foundationalist feminism, as briefly outlined above, offers a counterposition to many of the tenets of positivism. The way in which it presents a challenge to these tenets is now explored in some detail (with reference to the arguments of various authors).

Maynard approaches the positivist tenet of *scientism* by considering the implications of this for the way in which quantitative and qualitative methods have been treated within feminist discourse. She notes that for many feminists the positivist (scientistic) suggestion of the unity of the natural and social sciences is associated with the imposing of "a 'masculanist' form of knowing, where the emphasis [is] on the detachment of the researcher and the collection and measurement of 'objective' social facts" (1994, p. 11). These feminist researchers have challenged the quantitative, measurement-orientated approach in social inquiry that they associate with positivism. They have highlighted instead the importance of methods that can be used to "explore experience rather than impose externally defined structures [categories] on women's lives" (1994, p. 12). Maynard puts this argument in historical perspective. She suggests that there was a time

> when feminist research was in its infancy and when women's lives and experiences were still largely invisible. What was most usefully required then was an approach to research which maximized the ability to explore experience .... (1994, p. 12)

Maynard notes that springing from this historical legacy, there still exists a tendency within feminism to "equate feminist work with a qualitative perspective" (1994, p. 12). Of course, Maynard notes, feminism (of whatever kind) rightfully challenges scientism by "focusing on the fact that the subjects of social research are conscious, language speaking and meaning-creating" (1994, p. 13). But this recognition in itself does not imply that "quantitative" methods are to be eschewed altogether. More important, for Maynard, is to recognize that "it is not so much quantification *per se* as naive quantification [or any naive research practice] which is the problem" (1994, p. 13).

Maynard's suggestion is that the feminist rejection of the positivist tenet of scientism can be used to open up a discussion on the way in which data — of whatever kind — are constructed in the research process (1994, p. 13). This, in any case, is the point that an anti-foundationalist position makes through the rejection of the positivist tenet of scientism. The point can also be made through a reconsideration of the tenet of *phenomenalism*.

The positivist tenet of *phenomenalism* is based on the idea that "data" appear in their givenness to researchers, who can "soak them up" so that theorizing can be "absorbed" thereby (to use a metaphor provided by Maynard, 1994, p. 15). Maynard notes that the guideline often given in "textbooks" is for researchers to be

> emotionally detached, calculating and in control of the collection of data. Those researched are regarded ... as the passive givers of information, with the researcher acting as a sponge soaking up the details provided. (1994, p. 15)

The anti-foundationalist critique of the "sponge" model of the researcher is that it fails to recognize the ways in which data arise out of the *process of interaction* between "researchers" and "researched". As Maynard indicates, even when experiments and surveys are undertaken, as well as in processes of (qualitative) interviewing, a human relationship is established with "the researched". It is through this relationship that the "collected" data emerge. Both researchers and researched are active participants giving meaning to the interaction and thereby influencing the data that are developed through the interaction.

In challenging the positivist tenet of *phenomenalism,* anti-foundationalist feminism emphasizes that human experience is in any case never "raw" (Maynard, 1994, p. 23). Referring to poststructuralist thinking in this regard, Maynard points out that

> the very act of speaking about experience is to culturally and discursively constitute it. People's accounts of their lives are culturally embedded. Their descriptions are, at the same time, a construction of the events that occurred, together with an interpretation of them. The researcher is also, of course, involved in interpretation. (1994, p. 23)

Maynard remarks that feminists have therefore raised as an issue the question of "how to produce scientific knowledge about meanings and social relationships when people understand and experience these differently" (Maynard, 1994). Maynard's own suggestion is that researchers cannot shy from creating attempts to link "experience to understanding" (1994, p. 24). In organizing these links, however, they should take account of the possible effects of their understandings in society. They should consider as relevant to their inquiry the possible impact of their way of organizing interpretations "on the lives of those included" in the study. And they should also consider that their inquiry "may be important for the category of persons they are taken to represent" (1994, p. 17). They cannot properly conduct their research without taking cognizance of the way it may be seen by others.

As far as the positivist tenet of *empiricism* is concerned, the question that is addressed by anti-foundationalist feminism is *not* how to use logic combined with experience in order to advance knowledge of external realities. The question is how one can construct a science "after truth" (namely, without believing that scientific practice must be geared to seeking "the truth" about reality — see Lather, 1993, p. 673). Part of the aim of (anti-foundational) research must be to "unsettle conventional notions of the real" that underpin most scientific work (Lather, 1993, p. 685). This implies that inquiry is directed toward "dissolving interpretations by marking them as temporary, partial, and invested" (with personal meaning) (1993, p. 681). The positivist *empiricist* tenet that rests on the idea that there is "a world" to be found through the scientific research process, is replaced in anti-foundationalist feminism with a celebration of the multiplicity of discourses (Lather, 1993, p. 680).

The positivist tenet of *value freedom* is also challenged within anti-foundationalist feminism on the grounds that belief in this tenet by scientists and by others, is what allows the products of scientific inquiry to be valorized as "knowledge" in society. The positivist suggestion that scientific research is aimed at providing a representation of external reality that is (as far as possible) free of researcher values, means that it becomes accorded with cognitive authority in society above other kinds of knowledge claims. Its authority derives from the perpetuation of the belief that in striving for value freedom the research community is indeed able to make advancements in the (scientific) representation of external reality.

However, Lather proposes that it is both possible and desirable for researchers to "ironically use researcher power to undercut practices of representation" (1993, p. 678). This can be done by, for instance, creating a research text that "is dense with the absence of referential finalities" (1993, 678). Such texts become devices to "implode controlling codes", rather than devices to tell a (more or less) true story about externally posited realities. The value behind anti-foundationalist research efforts, can be conceived as "refin[ing] our sensitivity to difference and reinforc[ing] our ability to tolerate the incommensurable" (Fritzman, 1990, as cited in Lather, 1993, p. 679). Otherwise put, the goal is to "foster differences and let contradictions remain in tension" (Lather, 1993, p. 679). Or, as Young indicates, social inquiry can be aimed at fostering the value of a "principled call for heterogeneity" (1990b, p. 312).

Objections to the positivist tenet of *instrumental knowledge* are made within anti-foundationalist feminism by challenging the distinction between knowledge and its application. It is argued that the process of social inquiry itself is shot through with social implications. As "respondents" address the questions asked and the issues raised (by researchers), so their attitudes toward, and conceptions of, the issues, may be impacted upon. Furthermore, the reports "about" reality as ordered and understood through the theoretical investments of researchers, are part of this reality. To try to separate out the moment of research from the moment of its application, is to fail to take sufficient account of what Lather calls "the rules and effects of [researcher] investments" (1993, p. 680).

2.7.1 Considering Accountabilities in Terms of an Anti-foundationalist Feminist Position

Lather directs her efforts at showing how knowledge might be legitimated (accounted for) "in contemporary postpositivism" (1993, p. 673). Lather concentrates on considering possible meanings of "validity" using an "antifoundational problematic" (1993, p. 674). Her proposals for validating research via this problematic (or approach) are discussed in this section, along with those of certain other authors concentrating on the issue of accountability. Their views are examined in relation to positivism, non-foundationalism, scientific realism, interpretivism, critical theory, and discursively-oriented constructivism.

2.7.1.1 Positioning Anti-foundationalist Feminism in Relation to Positivist Argumentation

As indicated in Section 2.7 above, Lather attempts to develop ways of seeing validity that move outside of the framework provided by positivism. She notes that within positivist thought, validity of knowing efforts is defined by success in "finding" the world. Positivist conceptions of validity rest on the premise that scientific propositions can be justified as long as they are reality-bound. An anti-foundationalist conception of validity, however, "shifts our sense of the real to discourses of the real" (1993, p. 675). Instead of lamenting the difficulties of "finding" the world, and instead of trying to improve practices to seek validity as representation thereof, it is possible to "celebrate the felt loss of found worlds" (1993, p. 675). One can celebrate this by interpreting as salutary the "political possibilities that open up when truth is positioned as made by humans via very specific ... practices" (1993, p. 675).

Lather suggests that the standard (positivist) way of defining validity, sets up a power relationship between those considered as having a route to gaining (more or less valid) knowledge and those not taking that path. She suggests (following McGowan, 1991) that it is possible to treat as a "site of political action" the way in which the legitimation of notions of validity is constructed (Lather, 1993, p. 676). The notion of validity can then become an entry point for thinking the as-yet "unthought in our thought" (1993, p. 676).

Positivism confines our thinking about accountability in social research, to thinking about how practices of generating validity can be improved. This thinking is directed toward seeing validity as a "technical problem" (1993, p. 675). The problem is, firstly, how to organize the research such that respondents' behavior and attitudes can be understood with the least possible "reactivity effect" (through their responding to the effect of being observed), and, secondly, how an analysis of results can be achieved so that conclusions can be justified (as far as possible) with reference to "the evidence". According to a positivist self-understanding of science, researchers become the more accountable (to colleagues and others in society) the more they can improve their techniques of unobtrusive observation and the more they can justify their drawing of conclusions with respect to observed evidence. But Lather tries to revise the positivist notion of validity (and attendant notions of researcher accountability) so that the social relationship between those involved in the research process can be foregrounded. Researchers then have to account for how they conduct their relationship with "respondents" such that they do not use the process to reinforce the belief in society that "doing science" offers some privileged way of accessing reality. One way of "unlearning privilege", is to develop the research process as an opportunity for all parties involved in the study, including the "professional" researcher, to explore and reconsider their attitudes and practices. Another way is to pay attention to the style in which textual reports of the study are presented, and to try to organize these in such a way that clear-cut referential language is foregone (Lather, 1993, p. 677).

The following bullet points provide a rendition of the anti-foundationalist feminist argument positioned in relation to positivism.

- Whereas positivism locates the accountability of researchers in practices aimed at improving access to "the truth", anti-foundationalist feminism sees the research process as an opportunity for people to explore what it might mean to work with difference.
- In positivism, accountability of researchers is linked to their efforts to gain valid findings, where validity is defined as an achieved correspondence of scientific statements with reality.
- In anti-foundationalist feminism, *accountability is linked to efforts made in the knowing process to create the space for people (including "professional" researchers) to reconsider their attitudes to knowing and living.*

2.7.1.2 Positioning Anti-foundationalist Feminism in Relation to Non-foundationalist Argumentation

The conception of accountability offered within the non-foundationalist position as expressed by Hammersley and Gomm (1997a) and discussed in Section 2.3.1, is criticized by anti-foundationalist feminists on various scores. Temple provides a good indication of the types of criticism that can be leveled from an anti-foundationalist point of view.

Temple points out that in developing their non-foundationalist position, Hammersley and Gomm assign special importance to the research community (Temple, 1997, paragraph 2.1). The scientific community has a crucial role to play "in subjecting knowledge claims to assessment on the basis of criteria of plausibility and credibility that are generally more skeptical than those operating in other areas of social life" (Hammersley & Gomm, 1997a, as cited in Temple, 1997, paragraph 2.1). Temple argues that what is of concern to feminists when scientists with particular criteria of assessment are seen as the source of judgments about the status of research undertaken, is that questions of "how knowledge is produced and who is allowed to speak in the name of the research community" are glossed over (Temple, 1997, paragraph 2.7). She points out that in any case within what is called the research community, "there is ... no single approach to science and researchers hold different perspectives" (1997, paragraph 3.4). For example, she notes that "feminist critics of traditional science question a model of scientific research which is hierarchical and which uses binaries such as emotion/reason, subjective/objective, subjects and objects of research" (1997, paragraph 3.4). The feminist complaint is that notions of what is "rational" and "scientific" are used to exclude other views on what may be involved in doing (acceptable) research.

According to Hammersley and Gomm's (non-foundationalist) view of accountability, scientists need to be prepared to engage in argument in making a case for the credibility of findings produced (as offering tentative conclusions that are likely to bear some relationship with reality). While at face value this may sound a feasible way of defining accountable inquiry, on closer examination it can be challenged. Temple challenges the non-foundationalist view of the possibility of invoking "reason" to persuade others of the reasonableness of one's work, by referring to the inquiry undertaken by Mykhalovskiy on masculinity. She indicates that in the case of this research, the reviewer of his work took the line of arguing that Mykhalovskiy needed to show evidence of an "accountability to the subject matter". But Mykhalovskiy himself remarks that in practice this meant that the assessment of the work became based "on the principles held by the reviewer as to the nature of the subject matter and how it should be tackled" (Mykhalovskiy, 1997, as cited in Temple, 1997, paragraph 4.1).

Temple notes that in conducting his research and writing it up, Mykhalovskiy chose to

> analyze how his views differ from those others [participants] at the table and in the group and does not attempt a final resolution [or even provisional conclusion] to his research questions. For his reviewer, he argues, this is in itself problematic in that as the author he refuses to finish his work with the definite conclusion traditionally expected from a researcher. (Temple, 1997, paragraph 4.3)

Temple sides with Mykhalovskiy's decision to leave fundamentally open his conclusions. He does not attempt to finish off the study with some referential vision that he can present as "findings" to others (those participating in the study as well as other audiences). While a non-foundationalist stance would suggest that he should at least offer some provisional statements (albeit not as fully corroborated), Temple does not see the "reason" in trying to offer a referential account of realities via such statements. She takes the anti-foundationalist stance that views held by researchers should not become the "norm against which others' views can be judged" (1997, paragraph 5.2). She prefers that the research process be seen as an opportunity for people (including the researcher) to learn that "'the truth' is not the same for everyone" (1997, paragraph 5.2). She elucidates her argument:

> Rather than suggest any particular group as in a most favored position for the job of judge and jury, I argue that a more open and honest move is to declare your hand: these are my views on the research subject. You are welcome to compare them to your own but not to those of some mysterious disembodied entity that is assumed to speak for everyone. (1997, paragraph 5.3)

Temple here, in line with Lather (1993), challenges the idea that accounts generated by researchers should be judged for their validity by comparing them with some notion of what "the truth" may be. They can be compared only with other accounts as people "listen and learn from other people's experiences" (Temple, 1997, paragraph 5.2). The process of "science" can be used to cultivate dialogue across different visions, without reason being seen as having the (abstract) power to reconcile these differences (1997, paragraph 4.2).

The following bullet points provide a rendition of the anti-foundationalist feminist argument positioned in relation to non-foundationalism.

- Anti-foundationalist feminism challenges the non-foundational view that the scientific experience of reality is to be rendered plausible through the way reasonable argument is developed in the scientific community. Anti-foundationalist feminism suggests that it may be unreasonable (and overly controlling) to suggest that science should be geared toward seeking referential accounts of realities.
- In non-foundationalism, accountability of researchers implies an orientation on their part to using their membership in the scientific community in a reasonable effort to minimize their own and others' errors in the search for knowledge.
- In anti-foundationalist feminism, *accountability implies introducing a style of research that cultivates people's capacities to learn that "the truth" is not the same for everyone, and to work with difference in the light of this learning.*

2.7.1.3 Positioning Anti-foundationalist Feminism in Relation to Scientific Realist Argumentation

Anti-foundationalist feminists criticize all forms of realism, including scientific realism, for holding on to the assumption that "the real" can become the object of scientific inquiry. Stanley and Wise express a criticism of realism thus conceived:

> Different states of consciousness aren't just different states of interpreting the social world. We don't accept that there is something "really" there for these interpretations to be interpretations of. Our differing states of consciousness lead us into constructing different social worlds. (1983, pp. 130-131)

They continue to express their argument concerning the human experience of the social world:

> Obviously we recognize that everyday life occurs around the assumption that an objective reality exists and because of this it is possible to make assessments of the validity and invalidity of experience. It has certainly been our experience that most of the reactions to our interpretation ... have been based on the belief that we are out of touch with the "real reality" that the reactor perceived. (1991, p. 279)

Here they make the point that the charge that a person is out of touch with "real reality" can be translated as a charge that they are out of touch with the reality as perceived by some "reactor" (that is, someone reacting to the interpretations). They argue that as soon as the research process is seen as oriented to gaining a "true" understanding of some "real reality", an assault on other experiences of reality is likely to come into play. Hence, instead of basing their own research on the assumption that they are seeking to find out about some "real reality", they suggest that it is as feasible (and more desirable as far as they see it) to work with the assumption that "oppressed people of all kinds see and experience social reality in uniquely different and interesting ways. Reality *is* contradictory, realities do co-exist and overlap, and conflict" (1993, p. 169). Having said this about the differential experience of "oppression" (and other social experiences), they indicate that they do not wish to present any particular theoretical account as providing an explanation of "why women are oppressed" (1993, p. 170). They are cautious of presenting such an explanation, because to do so would be to perpetuate through the research process a power relationship that they regard as itself oppressive. So they reiterate that the purpose of their theorizing is *not* to provide

> a view which would be imposed on other people's experiences during the conduct of research. Having at its heart the belief that many "objective realities" exist, it takes as its task the exploration of these, not their obliteration, their dismissal, as "false" or "inadequate". (1993, p. 171)

According to Stanley and Wise, the (realist) suggestion that the aim of science is to provide explanations supposedly rooted in some "real reality", means that knowing becomes defined as a process of sifting truth from falsehood based on the idea that some people (those involved in proper scientific research) have a better way of accessing reality. Against the claims of those (including scientific realists) who propose that science must, by definition, be oriented to painting a picture of "real reality", Stanley and Wise argue that this conception of science is based on an understanding of the character of social "reality" that differs from theirs (1993, p. 171).

Stanley and Wise are aware that realist-oriented social scientists reject their argument about "reality" for its being "out of touch with reality" and for being "unrealistic". But for them, it is not unrealistic to posit that there are many "objective" realities — because reality in any case (as they see it) is defined through the *experience of participating in reality-construction.*

Lather makes a similar point when she notes that "the bugaboo of relativism [can be] displaced, positioned as a foundationalist concern" (1993, p. 682). She suggests that instead of grounding the authority of statements with respect to some supposed reference to an independently existing reality, research can be legitimated through self-reflexivity (Lather, 1993, p. 682). She gives the example of Richardson's way of treating her research goal (in interviews with unmarried mothers) as an opportunity also to probe her own lived experience (1993, p. 682). As she proceeded, she tried to "create a position for experiencing the self as a sociological knower/constructor — not just talking about it, but doing it" (Richardson, 1992, as cited in Lather, 1993, p. 683). Lather indicates that Richardson's study can be seen as creating a dynamic of "opening up spaces [in reality-construction] in which no-one is as yet the master" (1993, p. 684). This is not merely a matter of admitting (as would scientific realism) the uncertainty of attempts to arrive at knowledge of reality. It points rather to the *possibility of a way of life where people can forego the attempt to assimilate different experiences of reality into some integrated picture.* In order to keep alive a discourse about this possibility, Lather, in propounding various forms of "transgressive" validity, suggests that researchers could, for example, choose to

> foreground a suggestive tension regarding the referent and its creation as an object of inquiry; ... foster differences and heterogeneity via the search for "fruitful interruptions"; ... work against constraints of authority via ... multiple openings; ... [and] create a questioning text that is bounded and unbounded, closed and opened. (1993, pp. 685-686)

These kinds of suggestions that Lather offers for practicing social inquiry offer some possibility (though not an exhaustive list) for generating "counter-practices of authority" in the process of research (Lather, 1993, p. 687). Researchers can contribute, through their research, to a recognition in society that living as if there is only one "reality" to know, is not the only way to live. As part of their contribution they can point to, and try to practice, an alternative style of living.

The following bullet points provide a rendition of the anti-foundationalist feminist argument positioned in relation to scientific realism.

- While scientific realism embraces the assumption that the proper object of scientific inquiry is a reality existing outside of the knowing process, anti-foundationalist feminism posits that research as a social practice can point to the possibility of foregoing the attempt to access "real reality".
- In scientific realism, accountable scientific inquiry requires that researchers enter into debate around ways of theorizing about reality in an effort to increase the chances of developing an integrated picture thereof.

- For anti-foundationalist feminism, *accountable research practice involves creating interruptions in univocal expressions of (experienced) reality — so as to point to a possible way of living where difference and heterogeneity can be incorporated in an unassimilated co-existence.*

2.7.1.4 Positioning Anti-foundationalist Feminism in Relation to Interpretivist Argumentation

As noted by Maynard (1994) and by Oakley (1998), many feminist arguments have revolved around the insistence on the need to develop a research space to explore the (subjective) experiences of women. Some form of ethnographic research, committed to "construct a sociology respectful of women's subjectivity", is seen as an alternative to the "dominant [research] frame" — in which such subjectivity has traditionally been sidelined (Oakley, 1998, p. 713).

However within feminist discourse, argumentation on what is involved in exploring the realm of subjective experience is not settled (and nor is the question of how this may relate to more "quantitatively" directed research). While some feminists have tried to legitimate "softer" inquiry on the basis that it can better allow for the expression of the "unmediated views of participants", others have questioned whether this is possible (Oakley, 1998, p. 715).

From an interpretive (Weberian) position, the aim of social science would be to reflect as far as possible the views of participants (their ways of assigning meaning), and to check the plausibility of interpretations with reference to observations of patterns of conduct. Oakley indicates that this implies some preference for multi-method studies, which would open up possibilities for checking the researcher's interpretation of participants' meaning-making with other collected data. But she notes that a problem arises when interpretations of the different data "are at odds with one another" (1998, p. 715). She cites the case of Sidell's research on the health status of women, which was undertaken using a variety of "data collection" methods (national statistics, sample surveys and interviews). The result of the research was a "mass of paradox and downright contradictory evidence" (Sidell, 1993, as cited in Oakley, 1998, p. 715). For Oakley, this implies that we should recognize that "different types of data may yield very different conclusions" (1998, p. 715). How one decides to approach the research process has an effect on the way findings are generated.

One way of dealing with a recognition of this that has been suggested by certain authors is to treat the research process as an opportunity for exchange of perspectives — so that both "researchers" and "researched" can engage in rethinking their attitudes and practices (Lather, 1993, p. 681; Temple, 1997, paragraph 5.2). This is a way of transfiguring what Lather and others see as the control effect of researchers' specific way of posing questions, interpreting answers, and uniting these with other forms of data collection.

Josselson indicates similarly that many "narrative researchers" (researching others' lives) "have begun to explore the ways in which our exchange with participants in the interactional phase of our research may affect those who share their lives with us" (1996, p. 61). The (feminist) argument here is that the potential of the researchers' presence to make a difference to people's lives, cannot be dismissed as being outside of the range of their concern as researchers. Josselson also argues that researchers need to take seriously the way in which their write-up of others lives may be felt by those others. She suggests that although narrative researchers

> recognize that our hypotheses and our conclusions about people originate in our own complex conceptual processes, we often lose sight of the additional authority our words and ideas carry when transferred to the permanence of print. (1996, p. 61)

In line with feminist argumentation concerning the need to deal with the issue of researcher "control", she uses the metaphor of an oracle to indicate how participants might turn to researchers "to tell [them] what it all meant" (1996, p. 67). She continues: "I don't think we can underestimate the projected, imagined powers our apparent authority, which rests on our access to print, invokes" (1996, p. 67). Her solution to this is to remind us (and herself) that "no matter how gentle and sensitive our touch, we still entangle ourselves in others' intricately woven ... tapestries" (1996, p. 70). Instead of treating as irrelevant to her role as researcher the issue of her effect on others' lives, she suggests that she "would worry most if [she] ever stopped worrying" about this (1996, p. 70). By continuing to "worry", she can at least consider that a concern with this effect is part of her research remit.

The following bullet points provide a rendition of the anti-foundationalist feminist argument positioned in relation to interpretivism.

- While interpretivism sees the research process as a forum to create accounts of the meaningful character of social life, anti-foundationalist feminism concentrates on how researchers can account for the possible effects of their presence on the lives of those studied and of others.
- In interpretivism, accountability of social researchers implies that they direct their attention to creating what might be considered to be (and are argued to be) plausible interpretations of aspects of social life that are chosen for study.
- In anti-foundationalist feminism, *accountability implies that researchers develop a prime concern with the way in which social research might impact on the lives of participants and other audiences as meanings are created and developed through the research process.*

2.7.1.5 Positioning Anti-foundationalist Feminism in Relation to Critical Theoretical Argumentation

Lather presents her anti-foundationalist feminist stance by indicating its ties to postmodernist/poststructuralist discourse, which, as she notes, "displaces both the criterion of efficiency and the Habermasian drive for consensus" (1993, p. 679). While agreeing with Habermas's concern with the way in which the drive for efficiency in science and society can function to close the terms of social discussion, Lather feels that Habermas's posited drive for consensus can lead to other forms of control. To make this point she draws, inter alia, on writings of Lyotard (1984) and McGowan (1991), whom she indicates are "opposed to the recuperation of the other into the same" (1993, p. 679).

For Habermas, human (communicative) rationality is defined through people's orientation toward the possibility of making judgments that can be consensually validated. Lyotard's postmodernist argument challenges Habermas's assertion regarding the human need to reach toward consensus, that is, to generate "sameness" of vision. For instance, Lyotard expresses concern about the potential terrorization of "all those who do not participate in saying 'we'" (Readings, 1991, p. xxv, commenting on Lyotard's postmodernist position). Habermas's procedures for sound rational argumentation, which he sees as written into human speech, are regarded with caution within anti-foundationalist discourse — in that reliance on some idealized form of argument can exclude or mute people who are not prepared to enter such forms of exchange.

Anti-foundationalist feminists nevertheless appreciate Habermas's efforts to find a way of theorizing that allows people to rethink their practices of human sociation. Lather notes that her own anti-foundationalism accepts that there will be both a rapprochement toward, and a collision with, emancipatory discourses that endorse an investment in "transformative praxis" (1993, p. 681). In some sense she can find a meeting point with Habermas's efforts to "invest" in social transformation.

Postmodernist/poststructuralist thinking has been interpreted by some theorists (including some feminists) as being too vacuous in its conceptualization of power to be able to express any vision of a better way of life in society. Hoy has argued that the notion of power that informs Foucault's (poststructuralist) historical studies, for instance, may be too broad: it becomes difficult in terms hereof "to speculate about whether future power configurations might be better" (1986, pp. 136-137).[12] The charge of broadness (vacuity?) of Foucault's notion of power is also leveled by Hartsock, who argues that "a theory of power for women, for the oppressed, is not one that leads to a turning away from engagement but rather one that is a call for change and participation in altering power relations" (1990, p. 172).

Seen from this angle, Habermas's theoretical discussion of communicative rationality, and his attempts to substantiate a theoretical vision of communicative power as a basis for generating legitimate use of power, can be regarded with some appreciation within anti-foundationalist feminist argumentation. However, Young (1990b), as Lather (1991, 1993) argues that it is a "political vision of inexhaustible heterogeneity" — rather than one of moving toward consensual forms of human sociation — to which she, as many feminists, commits her energies (1990b, p. 301).

---

[12]Simons asserts, though, that Foucault's conceptualization of power relations does implicitly include a principle for the assessment of political regimes. A regime, she notes, "is judged [by Foucault] unfavorably as dominative if it minimizes the possibilities for strategic reversal and thereby confines practices of liberty. ... . The regulative aim should be to pursue games of power ... played with a minimum of domination" (Simons, 1995, p. 94). Although Foucault is wary of generalizing about what possibilities exist for playing such games, Simons interprets his position as implying that inventions can be made possible within power relations.

Young prefers to build up a vision of a better form of human sociation by imagining as a political ideal "the unoppressive city" (1990b, p. 319). City life is her guiding metaphor for thinking about new forms of sociation based on "openness to difference" (1990b, p. 319). She indicates that she presents this vision in order to point to, and invoke, forces that can work against unity hence creating difference. She conceptualizes this as a vision of justice, in that it hopes to make room for a "general will", where the general will is to protect differences, which is for the common good (1990a, p. 97). She utilizes the metaphor of city life as follows:

> In the city strangers live side by side in public places, giving to and receiving from one another social and aesthetic products, often mediated by a huge chain of interactions ... . The social differentiation of the city provides a positive inexhaustibility of human relations. The possibility always exists of becoming acquainted with new and different people, with different cultural and social experiences; the possibility always exists for new groups to form or emerge around specific interests. (1990b, p. 319)

Young admits that her account of the city as pointing to forms of human sociation based on "openness to unassimilated otherness" is a vision that is not yet instantiated. She emphasizes that "we do not have such openness to difference in our current social relations" (1990b, p. 319). But, in similar fashion to Habermas, she points to the need for a type of social theorizing created with the normative intent to pose a "better" possibility.

Habermas suggests that his own social theorizing can be accounted for insofar as it puts itself up for discursive validity checking through the same kinds of discursive exchange that he posits as rooted in all human speech. Anti-foundationalist feminists do not, however, accord with this manner of posing the possibilities for "validation" of visions advanced. According to them, views are to be exchanged as a way of enriching the sensitivities of all those involved in the exchange, without the parties necessarily orienting themselves toward developing a consensus on the matters under consideration.

The following bullet points provide a rendition of the anti-foundationalist feminist argument positioned in relation to critical theory.

- While appreciating Habermas's efforts to engage in social theorizing with normative intent, anti-foundationalist feminism is cautious of his suggested way of treating the validation of claims created in this way of knowing/arguing.
- In (Habermasian) critical theory, the accountability of social inquirers is linked to their partaking in debates in the public sphere, utilizing (as far as possible) the procedures for rational argumentation written into the structures of human speech.
- In anti-foundationalist feminism, *accountable social inquiry is linked to the commitment made to the exploration of possibilities for developing less controlling forms of human sociation than (argued to be) embedded in conventional forms of inquiry.*

2.7.1.6 Positioning Anti-foundationalist Feminism in Relation to Discursively-oriented Constructivist Argumentation

Anti-foundationalist feminism develops the constructivist position to the point of suggesting that researchers, no less than others in society, are engaged in creating worldviews as part of the process of living. Oakley indicates that feminist argumentation takes up and extends Kuhn's efforts to demystify "popular conceptions of science" that portray scientists as "reasonable men" geared to finding out about "'objective' reality" (1998, p. 717). She notes that in Kuhn's *The Structure of Scientific Revolutions* "the actual practices of science were revealed as things that depend on experiences, on feelings and intuition and on the operation of ordinary human events. Science is a socially embedded activity" (1998, p. 718). Oakley suggests that Kuhn's work helps to demystify conceptions of science (including social science) as offering a means to "find out" about an independently existing world (unmediated by human experience).

Maynard concentrates on considering the status that should be accorded to various attempts to develop "knowledge" in society. She argues that a focus of attention on the way in which realities are constructed in the knowing process, needs to be accompanied by suggestions for assessing the quality of competing constructions. In regard to the issue of how quality might be defined, she proposes that "rigor involves being clear about one's theoretical assumptions, the nature of the research process, the criteria against which 'good' knowledge can be judged and the strategies for interpretation and analysis" (1994, p. 25). She suggests that all of these things can be made "available for scrutiny" (1994, p. 25). In the process of their being subjected to scrutiny, evaluations and judgments will be made. But Maynard also notes that the debate within feminism about how to address competing accounts (as constructions) is still ongoing.

Temple indicates that in terms of her interpretation of constructivism, researchers are encouraged to extend their understanding of their accountabilities beyond trying to create findings that will become "condoned by a group [of scientists] with particular standards of assessment" (1997, paragraph 2.6). She suggests that once researchers recognize how they are involved in creating constructions, it becomes easier for them to recognize that "being accountable to different kinds of audiences can enrich research" (1997, paragraph 3.9). They can then organize the research process and its write-up in a manner that opens the space for others (academic and other audiences, including participants in the study) to enter the process of reality-construction. Lather makes a similar point when she suggests that research in society should aim to be "accountable to people's struggle for self-determination" (1995, p. 42). That is, the research process can be used to widen people's sensitivities to alternative choices for envisaging, and acting in, their (experienced) situation.

The following bullet points provide a rendition of the anti-foundationalist feminist argument positioned in relation to discursively-oriented constructivism.

- Anti-foundationalist feminism has an affinity with constructivist thought that highlights the way in which people (including "professional" researchers) are involved in world-construction.

- Anti-foundationalist feminism concentrates on excavating how constructivist thinking can be linked to an intent to use the research process as a site for cultivating people's sense of their participation in creating ways of living.
- While a discursively-oriented constructivism suggests that accountability is assessed in discourse as social activity by those party to the (discursive) interaction, *anti-foundationalist feminism emphasizes that accountability requires an openness to explore the criteria in terms of which "knowing" in society is to be judged — so that it can become a site for developing a politics of difference.*

2.8 CONCLUSION

In this chapter, I offered an introduction to positivist argumentation, and an attendant view of accountable research practice, in order to provide a backdrop for discussing a number of alternative positions. Five alternatives to positivism were explored in detail in the chapter, namely: critical rationalism/non-foundationalism; scientific realism; interpretivism; critical theory; and anti-foundationalist feminism. I structured the chapter so that the various views on accountability in relation to other ones could be excavated.

I did not explore what I call discursively-oriented constructivism in this chapter, except through showing how the other five positions might relate to it. The brief discussion in Chapter 1 should be sufficient to allow the reader to make some sense at this point of the various responses proffered in this chapter to constructivist argumentation. (I elucidate and extend the position in Chapter 3, toward what I call a trusting constructivism.)

The elaboration of the arguments of critical rationalism/non-foundationalism, scientific realism, and interpretivism as provided in this chapter indicate that according to all three of these positions, researchers are required to be self-critical of any statements that they might make in relation to "the evidence", as well as critical of others' ways of seeing and using evidence. Researchers are accountable in the sense of indicating a preparedness to engage in argument around the use of evidence to support (provisional) statements made about realities being investigated. What is emphasized in interpretivism, though, is that social scientists need to pay specific attention also to considering the reality of human meaning-making, as a basis for offering an understanding of the constitution of social reality.

Non-foundationalism, scientific realism, and interpretivism are in accord with the view that the discourse of scientists (including social scientists) must be oriented toward accessing, as far as is reasonably possible at any point in time, the phenomena being studied. Science cannot be treated merely as a process of discourse aimed at creating statements that bear no relation to these phenomena. If scientific discourse were to be treated as such (by scientists and others), there would be no justification to warrant society's support for scientific activity and its grant of legitimacy to such activity. The only goal that warrants the financial and intellectual commitment that is required to maintain the institution of science in society, is the goal of advancing knowledge of reality. Where these positions take some distance from the positivist view of the role of science in society is in their suggestion that extending *provisional claims* about the realities under study should replace any hope of reaching certainty in knowledge claims. Nonetheless, all of the positions suggest that to continue to warrant the grant of credibility in society, scientists need to consider that their accountabilities rest on following the path that is most likely to lead closer to the truth. Funders of research in society, and other audiences, expect this of social scientists, as of all scientists.

Scientific realists have cautioned, however, that scientists should be wary of confining themselves to the remit of research that some (many) funders of research in society might prefer. They point out that funding bodies may be party to the perpetuation of theoretical thinking in social science that in fact *lacks theoretical depth*. To allow for theorizing beyond the "middle range", scientists might need to be critical of the expectations of certain funding bodies, and at least try to combine their suggested remit with other forms of theorizing. In this way social science can better fulfill its goal of offering advancements in knowledge.

Scientific realists thus criticize both non-foundationalist and interpretivist positions for tending to support too narrow a definition of what theorizing involves. According to them, it is crucial that theorizing about social structural entities and their properties is not excluded from the goal of social scientific inquiry. The non-foundationalist response to this scientific realist suggestion is that in order to continue to maintain its credibility in society, the scientific community must base its theorizing *as firmly as possible* in some logical connection to empirically observable phenomena. Interpretivists, for their part, suggest that any theorizing about social entities should be rooted in the way in which they are conceptualized by actors: *actors' meanings and the outcomes hereof* must remain the proper subject of social scientific inquiry.

If we wish to concentrate on exploring commonalities in these arguments, we can point to the (common) suggestion that science can rightfully be accorded its special cognitive status in society insofar as scientists orient their activities to finding out about the world and are accountable to others for their efforts in this regard. According to all these views, living in the world means that we need to be able to foreclose certain cognitive alternatives in order to be able to create a more or less sound (realistic) platform for acting. Thanks to the knowledge (albeit provisional) afforded by science, people are aided in their efforts to develop (more) reasonable cognitive closures. Unless scientific knowledge is indeed about reality, rather than sealed in its own discursive processes, it cannot perform this function.

In each of these positions it is argued that scientists can choose topics to study that fit in with their own conception of what is valuable to study. Topics about which information is likely to be abused later (as a specific scientist sees it), might therefore be kept off the agenda of their research. But having chosen the topic, the important requirement for researchers is to orient themselves toward developing knowledge of the realities under consideration. When scientific activity is directed toward serving goals other than the goal of advancing knowledge about reality, there is good reason for people to be suspicious of the accounts produced. Conversely, when scientific activity is directed toward advancing knowledge, then even if it has effects on others' understanding of their realities (and hence on their actions), this can be regarded as all part of the process of people rightfully drawing on scientifically informed understandings as they act in the world.

Critical theory, anti-foundationalist feminism, and a discursively-oriented constructivism, however, raise concerns about the relationships between knowers that become established in society when knowing is defined as a matter of representing (natural or social) realities. From the perspective of these positions, it is important to focus on the effects that this definition of knowledge itself might have on discourse within the "scientific community" as well as between this and the wider society.

Critical theory suggests that the theoretical discourse used in social science should allow openings for people (scientists and others) to reflect upon the goal direction being pursued in both science and society. The concern is that owing to the prominence of definitions of science as ideally screening out normative reasoning, theorizing that tries to *reflect on its own normative basis* is given little credibility in the scientific and wider public community. Critical theorists develop a mode of theorizing (and a way of accounting for it) that appeals to people to join a process of discourse around the normative standards governing the practice of science in society. They suggest that public discussion around the role of science in terms of its relation to society, can and should become part of scientific theoretical discourse. In this way they provide for the "justification" of science in communicative terms.

Anti-foundationalist feminists have opted to concentrate attention on what they see as the controlling character of a research process in which scientists are geared to creating valid scientific accounts — where validity in turn is defined in terms of researchers' success in finding out about reality. The concern of such feminists is that the orientation of scientists toward trying to access reality, goes hand in hand with the presumption that the path to knowledge chosen within (some part of) the research community does indeed offer a better chance of accessing the posited reality. However, this presumption carries *cognitive as well as political implications* that require further consideration. Anti-foundationalist feminists prefer that the definition of knowing becomes revised to cater for the (ethical) idea that no-one has a right to claim a better way of coming to grips with "real reality".

Although the first three positions discussed above (non-foundationalism, scientific realism, and interpretivism) admit that human reason is fallible, it is still argued that somehow scientific reason can be used to progressively sift out unreasonable conclusions. As long as people are prepared to be reasonable (and are accountable to their scientific colleagues for their reasoning), then it is assumed within these positions that the scientific community can function to create conclusions that are more reality-bound than are likely to be created by "non-science". But the concern raised (in particular) by anti-foundationalist feminists is that "scientific" knowing — as defined by some set of people — becomes a means of sorting out apparently more credible accounts of reality from those less credible. This sorting activity is itself based on a tenuous starting position, which appears to be firmer than it can ever be. Anti-foundationalist feminists highlight the tenuousness of the belief that one can find a way of accessing, or even getting close to accessing, "real reality" through choosing some defined route argued to be the "path to knowledge". In the light hereof, they suggest that it is not unreasonable to call for an epistemological and at the same time ethical position proposing a way of life in society where people can orient themselves to the idea that "real reality" cannot be accessed. While realist-oriented epistemologies suggest that people simply cannot live unless they attempt to ground their ideas about reality in "real reality", anti-foundationalist feminists argue that there is no need for people to live in terms of such conventional notions of "the real".

Realist-oriented authors contend that any form of nonrealism (which they see as moving toward relativism) is unsustainable as an epistemology — in that proponents end up developing an inconsistent position. For instance, when it is averred that there "are" many realities to be appreciated, realist-oriented authors suggest that a claim is being made about something existing outside of the knower's consciousness — namely the fact that there are many realities. This claim, it is argued, is presented as a claim to truth — that is, a claim that depends on a reference to what is actually the case. But the rejoinder to this from a constructivist/anti-foundationalist feminist stance is that a statement to the effect that there "are" many realities, *expresses a suggestion concerning a way of orienting ourselves in our relationship with (our experience of) others' worlds.* Nevertheless, the question that any constructivist-oriented position is left to consider is how, in social living together, competing accounts presented by communicators might be assessed. The next chapter takes up this question.

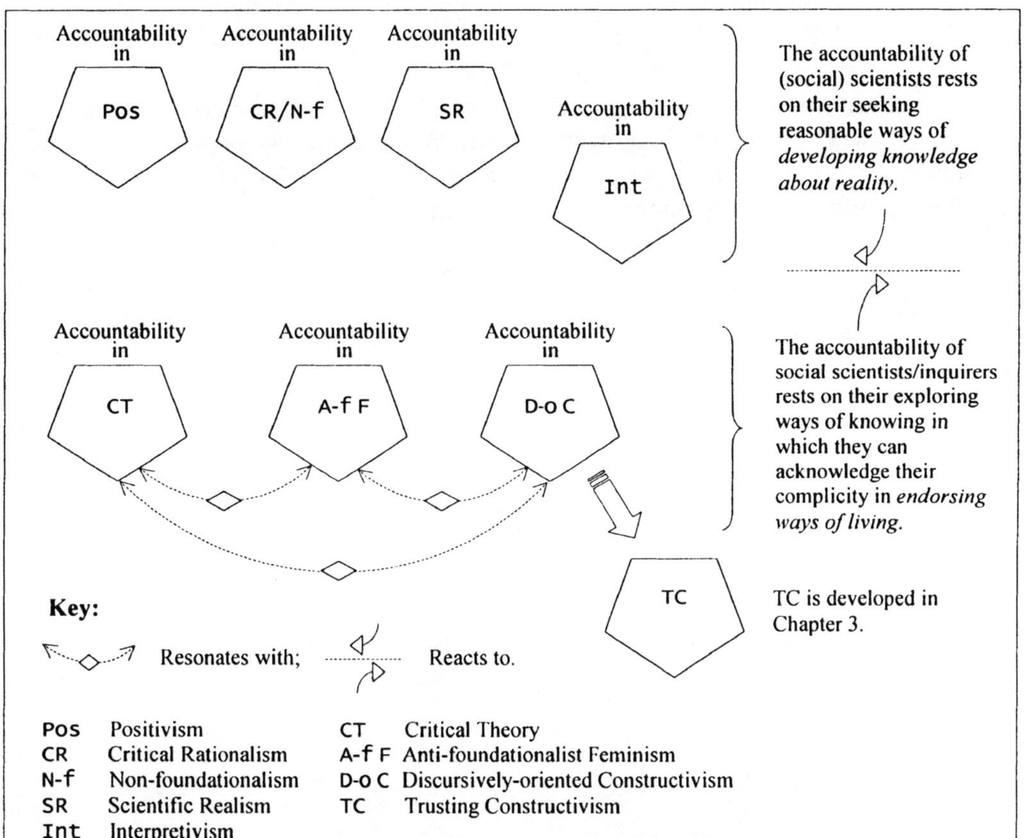

Figure 2: A locating of arguments on accountability.

Figure 2 depicts a way of locating the various arguments on accountability that have been discussed thus far. It illustrates a resonance between critical theoretical, anti-foundationalist feminist, and constructivist proposals for addressing researcher accountability in the context of considering ways of living in society. I have not tried via the figure to express connections between positivism, critical rationalism/non-foundationalism, scientific realism, and interpretivism. But I have tried to indicate that they all react to (and problematize) critical theoretical, anti-foundationalist feminist, and constructivist pleas to extend the remit of social researchers beyond the primary goal of seeking knowledge of reality. And I have also pointed out that critical theory, anti-foundationalist feminism, and constructivism introduce different concerns as they react to positivism, critical rationalism/non-foundationalism, scientific realism, and interpretivism in various ways.

It should be noted that in Figure 2 the interpretivist position has been shown as offset from aligning with positivism, critical rationalism/non-foundationalism, and scientific realism. This is in order to take account of the interpretivist suggestion that the human experience of reality needs to be given specific attention in the process of studying the social world. However, notwithstanding the interpretivist insistence on approaching the study of the social world as an enterprise that is fundamentally different in kind from natural scientific inquiry, it is still considered the task of scientists to seek to develop knowledge about, in this case, social reality. Their accountabilities are judged accordingly (as illustrated in the figure).

In the next chapter, I develop (trusting) constructivist arguments concerning the practice of social inquiry (and researchers' accountabilities); and I spell out how such argumentation can be seen as linked to options for ways of living in society. I also undertake another review of points of contention in the debates pursued in the book.

3

A Reconsideration of Constructivism: Discursive Accountability Explored

3.1 INTRODUCTION

In order to develop the constructivist argument in this chapter, I begin by referring to Kuhn's (1970) conception of the practice of natural and social science. I show how he draws attention to the construction within scientific communities of ways of seeing and defining problems that are located for scrutiny. I consider how his position is regarded by Feyerabend (1993) — who proposes a relationship between science and society differing from that of Kuhn. I then refer to Edwards' (1997) definition of discourse as social activity, showing how it can be used to envisage processes of discourse within scientific and other communities.

I consider Kuhn's, Feyerabend's, and Edwards' arguments, as expressions of what I call a discursively-oriented constructivist approach to the practice of science. Having briefly outlined their arguments in Section 3.2, I proceed to examine them in relation to the tenets of positivism (Section 3.2.1). When discussing these tenets, I consider implications of a constructivist position for conceiving the development of *discursive accountability* as a way for people to express their (accountable) involvement in society. I relate this to a discussion of possibilities for trust earning and trust awarding in processes of social discourse. And I suggest that this is the focus of what I call a trusting constructivist position. To organize the discussion hereof I draw on and extend a number of authors' arguments, including that of Edwards.

This leads me onto a consideration of the accountabilities of so-called "professional" inquirers engaged in social research. I show (drawing, inter alia, on Feyerabend's deliberations) that although the boundary between professional and other forms of knowing can be regarded as blurred, it is still possible to isolate certain expectations for the practice of accountable research. In Section 3.2.2, I locate two requirements that have sometimes been cited as marking off scientific inquiry from other discourses (namely, the requirements to render processes of inquiry recoverable in defending them, and to make public in some form the results of such processes). In Sections 3.2.2.1 and 3.2.2.2, I consider how these requirements might be treated from a (trusting) constructivist point of view. Having in this way outlined what I consider to be a workable constructivist position, I show (in Section 3.2.3) how it can be developed in relationship to the alternatives discussed in Chapter 2. In Section 3.3, I bring together my discussions from Chapters 1-3 by reviewing and rounding up the debates therein.

My comparison between arguments as presented in Section 3.3, forms the basis of my discourse around the various research examples that I present in following chapters.

3.2 DISCURSIVELY-ORIENTED CONSTRUCTIVISM AS A WAY OF DEFINING THE PRACTICE OF SCIENCE

As indicated in Chapter 1, Kuhn is sometimes considered as supporting a constructivist position because of his emphasis on the role of the scientific community in the construction of science. Kuhn (1970) explores the way in which scientific inquiry proceeds within the scientific community in periods of both normal and revolutionary science. During times of normal science — which he sees as characterizing most periods of natural scientific endeavor — the scientific community works on solving problems or puzzles set within a particular way of defining its field of study. It develops "efficient instruments for solving the puzzles that its paradigms define" (1970, p. 166). It makes progress in developing this efficiency. Its approach to, and definition of, reality is largely unquestioned — and it is indeed for this reason that scientists are able to explore in depth their defined puzzles. Kuhn explains that,

> once the reception of a common paradigm has freed the scientific community from the need constantly to re-examine its first principles, the members of that community can concentrate exclusively upon the subtlest and most esoteric of the phenomena that concern it. (1970, pp. 163-164)

Kuhn notes that this exclusiveness of concentration implies a certain rigidity of perspective. When some "young man [sic] or one new to the field" proposes "a new candidate for paradigm" it is likely that the person's argument will not be given due attention within the rest of the community (1970, p. 166). But Kuhn states that the "loss due to rigidity accrues only to the individual" (1970, p. 166). By this he means that although individuals with a "fresh approach" are often not accommodated within the scientific community, there are times — "when the occasion demands" — that the "community can switch from paradigm to paradigm" (1970, p. 166). A revolutionary switch to a new way of addressing puzzles and problems then takes place within the scientific community.

In discussing the relationship between science and society, Kuhn argues that "mature scientific communities" are characterized by being insulated "from the demands of the laity and from everyday life" (1970, p. 164). The insulation from societal pressures allows the scientist to concentrate attention on problems "that he [sic] has good reason to believe he will be able to solve" (1970, p. 164). Kuhn refers to the difference between the natural and social sciences in this regard. He suggests that in the social sciences, scientists seem to be required to "defend their choice of a research problem — e.g. the effects of racial discrimination or the causes of the business cycle — chiefly in terms of the social importance of achieving a solution" (1970, p. 164). Kuhn also notes that in the social sciences, members of the research community are made aware of the "immense variety of problems" that in the course of time they will be expected (by colleagues and by others in society) to address. At the same time, they become aware of "a number of competing and incommensurable solutions to these problems" (1970, p. 165). Kuhn does not offer detail on how social scientists address this awareness of incommensurability; but he does suggest that ultimately they have to evaluate for themselves how to proceed (1970, p. 165).

Feyerabend takes issue with Kuhn on a number of points when developing his own account of the practice of (natural and social) science. To start with, he is more concerned than Kuhn is about what Kuhn calls the "losses" accruing to the individual due to the rigidity of perspective that is perpetuated in times of "normal science". Kuhn's disregard for the effects on individual scientists is not appreciated by Feyerabend. Nor does he concur with Kuhn's suggestion that a mature scientific community operating in "normal" periods needs to practice science in terms of supposedly commonly accepted standards in order to operate efficiently. He argues the case for pluralism as follows: "Pluralism of theories and metaphysical views is not only important for methodology, it is also an important part of a humanitarian outlook" (1993, p. 38). Feyerabend thus calls for a scientific outlook that admits of pluralism of ways of addressing issues.

In addition, Feyerabend questions Kuhn's conception of the autonomy of science from social pressures. Feyerabend indicates that while he can join Kuhn in questioning the grounding of science in abstract appeals to scientific rationality, he differs from him "by opposing the political autonomy of science" (1993, p. 213). Feyerabend suggests that some kind of *open exchange* can and should be organized between scientists and others in society: within such an exchange one should not need to "play the game of the intellectuals" (1993, p. 227). The exchange that currently takes place is asymmetrical — favoring only discourse within the framework of the terms set by scientists (1993, p. 228). The exchange is based on "intellectuals trying to convince their fellow citizens that the money paid to them is well spent and that their ideology should continue to assume the central position it now has" (1993, p. 228). Feyerabend proposes an alternative arrangement for the relation between science and society, where "the debates settling the structure of a free society are open debates not guided debates" (1993, p. 229). That is, the debates should not be guided by the central position accorded to scientific knowing in society.

Edwards (1997) similarly approaches the question of how science in society should be understood, by focusing on the discursive practices of those engaged in (self-named) "scientific activity", and by challenging the time-honored conception of science as "revealing the nature of the world" (1997, pp. 249-250). Edwards explores the discursive processes through which within the scientific community the faith in science as offering a way of accessing reality is perpetuated. He suggests that the manner in which the term "error" is used, creates the impression that errors are to be seen as contingently produced within a scientific endeavor that is nevertheless still able to move closer toward truth-development. The term "error" is often used by scientists to discredit colleagues' ways of proceeding, by referring to their

> personal motives and thoughts, insights and biases, social settings and commitments, a realm in which speculative guesses and intuitions can operate, and

> where conclusions and theory choice may give rise to, rather than follow from, the empirical work that supports them. (1997, p. 58)

In locating others' errors, a "contingent repertoire" is drawn upon by scientists in order to "account for how and when things go wrong, particularly in rival laboratories, and with regard to discredited findings" (1997, p. 58). The reference to the contingent repertoire is accompanied by the assumption that this realm of would-be science is an "informal side-show to the proper arena of scientific discourse". Edwards notes that "one of the things that has to be done" in the human practice of science "is resolving disagreements or contrary claims" (1997, p. 58). References to "stubbornness, interests, rivalries, and misguided allegiances" serve to explain how errors can be produced, and at the same time to suggest that the search for truth (as a process of minimizing the sources of error) is possible, as long as people are prepared to be more rational. But Edwards points out that as soon as scientists insist on defining others' "errors" in this way, the consequence is that "there is no legitimate room for debate" (1997, p. 249). The belief is that somehow "reality" can be appealed to "as a criterion for ending their debate" (1997, p. 249). However, as Edwards indicates, whether recourse to this "reality" can be invoked, is precisely what is under debate (1997, p. 249).[13]

Edwards contends that the metaphor of communication underlying the view that knowing is a process aimed at grasping external realities, is unhelpful in forwarding this debate. Edwards therefore prefers to concentrate on highlighting the character of discourse as a social activity in which people engage as they "become participants in event construction, offer ... versions of things, [and] choose amongst accounts" (1997, p. 16). The question still remains, though, as to how the choice amongst accounts to which Edwards refers, can be rendered accountable through processes of (scientific or other) discourse. This question is returned to later in the chapter.

3.2.1 Discursively-oriented Constructivism Conceived in Relation to the Tenets of Positivism

Kuhn challenges the positivist suggestion that natural scientific inquiry involves making progress in reaching "a permanent fixed scientific truth" (1970, p. 173). But he does state that the natural scientific community is efficient in being able to solve the problems that it sets for itself; and he does see this as evidence of a "mature"

---

[13]In relation to social scientific activity, many other authors have made similar remarks (albeit with less of a focus on discourse as social activity in the way Edwards conceives it). See, for example, Gouldner (1975, 1976, 1980); Knorr-Cetina (1981); Romm (1990, 1994); McKay & Romm (1992); Guba & Lincoln (1994); Deetz (1996); Weil (1996, 1997); Barry & Elmes (1997a); and Jacobson & Jacques (1997). Davis (1997) provides a detailed elucidation also of how the notion of moving closer to "the truth" can be challenged in regard to the achievements of the natural sciences. He offers an argument for interpreting natural science as an enterprise that creates and shapes reality. Fisher points out that epistemological arguments of this kind have now penetrated nearly all the disciplines, and can be seen as "transdisciplinary" (1996, pp. 161-162). (See also Midgley, 1996a, 2000.)

community (1970, p. 167). This implies that he sees the processes involved in doing natural science as in some way exemplary of scientific activity proper. Feyerabend takes issue with Kuhn in this regard. He avers that a recognition of the way in which the natural scientific community tends "normally" to exclude dissidents, should be taken as a sign to revise our understanding, and attendant practice, of both natural and social scientific activity. Instead of endorsing the idea of modeling social scientific activity in *scientistic* fashion on the Kuhnian vision of the natural scientific community, Feyerabend pleads otherwise.

As far as the tenet of *phenomenalism* is concerned, Kuhn suggests that perception takes place in terms of certain paradigmatic commitments: to "see" phenomena, is to see them from the perspective of some paradigm. He suggests that "observation and conceptualization, fact and assimilation to theory, are inseparably linked in any discovery" (1970, p. 55). Without the assimilation of facts in terms of certain conceptual categories, they cannot be said to be "observed". Feyerabend sharpens this point when he suggests that: "it suffices to remember that observational reports, experimental results, 'factual' statements, either *contain* theoretical assumptions or *assert* them by the manner in which they are used" (1993, p. 22). Feyerabend points out that usually we are not aware of the way in which our observations are shaped by theoretical assumptions. We recognize the effects of these assumptions "only when we encounter an entirely different cosmology ... not by analysis" (1993, p. 22). This feature of everyday observation is paralleled in the process of scientific observation: "The material which the *scientist* has at his [sic] disposal, his most sublime theories and his most sophisticated techniques included, is structured in exactly the same way" (1993, p. 22). Thus inventing new ways of seeing "the world" means that:

> We must invent a new conceptual system that suspends, or clashes with, the most carefully established observational results, confounds the most plausible theoretical principles, and introduces perceptions that cannot form part of the existing perceptual world. (1993, pp. 22-23)

In this way, Feyerabend concurs with Kuhn's suggestion that the perceptual world is a product of the application of a particular cosmology (way of defining the world).

Edwards takes a similar line of argument when he notes that disputes involving differing accounts of the world cannot be resolved with reference to an appeal to extra-linguistic realities. The discursive practices of people (including scientists) produce "versions of the world" that cannot be checked against "the world". So he states:

> In discursive psychology [the position he adopts] there can be no process of "reality checking" that is independent of descriptions, except as a rhetorical device in itself, deployed both in science and in ordinary talk, for "externalizing" descriptions ... . (1997, p. 253)

People may try to externalize their descriptions by arguing that they refer to a reality that exists outside of the descriptions — but there is no manner in which a reality check can be accomplished by people engaged in disputes (as is implied by the phenomenalist tenet of positivism).

Following the constructivist reconsideration of the tenet of phenomenalism is a reconsideration of the tenet of *empiricism*. The positivist tenet of empiricism is linked to the view that striving for truth, via recourse to the evidence of reality, is the goal toward which scientific inquiry is properly directed. However, Kuhn, in contrast, suggests that "we may ... have to relinquish the notion, explicit or implicit, that changes of paradigm carry scientists and those who learn from them [their audiences] closer and closer to the truth" (1970, p. 170). Feyerabend extends these deliberations of Kuhn. He criticizes the positivist suggestion that induction, combined with deduction, is the logic by which scientists proceed to verify certain views of the world (while discounting others as false). He suggests that even taking the positivist view that science can be used to make advancements in knowledge, it should be recognized that scientists have in the past often proceeded in terms of procedures that, at the time, could have been, and were, argued to be unreasonable (1993, p. 117). For instance, Galileo could be argued to have transgressed the protocols of scientific inquiry that defined what was reasonable scientific behavior in his time. Feyerabend remarks that a modern scientist cannot

> cling to his own very strict standards and at the same time praise Galileo for defending Copernicus ... . The situation becomes even more complex when we consider that the Copernicans changed not only views, but also standards for judging views. (1993, p. 129)

Feyerabend makes the further point that there is no reason to believe that science, in whatever way it proceeds, can make advancements in developing "truths" that should be regarded as sacrosanct. He indicates in this regard that the Church in Galileo's time had wanted him to "teach Copernicus as a *hypothesis*; he was forbidden to teach it *as a truth*" (1993, p. 132). Feyerabend sympathizes with the Church in wanting Galileo to admit the fundamentally hypothetical character of the statements that he made: the aim of the Church was to "protect people from the machinations of specialists" (1993, p. 133). The Church's intention in this regard, as far as Feyerabend sees it, was correct: it had the "right social intention" (1993, p. 133).

Edwards likewise suggests that if we lived in a world where "descriptions would be definitive and accurate, explanations would be correct, and knowledge would be assured" then the (positivist-inspired) search for organizing a "one-to-one mapping between language, mind, and reality" could be accepted. But, he notes, "things are not like that, and it is time to count it not as a nuisance or inconvenience ... but as a blessing" (1997, p. 18). He argues that in scientific as in everyday discourse, the inability of people to find a way of mapping statements onto extra-linguistic reality should not be regarded as a nuisance. Instead of concentrating attention on determining a scientific route that supposedly offers us better access to "reality" as such, Edwards suggests that we should shift our focus (as inquirers) to "how specific versions of the world are put into place, undermined, and defended" (1997, p. 19).

Positivist argumentation suggests that the ideal of *value freedom* should in principle guide scientific activity; and this can render it distinct from "ordinary" inquiry (undertaken as part of everyday life). But Feyerabend undercuts the positivist tenet of value freedom by expressing his concern that science — natural or social — might contain ways of seeing the world that *come to exclude other possible and valuable relationships with "the world"*. He gives an example of the way in which a science aiming for objectivity can end up (arguably) imposing a worldview that is not free of values. With reference to the application/imposition of a Western scientific approach in the (so-called) Third World, he indicates that:

> Both human decency and some appreciation of the many ways in which humans can live with nature prompted agents of development and public health to think in more complex or, as some would say, more "relativistic" ways. (1993, p. 248)

The term "relativism" here refers for Feyerabend to a recognition that science need not be defined as offering information that can be used for guiding life decisions. As Feyerabend argues, "the search for objective guidance is in conflict with the idea of individual responsibility" (1993, p. 249). For Feyerabend, the recognition that scientific "experts" are fallible is what allows (or should allow) people to treat the scientific enterprise as one amongst other possible ways of knowing that might be adopted (1993, p. 251). If people assign a central place to scientifically derived information, there is the danger that this will be turned to for "objective" guidance — and will impact unduly on their choices (1993, p. 251). Feyerabend argues that in the light of Kuhn's contribution to the sociology of science (in exploring the social practices involved in doing science) it can be shown that "a general appeal to the authority of science runs into contradictions" (1993, p. 271). This appeal can therefore be "undermined" (1993, p. 271).

Feyerabend's argument in relation to the authority of science also implies that we can no longer (after Kuhn) accept the positivist tenet of *instrumental knowledge*. This tenet is based on the presumption that if science is practiced properly (utilizing reasoned argumentation in relation to "the evidence") it offers a route to generating some knowledge about (efficient) ways to pursue desired goals. But Feyerabend suggests that science cannot, and should not, perform in society the function of exercising an authoritative voice in this regard. Feyerabend notes, nevertheless, that by rejecting the general appeal to the authority of science, we still are left with the issue of "how science should now be interpreted and used" (1993, p. 271). He argues that it is not for him to try to answer fully these questions. Rather, they "have to be answered by the interested parties themselves, according to their standards, conceptions, [and] cultural commitments" (1993, p. 271).

Edwards' (1997) focus on discourse as social activity provides some entry point for (re)considering this issue as raised by Feyerabend. Edwards suggests that it might be fruitful to explore ways in which people — scientists and others — offer versions of "things" that they add to ongoing debates, and create their own choices (amongst accounts) at points in time. By shifting the focus of his analysis away from a concern with excavating "what the world is like" (1997, p. 16), Edwards may be seen as endorsing the development of climates of human trust based on people's recognition that, as Salipante and Bouwen put it, "the interplay of differing perspectives will be ongoing and will continue in future conflicts" (1995, p. 93). Salipante and Bouwen suggest that it is important to cultivate (rather than destroy) "the value and legitimacy of multiple perspectives [so as to] permit the examination of a broader set of interpretations and alternative actions" (1995, p. 93). For them, human trust can just as well be built on this basis than on people's expectation that their conflicts of perspective can somehow be resolved.

But if climates of trust are to be built up amidst recognition of the value of a multiplicity of ways of seeing, how may people's accountabilities be directed? One way of conceiving this is to highlight the expectation that people will be sufficiently sensitive in their engagement with others to *take into account a range of considerations springing from debate with alternative perspectives and values.* (See also Romm, 1996b, p. 34, 1996c, pp. 184-185, and 1998a, pp. 69-70.) Of course, in working with such a view of (discursive) accountability, those holding other people to account can never know whether indeed someone, or some group, has taken seriously enough alternative positions, or whether these have been sufficiently brought to bear in their thinking. All that can be said — in terms of this argument — is that if some party(ies) feel(s) that others have not been sufficiently sensitive in their considerations, then they would need to re-enter alternative concerns into the arena, with the aim of developing the trust relationship. An implication of this, though, is that people should be able to earn trust other than by trying to defend positions through claiming recourse to an understanding of extra-linguistic realities. Conversely, trust awarding becomes a matter of allowing people to make choices — defended in discursive engagement with alternatives — about both how to see and act.

This is not to say that trust awarding need be regarded as set within a social relationship once-and-for-all. The according of trust requires some "leap of faith" that the trusting party will not be disappointed (Wicks, Berman, & Jones, 1999, p. 100). The workability of such faith can be assessed only over time. Trust awarding is therefore not a static feature of a human relationship. As suggested by Wicks, Berman, and Jones, "relationships unfold so that individuals continually update their information base and their decisions to trust" (1999, p. 101).

What is proposed in terms of the argument offered here — toward a trusting constructivist position — is that people's decisions to trust others as they create and enact ways of seeing at points in time, can be focused around considerations of whether it is felt that they have expressed a *discursively-accountable* orientation, as elucidated above. It need not be focused on requirements to ground "knowing" in some (supposed) recourse to extra-linguistic realities. Indeed, it may be argued, through justifying themselves in this way, people do not necessarily instill others' trust in their ways of seeing and acting — especially insofar as others believe that there are alternatives that have not been sufficiently accounted for in the decision-making. And, as Edwards (1997) notes, an appeal to "scientific" ways of knowing to settle the debates, is unhelpful in this regard — because the question of how people might justify the supposedly better access to reality is still under debate.

But if science cannot perform the function in society of foreclosing options appearing "unrealistic", is there any point in still separating out "scientific" from other forms of discourse in society? This question is explored below.

3.2.2 The Distinctiveness of "Professional" Inquiry

Scientific or professional inquiry is sometimes seen as expressing its distinctiveness from other discourses in society in terms of an explicit allegiance on the part of inquirers to both:

- render visible the conduct of the inquiry process — so that this is laid open for anyone wishing to appreciate/review the way in which the inquiries have been organized (Shipman, 1982, p. xiii; Gummesson, 1991, p. 159; Checkland & Holwell, 1998, p. 17); and
- disseminate, in appropriate ways, interpretation(s) of the results developed through the inquiry process — so that academic and other audiences (within and outside of the specific domain where the research has been set), can engage with the study (Gummesson, 1991, p. 161; Flood & Romm, 1996, p. 135; Hall, 1999, pp. 152-154).

These two points of (possible) distinction can be addressed from within a (trusting) constructivist position as follows.

3.2.2.1 Accounting for Strategies of Inquiry

A distinguishing mark of so-called professional inquiry (in line with expectations for professional conduct) is that attempts are made to make visible the way in which inquiries are conducted; the processes by which investigations are undertaken, need to be recovered for others to examine.

A trusting constructivist position would add to this requirement that in organizing, and accounting for, their manner of proceeding, some attention should be given by the inquirers to keeping alive a discussion around differing criteria that might be used to define "acceptable" research practice. (See also in this regard Gummesson, 1991, p. 159.) Feyerabend's arguments concerning the history of scientific practice are instructive here. He argues that "in the 19th century the idea of an elastic and historically informed methodology was a matter of course" (1993, p. 10). He treats Mach's statement that "the examples of great scientists are very suggestive", as implying that these examples serve to make the mind of scientists "nimble" — rather than to offer strict rules to follow (1993, p. 10). Feyerabend comments that scientists are now trained to believe that "logic" should condition their work in the scientific domain, and that they should "inhibit intuitions that might lead to the blurring of boundaries" in relation to other forms of knowing (1993, p. 11). They are trained so that their "imagination is restrained" to what is considered to be scientific ways of proceeding: even their "language ceases to be [their] own" (1993, p. 11).

Feyerabend indicates that from his interpretation of the history of science, scientists have not remained within the boundaries of traditions "defined in this narrow way" (1993, p. 11). He states in this respect that "inventing theories and contemplating them in a relaxed and 'artistic' fashion, scientists often make moves that are forbidden by methodological rules" (1993, p. 148). He argues that if people are called upon to defend their way of proceeding "against all those who will accept a view only if it is told in a certain [scientific] way and who *will trust it only if it contains certain magical phrases called 'observational reports'*", then an impoverishment of the inquiry results (1993, p. 17, my italics).

Following this line of argument toward a trusting constructivist position, a scientific approach to research would be seen as serving the function of allowing others to participate in the processes — including imaginative moves — that go to make up the research strategy (how ever emergent this may be).[14] The expectation that scientific inquirers make an effort to render this visible, suggests that to be labeled "professional", they do have to account for their manner of organizing their investigations (see also Sharma, 1997, pp. 764-765). But a certain trust also needs to be afforded to them — so that they do not feel compelled to have to iron out the areas of their work that seem not to fit some (supposedly accepted) protocol in order to defend themselves. Those engaged in organizing research can then earn trust, without having to earn it by denying the human ("subjective") character of the research effort. This can be seen to apply in the case of both natural and social scientific research. In the case of social scientific research, nevertheless, trust earning would have to involve an explicit discursive engagement with alternative visions that may be presented regarding the purpose of the "professional" inquiry in terms of its relationship to ordinary sense-making in social life.[15]

3.2.2.2 Creating Publicly Available Information

Scientific inquiry may be characterized also with respect to an intention to present publicly in some form — to academic and other audiences — the results, how ever tentative, created via the conduct of the inquiry. (The form(s) used to discuss results need to take into account what is considered to be respondents' right to remain, if

[14]I follow Layder's suggested use of the word "strategies" to point to (or invoke discussion around) "ways of doing research", and his distinction between this and techniques for data gathering (1993, p. 108). However, I prefer to regard the latter as devices for data *evocation*. It may also be noted that the term "method" is sometimes used by authors instead of the term "technique" to point to a style of collecting data; but because the term has traditionally been associated with the idea that there is a "scientific method" that can be followed in doing science, I have chosen to avoid it where possible. I use the term only seldom (in cases where authors of reports that I have cited have used it).

[15]Lincoln uses the term "trustworthy" to symbolize social research that shows an appreciation of the "political and ethically sensitive and complex" character of social scientific research (1995a, p. 277). A trusting constructivist position takes up this concern of Lincoln and considers what kind of (discursive) relationship is involved in people awarding others (including researchers) their trust.

desired, anonymous in reports involving them.) However, what a constructivist perspective emphasizes is that constructions can be fashioned without either authors or audiences regarding these as serving to advance knowledge defined as reflection of external realities. A good construction may be considered as one that is experienced as providing what Barry and Elmes call a "provocative optique, a view that opens up new trains of thought" (1997b, p. 432). While opening up new trains of thought, it can simultaneously invite further understanding and meaning-making (Romm, 1997a, paragraphs 7.3-7.7). Information provided can be worked with, and past, by people as they address its (possible) relevance in different contexts (Romm, 1997b). A constructivist position focuses on the requirement for researchers to indeed invite continuing discursive participation in the interpretation/review of what is "found" through some study.

In regard to the addressal of social research, Kilduff and Mehra suggest that "provocative research" can be guided by the intention to encourage creativity (1997, p. 454).[16] Readers can be invited to "analyze the work from their own perspectives" (1997, p. 469). In this way, it can serve to "ignite discourse" (1997, p. 469). Insofar as the inquiry leads to the provocation of discourse in relation to conflicts, tensions and dilemmas faced by people, it can be regarded as (potentially) meaningful. The task of social theorizing is thus not seen as an attempt to "capture" the realities of social life, but rather to make some contribution toward enriching the discourses through which the dynamics of human interaction are enacted. People can continue their engagements with issues toward which the inquiries draw attention — in the recognition that there is no one way of defining the realities in which they, with others, are involved.[17]

In summary, a trusting constructivism would see the arena of scientific/professional research as one where publicly available constructions (presented as constructions) are created, using strategies of inquiry that are accounted for with reference to an engagement in debates about criteria for judging "acceptable" conduct in this arena. The way in which a trusting constructivist view of accountability in social research can be positioned in relation to other positions is given attention in the next section.

---

[16]In laying out their argument, Kilduff and Mehra indicate that they subscribe to a form of postmodernism, as long as this is not understood as implying a call for "the death of all scientific inquiry; ... [and] a banishment into utter relativism wherein a clamor of fragmented and contentious voices reigns" (1997, p. 454). They suggest that "affirmative postmodernism" still implies the possibility of "making discriminations among competing interpretations" (1997, p. 455). However, as I noted in my discussion of anti-foundationalist feminism (Chapter 2, Section 2.7), the criteria for organizing such "discrimination" are reshaped in postmodernist discourse (in relation to other discourses).

[17]I have left in abeyance in this book a discussion of how the development of different forms of human interaction may be linked to a different sense of being in/with so-called "nature". It can be argued, as, for example, does Reason (1994), that a sense of our participation in reality-construction means that we do not experience our existence as apart from an experience in/with nature. (See also Fals-Borda, 1991.) Romm (1998b) offers an outline of a way of dealing with different options that may be experienced for nurturing sustainability in engagement with our worlds.

3.2.3 Trusting Constructivist Considerations on Researcher Accountability

3.2.3.1 Positioning Trusting Constructivism in Relation to Positivist Argumentation

The positivist tenet of *scientism* suggests that the accountabilities of social scientists are to be defined similarly to those of natural ones. In both types of research, scientists are supposed to be directed toward developing their theorizing around exploring possible connections between variables that have been isolated for attention. Davis, Schoorman, and Donaldson (1997) give an example of the kind of studies that would be endorsed within positivist-inspired research aimed at investigating human organizational realities. In this section, I use this example to indicate briefly how a trusting constructivist position would review it. (I use this example because Donaldson — as noted in Section 2.2.1 — labels himself as being positivistically-oriented.)

Davis, Schoorman, and Donaldson suggest that it is possible to set up investigations into what they call stewardship behavior in organizations. They put forward a number of propositions, to be tested through the protocols of scientific procedure. They indicate that their proposed theory of stewardship disputes some of the claims of "agency theory" regarding people's "individualistic utility motivations" in organizations (1997, p. 20). They propose that there exist motivations and behavior that cannot be explained in terms of this theory. According to the stewardship theory that they present as an alternative, the behavior of stewards in principal-steward relationships can be explained in terms of their operating pro-organizationally. One proposition that they proffer for testing this suggestion is as follows:

> People who have high identification with the organization are more likely to become stewards in principal-steward relationships than are people who have low identification with the organization. (1997, p. 30)

Referring to the term "identification" in the proposition, they suggest that identification occurs when people "define themselves in terms of their membership in a particular organization by accepting the organization's mission, vision, and objectives" (1997, p. 29).

A constructivist position, however, revisits the proposition, and the terms utilized therein, by calling attention to the way in which it already reflects back on Davis, Schoorman, and Donaldson's vision of relevant issues of concern. It seems that in terms of Davis, Schoorman, and Donaldson's understanding of organizations, people's possible expressions of doubt in regard to their sense of "identification" therewith, would be treated as indicators of a lack of pro-organizational commitment. It is not treated as a matter of concern in the inquiry that people might face dilemmas of integrity in deciding how to orient to "the organization". This issue is left out of the terms of the inquiry — being judged as irrelevant to it. But from an alternative point of view, it might be argued that felt tensions on the part of people can just as easily be considered as an expression of organizational commitment — understood as a concern with the way in which policies are being instantiated at some point in time. (See, for example, Salipante & Bouwen, 1995, p. 80-81; Gergen & Whitney, 1996, p. 350; Maclagan, 1998, p. 131; and Weil, 1998, p. 57.)[18]

From a constructivist perspective, the question of how to "indicate" or operationalize the terms used in the case of this inquiry, cannot be resolved by continuing to propound a conception of language as referring to external realities. Rather, the constructs utilized (and statements involving them) should be regarded as creating (contestable) frameworks for human sense-making.[19] Hence, instead of treating Davis, Schoorman, and Donaldson's hypothetical propositions as ones that might become corroborated if the evidence turns out (or is conceived as) supporting them, they would be seen as offering entry points for continuing addressal of issues raised. They would be seen as engendering a "provocative optique" — to use Barry and Elmes's language (1997b) — which opens up certain trains of thought. In whatever way attention is directed by those concerned (professional researchers and others), what is important is that the inquiry is recognized as *making an input* into the discourse about ways in which everyday conduct might be conceived and enacted.

A constructivist position suggests that scientists' accountabilities are diminished to the extent that they fail to take due account of the ways in which their *framing of realities* might have an influence in directing activities in social life. It appears that Davis, Schoorman, and Donaldson are satisfied to define their accountability as researchers in terms of whether they believe they can substantiate their claims with reference to "the evidence". But as Gummesson argues, "facts themselves are not that unambiguous" (1991, p. 123). In the case of this example, Davis, Schoorman, and Donaldson appear to have resolved the ambiguity around ways of treating people's "identification" with organizations in favor of a preferred way of defining what they call pro-organizational behavior. But, by their very decision to code phenomena this way, it can be argued that they set in motion a specific way of relating to participants (through setting questions for them to consider in regard to their organizational commitments). Also, when they later make statements about the results of their studies into pro-organizational behavior, this too can lead people to the point of deciding to address felt ambiguities in favor of David, Schoorman, and Donaldson's suggestion that pro-organizational behavior can be clearly defined.

---

[18]Davis, Schoorman, and Donaldson are aware that sometimes policies justified with reference to "organizational success" are seen (by others) as operating to the detriment of certain stakeholders. But they suggest that "a steward who successfully improves the performance of the organization generally satisfies most groups, because most stakeholder groups have interests that are well served by increasing organizational wealth" (1997, p. 25). Thus, according to them, it can be assumed that a steward operating with the intention of "maximizing organizational performance" operates in keeping with advancing a general social interest (1997, p. 25).

[19]For elaborations of this argument in relation to social scientific research, see also Knorr-Cetina (1981, p. 99); Heritage (1984, p.145); Cooperrider, Barrett, & Srivastva (1995, pp. 170-171); Dachler & Hosking (1995, p. 6); and Shotter (1993a, pp. 50-52). (Shotter uses the term "social constructionism" to focus specifically on the way in which constructs feature in our everyday existence.)

Maclagan raises similar concerns to those presented here by pleading (in the context of management science) for researchers to explicitly introduce moral considerations into their discourse. Taking the notion of "organizational goals" (a notion used in both scientific and everyday discourses) as a way of making his point, he argues that this notion is

> often ambiguous, and from an ethical standpoint other functions of an organization's activity, such as the provision of employment, service to dependent customers or pollution control, may be just as significant as return on investment for shareholders or benefits for local taxpayers in the case of a county council. (1998, p. 148)

Maclagan shows with reference to this and other examples that if social researchers try to sanitize moral considerations out of the language of their science, they fail to highlight for consideration the way in which people might develop their moral sensitivities in organizational and wider social existence. According to Maclagan, moral considerations are brought to bear in our thinking when we recognize "that there are several stories operating, not one objective story" (1998, p. 49). Part of the complexity of moral decision-making arises when we recognize that specific ways of framing "the realities" are themselves not without consequence. Or as Gummesson, citing Doctorow (1977), notes, "the world in which we live is still to be formed" (1991, p. 122). A trusting constructivist position suggests that it is preferable for people to indeed conceive "reality" as that which is in the process of being formed. Such a conception might encourage forms of discourse where people operate through a serious engagement with different "stories" as a way of expressing their involvement in social reality.[20]

The following bullet points provide a rendition of the trusting constructivist argument positioned in relation to positivism.

- In terms of a positivist conception of science, researchers' adherence to scientific protocol is presumed to be a matter of their reasoning on the basis of evidence amassed, toward improving their theorizing about reality.
- In positivism, the accountability of scientists is expressed through their attendance to scientific protocol in drawing conclusions from the available evidence.
- According to trusting constructivism, *the accountability of researchers is tied to their acknowledgment of the way in which knowing activities (including their own) can be rendered more discursively-oriented through taking into consideration a*

[20]Crotty (1998, pp. 10-11) indicates that one could use the term "ontology" to refer to a way of being-in-the-world. He indicates that although some authors speak of positivism as entailing an ontology (in its view of reality as made up of discrete and observable events), he believes that this term is not helpful in pointing to debates about ways of treating our (social) existence. He prefers in any case not to introduce the term "ontology" into his comparison of different arguments in the research literature. I have adopted a similar position in this book.

variety of arguments and concerns pertaining to the process and product of their research endeavors.

3.2.3.2 Positioning Trusting Constructivism in Relation to Non-foundationalist Argumentation

According to a non-foundationalist position (as expressed, for instance, by Hammersley and Gomm), researchers' accountabilities are linked to their "do[ing] their utmost to find and keep to the path which leads toward knowledge rather than error" (Hammersley & Gomm, 1997a, paragraph 4.3). Hammersley and Gomm argue that researchers can admit that they create constructions in the process of research — as long as this admission goes hand in hand with an attempt on their part to access phenomena existing outside of their knowing endeavors (1997a, paragraph 4.2). But the constructivist rejoinder to this argument is that the possible effect in society of a realist-oriented way of conducting and presenting inquiries can itself be called into consideration as an issue that needs to be confronted. The reliance of people in their discourses on some supposed reference to extra-linguistic reality — buttressed by science — can then be given due attention.

Of course, a non-foundationalist self-understanding of science can, and does, point out that people should not rely on scientists to offer the truth — as truth-seeking is, admittedly, subjected to errors of human judgment within the scientific community. The non-foundationalist position suggests that once people in society are made aware of this, then scientists are absolved of responsibility if people still insist on treating scientifically derived information as being a product of an infallible process of discovery. But a constructivist position retorts that if scientists are to prevent people assuming that somehow realities *can* be accessed through scientific inquiry, then they need to change the language used in accounting for their inquiries. If social science is to accept the "burden of the loss of certainty in our post-ideological age" — as Delanty (1997, p. 141) puts it — it cannot present itself as an expert discourse to which lay people need to turn in order to attain more or less accurate accounts of reality.

A constructivist position therefore shifts the focus of attention so as to enable both professional inquirers and others to consider the ways in which, as Keys indicates, any proffered analytic knowledge itself might become part of a "dynamic and emerging situation" (1997, p. 9). Constructivism extends Keys' considerations to the point of suggesting that it is crucial for social inquirers to take into account the possible social enactments that might be set in motion through the way in which their inquiries are undertaken and reported upon.

This focus has a number of implications. One implication is that the so-called "reactivity effect" is construed differently from the way in which it is conceived within a non-foundationalist position. For non-foundationalism, the reactivity effect (the effect that can be exerted on participants as they react to the presence of the researcher/observer) is an effect that should be recognized as possibly operative. Respondents' behavior and attitudes can then be understood as being in part a response to their being "under study". The reactivity effect can also be seen as something whose likelihood of occurring in different research situations can be investigated — so that researchers can explore what sorts of circumstances tend to produce greater or lesser of such effects. Knowledge of people's responses to varying circumstances (for example, varying research situations) can thus be advanced, and ways of minimizing the reactivity effect can also be suggested.

A constructivist approach follows the anti-foundationalist feminist one in arguing that the so-called reactivity effect can be treated differently. It can be regarded as an opportunity, for those initiating research processes, and others involved, to set up a relationship in which all parties are recognized to be active participants in making sense of the interactive process. What is then required is that the parties can account for their own presence in making a difference to the way in which the interaction proceeds, in such a way as to create spaces for one another to contribute to defining the meaning of the inquiry process. The "reactivity" of the parties to one another then becomes a matter of each trying to take seriously the manner in which the other is defining the research situation of which they are part. Constructions around ways of setting up the research situation can then be developed and accounted for in relationship to an appreciation of others' views of "the situation".

Furthermore, another implication of a constructivist approach is to counsel inquirers not to treat findings arising out of the research process as an account about reality that needs to be reported in these terms. It is emphasized that just because of this way of reporting, the possibility of self-fulfillment of reports might be increased. So, for instance, a constructivist position takes seriously Jervis's indication that, as he puts it,

> descriptions of the economic and social health of a neighborhood can be self-confirming because they depend in large measure on the characteristics, attitudes, and behavior of people who live there, which in turn are influenced by the descriptions people believe (and think others believe). Thus media spokespersons are incorrect to claim that they merely report what is happening ... . (1997, pp. 148-149)

Just as (journalistic) reporters arguably contribute to *creating* the news on which they "report", so the same can be said of scientists — whose power in society may indeed be strengthened through the realist language that they employ. This, it may be argued, is the import of Jackson's caution that we should not underestimate the power that scientific language can exert in society (1991, p. 212). This is also the import of Romm's (1995) account of the way in which knowing can become an act of intervention in the social network, and of Midgley's similar argument (1996b, pp. 19-20). Midgley explores this argument in the context of considering how inquirers make boundary judgments. As he notes, "the business of setting boundaries defines both the knowledge to be considered pertinent *and* the people who generate that knowledge (and who also have a stake in the results of any attempts to improve the system)" (1996b, p. 18). He proposes that the justification of boundary judgments requires alertness to the possible implications of setting them in a particular way. It requires an appreciation of what he calls "systemic logic" (1996b, p. 20). (See Midgley, 2000, for a full discussion hereof.)

In response to these kinds of qualms, Hammersley and Atkinson assert that they too are concerned about the practical effects that research activities might have in society — but that they do not want this concern to detract from research efforts toward truth-seeking. For them, the prime task of social researchers is to use the process of systematic inquiry in striving to make advancements in moving closer to the truth (1995, p. 17).

But in terms of constructivist argumentation, the non-foundationalist position is still too intent on preserving what Feyerabend calls the "chauvinism of science" (1993, p. 163) to appreciate that research need not be organized by attempting to follow the path that (supposedly) leads toward knowledge. Rather, it can be reconstructed and enacted as affording an opportunity for researchers, with others, to reflect on their possible impact in the world as part of their "knowing" endeavors (Romm, 1995, p. 137, and 1997a, paragraph 8.1). It is through these reflections that they can attempt to earn others' trust.

The following bullet points provide a rendition of the trusting constructivist argument positioned in relation to non-foundationalism.

- According to trusting constructivism, the possible reactivity effect (and attendant social enactments) created through the research process, as well as the possible self-fulfillment of presented findings (both of which are indeed acknowledged in non-foundationalism), need to be considered as part of the range of concern of researchers (as researchers).
- In non-foundationalism, the accountability of social researchers is directed toward the research remit of seeking truth, defined as representational knowledge of reality.
- In trusting constructivism, *accountability of social researchers requires their taking into account the possibility that their ways of organizing inquiries and presenting "results" might penetrate ways of seeing and enacting social life: provision therefore needs to be created (within the research process and through its dissemination) for responsible choice-making in terms of these considerations.*

3.2.3.3 Positioning Trusting Constructivism in Relation to Scientific Realist Argumentation

The scientific realist position follows the non-foundationalist one in suggesting that the scientific community can (and should) become organized to separate out better research from worse research. This implies that scientific activity is properly directed by researchers' trying to develop informed theorizing through some form of dialogical involvement in the scientific community. Layder suggests in this regard that the dialogue should not be

> completely open to all and sundry influences. The interchange and dialogue I envisage must be governed by the theoretical presuppositions implicit in the strategies and underpinning the overall perspective that I have employed [in the book]. (Layder, 1993, p. 209)

Appropriate dialogue in the scientific community should in other words for Layder be governed by the quest to develop knowledge of reality — knowledge being defined in terms of (scientific) realist presuppositions. When those engaging in scientific activity define scientific reasoning in this way, the realms of fact and value are supposed to be kept separate. But a constructivist position suggests that the endorsed form of argumentation already itself contains a value. This is the value of justifying "reasoning" with reference to the manner in which accounts are created using available evidence to make inferences about (underlying) realities. This definition of good reasoning within the scientific community may in turn have implications for styles of living in society: people become expected to justify their way of viewing situations in the social sites in which they are involved in terms of the likelihood that their constructions tell it (more or less) "like it is". A different way of organizing world-construction activities thus remains unexplored.

In response, a scientific realist position — as other forms of realism — suggests that there is no reason to believe that an apparently "different" way of organizing world-construction activities is either possible or desirable for human beings. It might be argued indeed that the likely explanation of why people in social life continue to adopt a realist conception of knowing — and therefore continue to turn to science as offering information about reality — is because this is what knowing is about. In addition, it could be argued that even if we might be able to redefine knowing along constructivist lines, there is little value in doing so. Why, it might be said, should the value of relinquishing the quest for knowledge of reality take precedence over the value of this quest?

A constructivist answer to this question would suggest that, in any case, a discussion about the purpose of science in society has to include considerations of the different ways of living that might be constructed and enacted through the way in which "scientific knowing" becomes defined. Alternative values need to be brought into confrontation as part of a discussion about the way in which the accountabilities of social researchers might be conceived. Decisions about how to *know* at the same time involve considerations about how to *live*. The (scientific realist) presupposition that knowing *must* be directed toward an understanding of extra-linguistic realities, fails to appreciate the way in which this definition of knowledge itself may contribute to constructing a lifestyle that needs to be accounted for. A trusting constructivist position offers as an entry into the debate around researcher accountabilities a conception of an alternative way of knowing and living.

The following bullet points provide a rendition of the trusting constructivist argument positioned in relation to scientific realism.

- Scientific realism, as all forms of realist epistemologies, treats the endeavor to comprehend reality as a separate activity from contributing to its unfolding.
- In scientific realism, accountability of social scientists is tied to their propensity to engage in dialogue with a variety of ways of theorizing in relation to the evidence available — with a view to developing a more or less accurate picture of social reality.

- In trusting constructivism, *accountability of social researchers requires that they take account of ways in which definitions of knowing themselves may affect the constitution of ways of living in society.*

3.2.3.4 Positioning Trusting Constructivism in Relation to Interpretivist Argumentation

Interpretivist-oriented argumentation regards the accountability of social scientists as lying in a propensity on their part to ground their accounts of social life with reference to the meaningful action of those being studied. But from a constructivist point of view, it is important to focus on the possibility that the construction of accounts supposedly "about" social reality might be part of the forming thereof. If reality is seen, as Gummesson (1991, p. 122) suggests it can be, as that which is in the process of being formed, then it is possible to take the interpretivist focus on its meaningful character to the point of concentrating on the way in which research itself is implicated in social meaning-making.

Instead of seeing researchers as ideally operating as knowing observers who can act as "change agents" through supplying knowledge of consequences of pursuing lines of action, they can, as Weil indicates, be seen as "implicated and embedded in the realities they [with others] are creating" (1998, p. 43). Constructivism draws attention to the possibility of what Weil calls "giving meaning to ... different epistemologies of practice as they are lived out in the behaviors and choices of people" (1998, p. 43). The Weberian epistemology, it can be suggested, still allows some "knower(s)" to define their remit as seeking information that can later supposedly be used by other people in society to manage change in accordance with goals set. But constructivism pays particular attention to the importance of acknowledging that as questions are posed and issues raised in research situations, so this already might affect the way in which "subjects" come to think about these issues. And as reports on findings are offered, these too can echo the initial manner of posing questions and interpreting answers as organized by those initiating the research. It is therefore deemed important that those defining themselves as "doing research" take some responsibility for the possibility that their own understandings, and ways of expressing and working with these, might affect the experience of reality for others.

The Kuhnian argument concerning differing "paradigmatic" assumptions that may underpin research endeavors, can thus be treated as an opening to *revisit the epistemological assumption* that scientific knowing must be directed toward knowing "about" (social) reality. According to a trusting constructivist position, then, the parameters of social scientific inquiry are still unnecessarily restrictive within the interpretivist position.

The following bullet points provide a rendition of the trusting constructivist argument positioned in relation to interpretivism.

- While conceding to the interpretivist suggestion that the social research process should take into account the meaningful constitution of social life, constructivism concentrates on how social research may itself become implicated in social meaning-making.

- In interpretivism, accountability of social researchers is tied to their propensity to develop theorizing about social reality with special reference to people's meaningful experience thereof.
- In trusting constructivism, *accountability of social researchers is tied to their propensity to account for the manner in which their epistemology in use may contribute to a way of defining their involvement in society.*

3.2.3.5 Positioning Trusting Constructivism in Relation to Critical Theoretical Argumentation

In a trusting constructivist position, as in critical theory, the fundamentally hypothetical character of any claims to validity, is considered as springing from the rootedness of human knowing in discursive encounter. Delanty cites Habermas (1996) as indicating that scientific activity, like everyday activity, is "anything but monolithic — it fragments into a number of competing viewpoints that are shot through with values" (1997, p. 141). This means, for Delanty, that it is not possible according to Habermas's schema for people to turn to science to resolve the "contentions" occurring in everyday life around definitions of reality. Delanty also points out that in Habermas's view, "the professionalized culture of social science does not in itself construct social problems from its own discourse but does so in response to public and media agenda setting". The social reality that enters into scientific discourse is a "constructed reality and one which is the product of contentious action" (Delanty, 1997, p. 140). Nevertheless, in Habermas's conception of social discourse, the fact that any validity claim is always open to continued contention, need not detract from people's efforts to seek consensual understanding. For the purposes of action co-ordination along communicative lines, people should participate in a discourse oriented toward consensus. But from a trusting constructivist perspective, the requirement for speakers engaged in discourse to orient themselves toward consensual understanding, is too strict and unnecessary a requirement. All that is required for people to address their intersubjective involvement in society is that they are able — as part of the development of trusting relationships — to *defend their arguments in discursive engagement with alternatives.*

Edwards suggests that in examining the way people operate with their descriptions, it is possible to explore the processes whereby they are defended in relation to other ones. He notes: "usually there is a very specific 'otherwise' [other way of seeing] that is at issue" (1997, p. 8). In other words, as he sees it, descriptions are advanced with reference to "alternatives, and sometimes specific counter-descriptions" (1997, p. 8). From a trusting constructivist perspective this suggests that speakers can orient themselves in relation to alternatives, without either they or others expecting that these need to be integrated in a consensual understanding — even in some indefinite future, as posited by Habermas (1993, p. 94). The focus in this perspective is on the way in which versions of the world can be accounted for by people (including "professional" inquirers) amidst a recognition that they cannot be defended with reference to objectified realities or with reference to the search for consensus.

The following bullet points provide a rendition of the trusting constructivist argument positioned in relation to critical theory.

- A trusting constructivist position does not concur with the critical theoretical argument that the justification of validity claims is dependent on the concerned parties orienting themselves toward developing consensual understanding.
- In critical theory, accountable social inquiry is oriented toward activating processes of discourse in the public sphere of society: contentious issues in this sphere are to be theorized in a manner that does justice to the human potential for communicative rationality.
- In trusting constructivism, *accountable social inquiry is oriented toward enabling a social discourse in which participants (including professional inquirers) can offer versions of "the world" as part of a social debate: in such a debate people are able to hold others to account while accepting their need for creating judgments in the light of fundamental differences between alternative (viable) constructions that can be created.*

3.2.3.6 Positioning Trusting Constructivism in Relation to Anti-foundationalist Feminist Argumentation

Anti-foundationalist feminists propose as a valuable endeavor the striving for social relationships that express the felt importance of respecting "that 'the truth' is not the same for everyone" (Temple, 1997, paragraph 5.2). They are committed to this value as one that — they admit — informs their research. They strive in their practice of research to make a contribution toward the invention of alternative ways of approaching knowing enterprises and therewith, the invention of different ways of living together in society. To make this contribution they incline toward constructively-oriented, rather than realist-oriented, approaches to human knowing.

The complaint of realist-oriented critics of anti-foundationalist feminism is that in the final analysis it becomes impossible to adjudicate between constructions proffered by people — whether by so-called professional inquirers or others. It seems as if all that can be achieved in social discourse (including "scientific" discourse) is to give voice to differing standpoints. In terms of realist thinking, anti-foundationalist feminism seems to offer no mechanisms for us to assess the quality of these standpoints.

In rejoinder to this complaint, anti-foundationalist feminists argue that one way of assessing the quality of views adopted, is to consider them in terms of the way in which they make provision for the implosion of their own codes. That is, the provision that they make for alternatives at the moment of their utterance, already serves as a sign of their quality (see, for instance, Lather, 1993, p. 674). Another, related, way of assessing them is to consider what provision they make for the self-determination of people (including authors and other audiences) in creating choices of vision and action (Lather, 1995, p. 42).

A trusting constructivist position takes up the anti-foundationalist feminist one at this point and works with it by focusing on the accountabilities of those engaged in processes of inquiry. Accountability implies (within this perspective) that as we "know" so we give credibility in our consciousness to a variety of possible options for seeing and acting — defined through discursive engagement with other people's visions and concerns. This means that inquirers accept that any way of seeing that they might develop at any point in time *arises as a choice (amongst other viable options) with which they, as others, might have to live.* But just because they are aware of their choicefulness of vision and of the possible impact that it might have on others, they try to organize their choices to open the space, as far as possible, for those of others. (See also Romm, 1997a, paragraph 5.2.) Insofar as this notion of responsible knowing is merited in society, inquirers can be called to account for their manner of addressing their choices in relation to alternatives.

The following bullet points provide a rendition of trusting constructivist argumentation positioned in relation to anti-foundationalist feminism.

- A trusting constructivist position appreciates the anti-foundationalist feminist shift away from a realist problematic toward considering opportunities to celebrate the uncertainty involved in the process of knowing.
- In anti-foundationalist feminism, accountable social inquiry is governed by the quest to instantiate a politics of difference and heterogeneity through the knowing process.
- In trusting constructivism, *social inquirers can be held to account for their ways of conducting inquiries and developing visions therefrom: trust earning on their part requires that choices of how to proceed and how to report on "findings" arise out of a serious engagement with differing arguments and concerns.*

3.3 A REVIEW OF THE DEBATES

In Chapter 1, I introduced some arguments that have been put forward regarding the meaning(s) to be attributed to research endeavors in society. Subsequently, I explored these arguments in much greater detail, paying special attention to foregrounding views on accountability. Following the detailed discussion of the various positions that I have elaborated, the arguments encapsulated in Table 1 (Chapter 1) can now be looked at again, with a focus on considering how they might be linked to proposing ways of conducting research (as social inquiry). Table 2 is structured around providing an inflection of the material presented in Table 1, with a view to outlining different visions of acceptable research practice. The table is aimed at generating a comparison between positions along these lines. So, for instance, it is shown how in positivism, the process of doing science requires creating a way of observing (measuring) phenomena and examining their possible causal connection; and it is also shown that the product sought is knowledge about reality. Likewise, for all the positions discussed, ways of organizing research are tabled. (The categories presented in Table 1 are carried through into Table 2, excepting that what was called critical rationalism/non-foundationalism has now become simply "non-foundationalism" following from the discussion hereof in Chapter 2, and discursively-oriented constructivism has been renamed trusting constructivism.)

Table 2: Comparing visions of acceptable research practice.

	Research Process			***Research Product***	
Positivism	*Scientism* Doing science	*Phenomenalism* Organizing measurement validity	*Empiricism* Seeking causal connections through data analysis	*Value freedom* Finding out about reality	*Instrumental knowledge* Offering useful knowledge
Non-foundationalism	*Scientism qualified* Attending to the effects of human perceptions in the social world	*Phenomenalism reviewed* Improving instruments for observation	*Empiricism reviewed* Creating theoretical explanations grounded in observations	*Value freedom* Seeking knowledge through operating self- and collegial criticism	*Instrumental knowledge* Providing opportunities for better-informed action
Scientific Realism	*Scientism qualified* Attending to the reality of generative mechanisms, including societal ones	*Phenomenalism reviewed* Developing acceptable observations	*Empiricism reviewed* Working at an analytic level to understand generative mechanisms in context	*Value freedom* Considering possible reformist allegiances and if necessary re-organizing one's focus	*Instrumental knowledge* Extending information about, inter alia, social structures
Interpretivism	*Scientism opposed* Doing *social* science	*Phenomenalism re-interpreted* Creating meaningful categories	*Empiricism re-interpreted* Developing understanding of the social world	*Value freedom* Paying attention to others' worlds	*Instrumental knowledge* Creating meaningful information relevant to actors
Critical Theory	*Scientism criticized* Avoiding the transference of an engineering approach to the study of social reality	*Phenomenalism reconsidered* Creating realities through the research process	*Empiricism reconsidered* "Discovering" social potentialities	*Value freedom undercut* Recognizing the value of human discourse	*Instrumental knowledge undercut* Furthering communicative understanding
Anti-foundationalist Feminism	*Scientism challenged* Overturning traditional research control	*Phenomenalism reconsidered* Working with ambiguity	*Empiricism reconsidered* Appreciating the situatedness of research	*Value freedom undercut* Appreciating the value of heterogeneity	*Instrumental knowledge undercut* Questioning the power to define realities
Trusting Constructivism	*Scientism revisited* Developing trusting relationships through the research process	*Phenomenalism reconsidered* Evoking and defending observations	*Empiricism reconsidered* Exploring alternative ways of seeing and using evidence	*Value freedom undercut* Encouraging discursive accountability as a value	*Instrumental knowledge undercut* Tempering the possible self-fulfillment of "results"

As depicted in the table, the first three columns (relating to arguments around the first three tenets of positivism), provide some advice on the way in which research as a *process* might be appropriately conducted; the last two columns (involving arguments around value freedom and instrumental knowledge) provide advice on the status of the research *product* to be sought (in terms of claims to knowledge). Of course, ways of addressing values (the fourth tenet) could be regarded as part of the research process. However, my focus in discussing this tenet is on the status of the product sought through the process. It is in terms of this focus that I have discussed the notion of value freedom and ways in which it can be undercut.

Table 2 highlights differing visions of acceptable research practice that may be invoked in judging the accountabilities of social researchers. The table can also be used to underscore certain similarities in some of the approaches. For example, connections can be created between critical theory, anti-foundationalist feminism, and trusting constructivism when one considers their grounds for challenging scientism. In all of these positions, the process of "science" is given meaning by explicating ways in which it can be regarded as embedded in social life. Responses to scientism in turn are linked to specific ways of reconsidering phenomenalism and empiricism, and to suggestions as to what this may mean in practice in terms of researchers' complicity in creating ways of living. (The responses to scientism given in critical theory, anti-foundationalist feminism, and trusting constructivism, constitute an opposition thereto — but on fairly different grounds from the interpretivist opposition, where doing social science is a matter of defining the study of human life in a particular way.)

It can also be noted from the table that in critical theory, anti-foundationalist feminism, and trusting constructivism, any research products are always considered (and assessed) in terms of the way in which they create opportunities for human discursive engagement. This manner of treating research products implies that the tenets of value freedom and instrumental knowledge are undercut. Hence, for instance, in critical theory, it is considered crucial that the value of human discourse is brought to bear in the way of creating research "products"; the aim is to use the research process to further communicative understanding. In anti-foundationalist feminism, a similar sentiment holds, except that in this case the value of forwarding ways of dealing with heterogeneity is espoused, and the power to define realities is examined/challenged in these terms. In trusting constructivism, it is the commitment to discursive accountability that is focused upon, and the way in which results can be treated accordingly (by those organizing, and touched by, the research).

Arising from my comparison of the various options for considering research as discussed above, I suggest that it is possible to group the arguments by positioning them specifically in relation to their ways of envisaging the connection between process and product in the doing of research. This "classification" of arguments regarding (accountable) research practice is given in Figure 3. Through the figure, I draw out the similarities between positivism, non-foundationalism, and scientific realism in terms of arguments about the relationship between process and product; I show how interpretivism may be seen as treating process in a distinct way (while still defining the status of products in terms of knowledge accumulation); and I point to the manner in which critical theory, anti-foundationalist feminism, and trusting constructivism introduce alternative concerns into the debate.

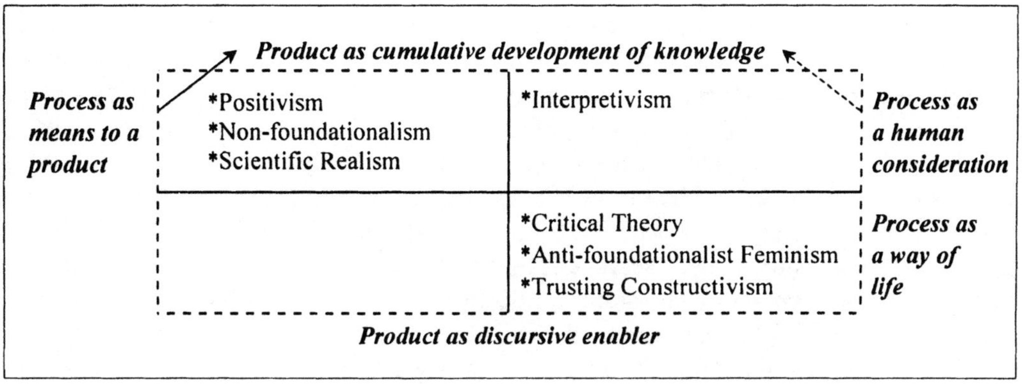

Figure 3: Accounting for social research in terms of process and product.

As indicated through the top left quadrant of the figure, non-foundationalism and scientific realism share with positivism the idea that the justification for scientists' manner of proceeding in their inquiries lies in the possibility of generating knowledge products through the scientific process. The scientific process is seen as a *means to the end of producing the product.* This is the purpose of the research endeavor. Non-foundationalism and scientific realism alike (more so than positivism) suggest that the process of generating knowledge is subject to human judgment (and error). But all three positions suggest that scientists are accountable to others in the sense that they must try to ensure that the processes they employ can be argued to function as, indeed, leading toward the advancement of knowledge of reality. Efforts must be made, as far as possible, to minimize sources of error. The information provided as scientists work with "the data" toward developing their theorizing, is then likely to be *more informative in relation to the realities than are uninformed opinions.*

As indicated through the top right quadrant of the figure, interpretivism suggests that in trying to generate knowledge products (in relation to social reality), it is important also to consider the way in which a relationship between researchers and participants/respondents is established in the inquiry process. The process of developing knowledge of social reality (as meaningfully constituted) needs to involve *treating people as meaning-givers in defining their worlds*; hence, the scientist's research agenda should not be used to pre-determine the way in which issues are discussed (and negotiated definitions pursued). However, what is produced through the research process, can be regarded similarly to the way in which research outcomes are seen in positivism, non-foundationalism, and scientific realism. That is, as a product, knowledge generated *derives its status from its ability to offer an insightful account of aspects of social life.*

As indicated through the bottom right quadrant of the figure, the trusting constructivist position, along with critical theory and anti-foundationalist feminism, sees the process of social research as ideally itself facilitating an experience of a way of life that can be regarded as ennobling for those involved. Research processes thus *need to be rendered accountable in these terms.* (In trusting constructivism, the focus is on the cultivation of trust through processes of discourse.) As far as the "product" of the research is concerned, *its status is seen to derive from the discursive possibilities that it serves to set up.* Critical theory, anti-foundationalist feminism, and a trusting constructivism do not necessarily concede on what may be implied in creating invitations to dialogue within the public sphere of society. But attempting to set the scene for such dialogue, in whatever way this is conceived, is regarded as being part of the accountabilities of social researchers. (The lack of an arrow from process to product in depicting these positions in Figure 3, illustrates that no route is proposed to advance representational knowledge of realities.)

The bottom left quadrant of the figure is left vacant, as it does not serve to elucidate any of the positions explored in this book. A case could be made, however, for placing certain forms of, say, postmodernist theorizing in this quadrant. Placing them here would suggest that whatever processes are used to develop theoretical arguments, audiences must be cautioned to treat the "products" developed in the process as discursive enablers at the point of their being appropriated/interpreted/revisited. (In this book aspects of postmodernist argumentation have been explicitly incorporated only in terms of their impact on, and relationship to, anti-foundationalist feminism. I have not otherwise delved into postmodernist thinking.)

3.4 CONCLUSION

This chapter was aimed primarily at exploring arguments that can be used in support of constructivism. To organize the discussion I drew on certain authors' arguments that could be considered as being constructivist in orientation, and developed these by focusing on issues of (discursive) accountability. In dealing with the question of whether there is any point (in terms of a constructivist position) in distinguishing professional inquiry from other discursive activities, I noted that the label of "professional inquiry" invokes the understanding that the inquiry process can be rendered recoverable for others to assess, and that ensuing information will be publicly available in some form. From a trusting constructivist perspective, the accounting for strategies of social inquiry implies the creation of an opportunity for others (academic and other audiences) to consider the inquirers' sensitivity in taking into account a range of arguments regarding the point of the inquiry. I argued that sensitive engagement in this regard is linked to the development of a way of living, and in this sense is not without consequence in society. I also pointed out that the rendering of results (or "products") public implies that they are presented as — in Barry and Elmes's words (1997b, p. 432) — stories "worth listening to", which can open up new trains of thought for those engaging with the issues raised by the study. They can be treated as opening up trains of thought, without having to pose as referring to realities unmediated by human discursive activity. However, the way in which discourse is invited, can be subjected to scrutiny by those assessing the (discursive) accountability of inquirers.

In the chapters that follow, I proceed to explore in some detail, from the various perspectives, some examples of research undertaken utilizing specific strategies and techniques of inquiry. (I use the word "strategy" to call up discussion around the approach — how ever emergent — for organizing the investigation; and I use the word "technique" to call up discussion around devices aimed at data gathering/evocation.) The way in which I structure my discussion in the following chapters is by drawing on all of the cells in Table 2 to create "headings" for judging the accountability of social researchers. I use the terms provided in Table 2 as a template to examine all of the examples from the perspectives of the various positions discussed in the book. Chapters 4-7 are thus aimed at continuing the debate between ways of conceiving research(er) accountability by locating arguments in relation to actual research projects. By organizing my comparison in this way, I hope to further the appreciation of ways in which differing definitions of acceptable research practice might be brought to bear in considering researchers' accountabilities. My discourse around the examples is aimed at lending more substance not only to the trusting constructivist position as explored in this chapter, but also to the other arguments that I have identified throughout the book.

4
Exploring Experimentation

4.1 INTRODUCTION

In this chapter, I generate a discussion around an example of what is labeled by the researchers involved (Dovidio et al., 1997) as an experiment. I do this by considering — from various perspectives — how they might be called to account for their approach adopted. The experiment amounted to what the researchers call a "manipulation" of certain conditions (the independent variable), organized so that the effects of this on other defined variables (hypothesized to be dependent ones) could be ascertained.

As noted by Gill and Johnson, experimentation in the social sciences in the "classical" sense of the term involves allocating people to different groups in such a way that they are matched as far as possible in terms of their starting positions (1991, p. 54). This matching can be organized by assigning people randomly to the different groups; or it can be done systematically by matching people in terms of a specific set of characteristics that researchers believe possibly may have a causal influence on the dependent variable(s) being examined. Or some combination of the two approaches can be used. The idea that groups must be matched — in some way — is in line with Comte's suggested definition of scientific experimentation. According to him, experimentation (in the natural or social sciences) can be defined by efforts made on the part of researchers to set up "parallel cases" (groups) — undisturbed by what can be considered as "irrelevant influences" — so that we can study the effects of introducing known changes (or influences) in one of the groups (1974, p. 254). The groups that are set up are sometimes called the experimental and control group, the former being the one where the experimenter manipulates the change in the independent variable (Gill & Johnson, 1991, p. 54). The aim of organizing the experimental manipulation is to investigate whether any consequent changes in (dependent) variable(s) can — with some confidence — be attributed to the independent variable. Statistical reasoning is used to determine the significance of any variation in the data from the experimental and control situations (see also Weaver, 1997, pp. 99-103).[21]

In the experiment discussed in this chapter, situations were indeed set up with the intention of investigating the effects of the manipulation of the independent variable on other variables that were isolated for attention. Taking the example of this experiment, I

---

[21]This is not to say that all studies classed under the banner of experimentation must follow this design. Bailey (1987, pp. 219-231) discusses ways in which the logic of experimentation can be preserved via a host of different designs.

examine it from the perspectives of all the positions that have been elucidated in Chapters 1-3, namely: positivism; non-foundationalism; scientific realism; interpretivism; critical theory; anti-foundationalist feminism; and a trusting constructivist position. I start by presenting an assessment from a positivist point of view of the experiment that was undertaken — using the tenets of positivism to create a way of judging aspects of the experimenters' approach in this case. This is followed by other assessments of the example. I present these assessments by positioning arguments through their ways of relating to — whether accepting, revising or reworking — positivist tenets. The headings used to structure the discussion (throughout Section 4.3) encapsulate the way in which these tenets are being treated at each point to create criteria for judging the proper conduct of social research. In this manner, the arguments developed in Chapters 1-3 are carried into this chapter.

The example that I present is of an experiment aimed at investigating the benefits of "recategorization" in intergroup relationships. In Section 4.2, I provide an outline of the conduct of the experiment (as I interpret it), explaining the operations undertaken within it and the presentation of results. I do this in what I regard as sufficient detail so that readers can follow my ensuing discussion in Section 4.3, where I provide assessments of the experiment from various angles.

4.2 THE BENEFITS OF RECATEGORIZATION (DOVIDIO ET AL., 1997)

4.2.1 The Organization of the Experiment

Dovidio et al. (1997) undertook an experiment that was organized around manipulating people's "intergroup contact". They were interested in exploring the effects that might be created when members of different groups were induced to "conceive of themselves more as a single group rather than as two separate groups" (1997, p. 401). The manipulation amounted to the experimenters' trying to instill an impression of people's belonging to one group (rather than two groups) in the experimental situation — so as to induce people to recategorize the situation (1997, p. 401).

Dovidio et al. state that the experiment was aimed at "investigating the effects of a manipulation designed to induce recategorization" (1997, p. 403). They indicate that they wished to examine the effects in terms of people's self-reported *evaluations* of others, and also in terms of their "forms of intergroup interaction" (1997, p. 404). Two types of interaction were considered. *Self-disclosure* was the first type. The researchers see this as "communicating a desire for intimacy and promoting closeness" (1997, p. 405). The second type of interaction investigated was *helping* orientations. This the researchers see as activity that is "co-ordinated to the needs of another" (1997, p. 404). Dovidio et al.'s expectation (or hypothesis) was that in people's *evaluations* of others, as well as in their *self-disclosure* and *helping* orientations, there would be lower bias scores (on average) for members from the situation where recategorization had been induced by the experimenters, as compared with the situation where it had not.

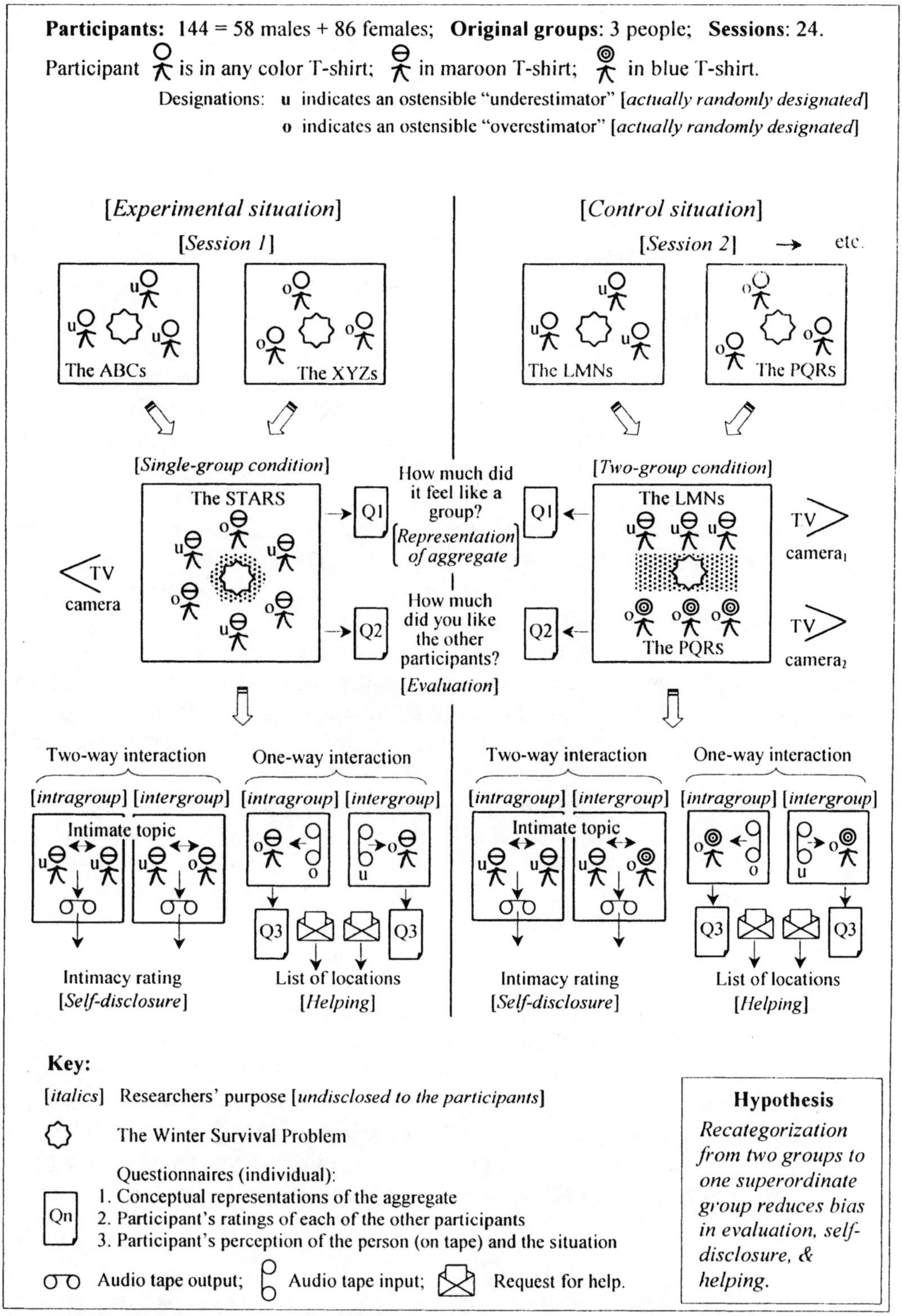

Figure 4: Dovidio et al.'s experiment.

Figure 4 (created for the purposes of the discussion in this book[22]) expresses in diagrammatic form the strategy utilized by Dovidio et al. to organize the investigation. The figure concentrates on the way in which the experiment was designed. (24 sessions with sets of six participants in each case were arranged.)

Participants were a group of 144 undergraduate psychology students. Dovidio et al. indicate that they "participated in the present study to fulfill one option of their introductory psychology course requirements" (1997, p. 406). In each session, participants met initially in three-person groups to which they were randomly assigned — with the proviso that each group had at least one woman and one man. They were told that their activities were being examined to explore the decision-making processes of "overestimators" (people who are inclined to overestimate the difficulties of tasks) or "underestimators" (those inclined to underestimate the difficulties). They were informed that they had been selected to participate in the study on the basis of the estimation task that they had previously completed. Their group behavior was ostensibly being examined with a view to exploring how they manage to interact as they "attempt to reach consensus" on a problem set for them (1997, p. 407). The problem in this case was the "Winter Survival Problem", which was discussed by all the groups involved.

The groups were all asked to choose their own names. These are designated as the ABCs, the XYZs, the LMNs, and the PQRs in Figure 4. (The labels have been made up here for the purpose of the illustration.) The groups were given some time to work on the Winter Survival Problem and then escorted to another set of rooms where they had been told a new interaction would take place (1997, p. 407).

At this point, the manipulation of intergroup contact was undertaken. As Dovidio et al. note, "the goal of the manipulation was to vary the participants' cognitive representations of the six participants as either one group or as two separate groups" (1997, p. 407). To induce the persons to have a *one-group* representation of the aggregate, instructors seated them in an "integrated seating pattern" and "emphasized their common identity as students at the same university". They were also instructed to wear the same color T-shirts (maroon), reflecting the school color. Furthermore, "an experimenter assigned the members of both groups a new single name to represent all six people and instructed them to use the new group name on all subsequent forms" (1997, p. 407). A TV recording of the session was arranged around the notion that the groups had merged to form the STARS (their group name).

In the *two-group* condition, a "segregated seating pattern" was arranged and each group wore a different color T-shirt — one group wore maroon, the other blue. Instructions "made no mention of their common university identity". The groups kept their separate three-person group names throughout the session. (The experimenter announced the session using the names of both groups, making no mention of any merger.) "Finally, each of the two TV cameras that recorded the session was aimed at a different three-person group" (1997, p. 407).

[22]The figure was drawn by Peter Adman, who read the text of Dovidio et al. (1997) and created a depiction of the design of their experiment on the basis of his understanding thereof. (My outline of the experiment in the chapter was in turn influenced by this figure.)

The task of the groups was again (as in the first task) to discuss and reach consensus on the same problem that they had been set earlier (namely, the Winter Survival Problem). They were given time for discussion and then each member was given questionnaires to fill in that asked about:

a) their conceptual representation of the group (Q1 in Figure 4); and

b) their rating of each participant (Q2 in Figure 4).

The first questionnaire asked whether it felt like one group or like two separate groups — and used a scale ranging from 1 (not at all) to 7 (very much) in order to establish the member's *conception of the aggregate*. The second questionnaire asked members to *evaluate each participant* with reference to how much they liked them, as well as how honest, cooperative, and valuable they found them. The rating again ranged from 1 (not at all) to 7 (very much), for each of the four items on the scale.[23] The ratings were organized using schematic figures to represent each participant's code and seating position, so that evaluations could be made with reference to these figures.

Once the forms were completed, "an experimenter prepared the participants for the next phase of the study, which involved helping and self-disclosure tasks" (1997, p. 407). Participants were chosen from the original subgroups and assigned into groups of pairs. The way in which they were chosen, provided the opportunity to assess both intragroup and intergroup *self-disclosure* and intragroup and intergroup *helping* for each six-person session (1997, p. 408). As far as self-disclosure was concerned (that is the two-way interaction in Figure 4), the experimenter chose either two overestimators or two underestimators for the first pair and one overestimator and one underestimator for the other pair. The example shown in Figure 4 is where two underestimators chosen in the first pair of participants represent intragroup self-disclosure, and one underestimator and one overestimator in the second pair represent intergroup self-disclosure. In the one-way interaction in this example, the overestimator interacting with the audio tape from another overestimator represents the opportunity for intragroup helping, whereas the participant being an overestimator interacting with an audio tape from an underestimator represents the opportunity for intergroup helping. This meant that the experimenters could assess opportunities for both intragroup and intergroup helping. The way in which the self-disclosure and helping tasks were designed is explained below.

The self-disclosure task was organized by the experimenters by firstly giving participants a list of potential discussion topics, and asking them to indicate which they would be most willing to discuss. Thereafter, they were all (each pair of participants) "assigned a moderately intimate topic to discuss: 'What are you most afraid of?'" (1997, p. 408). The definition of "moderate" was created through the fact that on a 1-10 scale, this item "was rated a 5.96 in a pretest of 20 students with no difference between the ratings of men and women" (1997, p. 408).

---

[23]These four items utilized in the questionnaire were considered as reflecting different dimensions of evaluative bias. See De Vaus (1996, p. 50) for a discussion of how concepts may become "indicated" with reference to their various dimensions; and see Foddy for a discussion of "the magic number 7" in designing social psychological rating scales (1995, pp. 165-166).

Participants were informed that they could decline to discuss the topic if they wished, but that if they agreed to participate, the discussion would be audio taped (while confidentiality would be kept). This task was ostensibly (that is, as far as participants were concerned) about "how people become acquainted and get to know one another". All participants agreed to complete this task (1997, p. 408).

The experimenter left the room while the participants discussed the topic. Later the audio tape output (see Figure 4) was coded for the amount and depth of information disclosed. The coding was organized through a rating scale by two independent judges. "The measure of intimacy of self-disclosure was the average of the judges' ratings" (1997, p. 408).

At the same time as the groups of pairs were meeting to discuss the topic, the other participants were "assigned to the helping conditions and escorted to separate cubicles" (1997, p. 408). During this task the participants were informed that they were in a one-way communication condition in which they would hear a recording of one of the earlier participants — whether an (ostensible) underestimator or overestimator. Participants in this task

> listened to a female student [ostensibly an earlier participant] describe how illness had prevented her from completing an important project — a survey of student life for a committee on which she served. After listening to the tape, participants completed questionnaires about their perceptions of the person and the situation ... and assessing their [own] emotional reactions. (1997, p. 408)[24]

After the experimenters had collected the questionnaires (Q3 in Figure 4), the participants were told that they had completed their responsibilities for the experiment. They were notified about how to obtain more information about the study and their research credit was recorded. Then each participant was handed an envelope containing a note with a written request for help — ostensibly from the female student — and a campus envelope. The experimenter informed them that the research team had agreed to distribute the note for her, since she had agreed to let them use her personal information. But the participants were also told that they were under no obligation to help. They were asked to please put their response in a box. The note contained

> an appeal to help by placing posters recruiting volunteers to participate in the student life survey in various locations across campus. To help, participants were asked to indicate their name and address and to identify the locations in which they would be willing to hang the posters. (1997, p. 408)

Dovidio et al. used as their measure of helping behavior "the number of locations indicated" (1997, p. 408).

---

[24]Empathetic concern was here measured using five items (compassionate, warm, moved, sympathetic, tender) on a 7-point (not at all, very much) scale.

4.2.2 Results of the Experiment

In discussing the results of the experiment, Dovidio et al. note that, "consistent with the hypothesis derived from the Common Ingroup Identity Model, creating conditions that foster a stronger one-group representation significantly reduced intergroup bias in both evaluations and behavior" (1997, p. 410). They indicate that the Common Ingroup Identity Model (as previously developed) is based on a social categorization perspective of intergroup behavior that "recognizes the central role of social categorization in ameliorating as well as in creating intergroup bias" (1997, p. 402).

Part of the hypothesis that guided this experiment, was that "the effect of the Group Manipulation on evaluations and behaviors would be mediated by representations of the aggregate" (1997, p. 411). In other words, it was hypothesized that the effect of the group manipulation as organized by the researchers, would be mediated by participants' representation of the aggregate in the different conditions. (See Q1 in Figure 4.) Statistical control was introduced to check the (separate) significance of members' group representations, in terms of their effect on evaluations, self-disclosure, and helping. In addition to this, tests of mediation using a multiple regression mediation approach (as a form of path analysis) were undertaken.[25]

Dovidio et al. indicate that mediation (via, in this case, the group representations) can be demonstrated by setting up a number of equations (1997, p. 411). In the first equation, the dependent variable is regressed upon the independent one and should indicate the extent to which the independent variable (that is, in this case the one- or two-group manipulated conditions) predicts the dependent one (in this case, the scores measuring evaluation, self-disclosure, and helping). In the second equation, the effect of the independent variable on the mediating variable (in this case, group representation) is examined. In the third one, the dependent variable is regressed on the independent variable together with the (potential) mediating one.[26] Mediation is demonstrated

> to the extent that the mediating variable predicts the dependent variable over and above the influence of the independent variable and to the extent that the effect of the independent variable is weaker than in the first equation in which it is considered alone. (1997, p. 412)

In the case of this study, Dovidio et al. note that consideration of the mediating effect of group representations (as measured through Q1) weakened the (separate) effect of the group manipulation created by the experimenters (the independent variable) for all the dependent variables. That is, participants' (negative) bias, for all the dependent variables, was demonstrated to have been mediated by group representations. This

---

[25]As indicated by De Vaus, the aim of path analysis is to "pinpoint the extent to which a variable's effect is direct or indirect" (1996, p. 225). It is therefore considered as providing information about the "path" of causality.

[26]See also De Vaus (1996, p. 229) for a summary of different ways of organizing regression analysis.

occurred most in the case of evaluative bias (as measured through Q2), next most for helping (as measured through the reported number of locations for putting up posters), and least for self-disclosure (as measured through the judges' intimacy ratings) (1997, p. 412). From the analysis it could be concluded that the mediating factor affected mostly participants' bias in their evaluations of others, and affected less the other bias scores (of helping and self-disclosure) (1997, p. 412). Dovidio et al. summarize the overall results:

> The manipulation of the intergroup contact situation that created stronger impressions of one group [rather than two groups] significantly reduced intergroup bias. Furthermore, ratings of the extent to which the six participants felt like one group mediated the reduction in bias. (1997, p. 413)

In response to the claim that participants may have been reacting to "implicit experimental demands to cooperate, which in turn produced the reduction in intergroup bias", Dovidio et al. contend that a previous study (Gaertner et al., 1990) "does not support this explanation" (1997, p. 413). Dovidio et al. point out that recategorization manipulation (as organized by the experimenters) was associated with people's seeing of themselves as belonging to one group, apart from the factor of group cooperation. The previous study indicated that both "intergroup cooperation and the recategorization manipulation independently increased one-group representations" (1997, p. 413). According to Dovidio et al., the Gaertner et al. study indicates that the recategorization manipulation "has its effects primarily by influencing intergroup representations rather than through implicit demands to cooperate or compete" (1997, p. 413). They state that:

> The results of the present study, taken together with those of Gaertner et al. (1990), thus offer continued support for the direct effectiveness of recategorization strategies for reducing evaluative bias in intergroup contact situations. (1997, p. 413)

They note that what their study has left open for continued exploration is whether there may be more proximal mediators of interpersonal cooperation (1997, p. 415). For example, they suggest that it is possible that creating a common group identity (through group manipulation) "produces higher levels of interpersonal trust, which, in turn, facilitates self-disclosure and other types of interpersonal cooperative behavior (Kramer & Brewer, 1984)" (1997, p. 415). They suggest that the degree to which the intergroup factors combine with interpersonal factors to facilitate such behaviors "may be a productive issue for further research" (1997, p. 415).

They suggest furthermore that a challenge for future research in regard to helping behavior is to

> find direct mediation of this sense of we-ness and common ingroup identity on helping, to determine when and how common ingroup identity may relate to interpersonal processes ... and to understand how intergroup and interpersonal processes can combine to influence helping. (1997, p. 416)

Generally, they propose that the advantage of the focus that they have put on

> intergroup as well as interpersonal processes in bias articulated in the Common Ingroup Identity Model permits not only a parsimonious theoretical explanation for the effects of diverse contextual factors, but it can also inform practical interventions designed to reduce bias. (1997, p. 416)

Indeed, they state that it has already been shown that "intergroup contact under certain prerequisite conditions ... promotes the development of more harmonious intergroup relations" (1997, p. 416). The Contact Hypothesis investigated by, for instance, Amir (1969) and Cook (1985) refers to the possibility of organizing contact as a way of generating more harmonious intergroup relationships. Dovidio et al. suggest that their identity model, developed through various experiments, including this one, adds the idea that "many of these features [of the results of Amir's and Cook's studies] operate through the common cognitive mechanism of producing a more inclusive, superordinate categorization of the memberships" (1997, p. 416). Thus they suggest that:

> Recategorization manipulations may also help to create interactions among group members in the contact situation that can activate other processes that can subsequently promote more positive intergroup behaviors and attitudes. (1997, p. 416)

The practical benefit of research in this direction is that it can be used by those wishing to promote such behaviors and attitudes. They conclude that

> the recategorization strategy proposed in our model and decategorization strategies, such as individuating (Wilder, 1984) and personalizing (Brewer & Miller, 1984) interactions, can potentially operate complementarily and sequentially to improve intergroup relations in lasting and meaningful ways. (1997, p. 417)

Recategorization as a strategy encourages members to employ a superordinate category to define their relationship with others as part of some (created) group. Because of the way in which they come to see one another under the umbrella of an inclusive (group) category, their "bias" is likely to be favorably inclined, allowing "personalized impressions" to "quickly develop" (1997, p. 417). As Dovidio et al. put it, "the newly formed positive bias is likely to encourage more open communication" (in which the decategorization processes of individuating and personalizing can set in) (1997, p. 417). Although these responses may not be automatic, the development of a common ingroup identity creates a "motivational foundation for constructive intergroup relations which can act as a catalyst for positive reciprocal interpersonal actions" (1997, p. 417). Dovidio et al. suggest that, taken together, both recategorization and decategorization can have the effect of "improving intergroup relations" (1997, p. 417).

4.3 EXPLORING ALTERNATIVE ASSESSMENTS OF THE EXAMPLE

4.3.1 A Positivist Assessment

4.3.1.1 Scientism: Doing Science

The positivist tenet of scientism suggests that a similar approach to scientific inquiry can be employed in the natural and social sciences. The experiment is regarded as a suitable procedure for investigating causal relationships between variables — whether one is studying natural or social phenomena — although it is considered as more easily administered in the natural sciences (cf. Smelser, 1984, p. 13; Bailey, 1987, p. 219; Remenyi et al., 1998, p. 56). It is suggested that via properly designed experiments, the causal influence of antecedent conditions in generating outcomes can be investigated.

In the case of the investigation organized by Dovidio et al., the experiment was guided by their hypothesis that creating a single-group condition would generate certain effects (in the form of self-reported evaluations, as well as self-disclosure and helping orientations toward others). The independent variable (the group situation) thus had to be manipulated (as it was done in this experiment) and effects on the dependent variables (evaluative bias and interactive behaviors) observed. Statistical analysis could be used to establish the impact of the hypothesized independent variable on the other variables. And statistical analysis could also be employed to consider the mediating influence of, in this case, people's representation of the aggregate (in terms of how this affected people's responses toward others in their evaluations and behavior).

According to the positivist tenet of scientism, there is no reason to assume that the complexity of the social world renders it impossible to study it via the same approach used in the natural sciences. Just as the natural world can be studied by attempting to discover relationships of causality operative therein, so too can social researchers attempt to study the social world. The experimental procedure — based on the logic of creating a change in some condition in order to examine hypothesized effects — was used for this purpose by Dovidio et al.

The researchers have left open certain questions for future research, such as, for instance, the way in which common ingroup identity relates to interpersonal processes. But they do not feel that in principle it is beyond the scope of social science to find a model that can explain "intergroup as well as interpersonal processes" (1997, p. 416). They have directed their inquiries and their defense thereof in a manner consistent with the (scientistic) tenet that it is possible to study human life by seeking relationships of causality between variables and building models on this basis.

4.3.1.2 Phenomenalism: Organizing Measurement Validity

The positivist tenet of phenomenalism suggests that the scientific observation of phenomena should be aimed at getting to grips with the factual data provided by reality. Bryman notes that emphasis on organizing measurement validity of concepts "can be seen as the transportation into social research of the [positivist] principle of phenomenalism and the doctrine of operationalism in particular" (1992, p. 22). He notes that the strict doctrine of operationalism that avers that concepts are to be seen as "synonymous with the measuring devices associated with them" does not have many adherents (1992, p. 22). Nevertheless, the insistence that concepts need to be rendered observable and must refer to the phenomena indicated (to be classed as valid), "has broad support among quantitative researchers" (1992, p. 23).

In the case of this example, Dovidio et al. show how each of the terms in their hypothesis can be related to what they take to be observable indicators thereof. However, Dovidio et al.'s way of rendering observable the variables with which the study was concerned, might be regarded as still not sufficiently well-validated from the viewpoint of the positivist tenet of phenomenalism. For instance, it might be suggested that the researchers need to better account for why they believe that the answer to the question about whether during the interaction it felt like one group or whether it felt like two groups, provides a good indication of people's sense of group identification (that is, their sense of "we-ness"). Or, as another example, the two judges' decisions (averaged) concerning the amount and depth of intimate information disclosed in the discussion around the intimate topic, only tells us that this is how the judges rated the participants' conversation. Although Dovidio et al. state that there was high interrator reliability here (1997, p. 408), a positivist position would require an indication of why this should be treated as a valid measure of participants' real orientation toward self-disclosure.[27] Likewise for, say, the observation of people's helping behavior, it may be asked whether adding up the number of locations that participants assert they would be prepared to put up posters, provides a well-validated measure of their real orientation to helping.

From a positivist point of view, it is important to offer some account (for colleagues and others) of why it is taken that the observations gathered during the research process refer to the phenomena being investigated. Otherwise, the statements being made about the relationship between the variables under study can be criticized for not being grounded in a sufficiently firm empirical basis. In terms of the positivist tenet of phenomenalism, Dovidio et al. could have better provided for this possible criticism.

4.3.1.3 Empiricism: Seeking Causal Connections through Data Analysis

Dovidio et al.'s experiment was guided by the hypothesis that recategorization of group situations (as one-group ones) reduces bias in evaluations and behaviors toward others.

---

[27]The term "reliability" is used in positivist discourse to indicate that observations can be made in similar fashion in different observation instances (see Slater, 1998, pp. 236-237), while "validity" of measures implies that the observations undertaken can be taken as *referring to the phenomena that are being investigated.* Kolakowski notes that according to positivism, to be considered valid, concepts should be created in such a way that, although being in some sense abstractions from the experience of individual facts, should serve as a "superior, more concise and more general description of empirical reality" (Kolakowski, 1993, p. 5). The difference between reliable and valid measurements (as treated within positivist and other discourses), is, however, not taken up as a point of discussion in this book. Suffice it to say that according to positivism it is important to provide an indication of how conceptual categories are being used to refer to the realities under consideration.

The hypothesis was tested using statistical reasoning to analyze the data ensuing from the experiment.

However, from a positivist point of view, the advance of knowledge about a topic cannot rest on the analysis of data from one experimental result (and nor do Dovidio et al. expect this to be so). Other forms of group manipulation and other contexts of group behavior need to be examined empirically to create more "external validity" for the results obtained. (In relation to experimentation, positivism defines "internal validity" as being a function of the way the experiment is designed so that causal connections between the variables being examined in the study can be located. The term "external validity" refers to the generalizability of results beyond the particular experiment undertaken.) A positivist outlook accepts that experimental research is weak as far as external validity is concerned. That is, it is not possible to generalize easily to different (other) contexts via the doing of particular experiments (see Gill & Johnson, 1991, p. 47). One way of strengthening the external validity of studies, according to positivism, is to organize experiments in a variety of contexts and to compare results therefrom. In the case of Dovidio et al.'s work, they have indicated what they think they have learned about intergroup behavior thus far, arising out of their various experiments (and arising out of their comparing their results with experiments undertaken by others too). And they have indicated appropriately that additional research is needed to explore further the relationship between the phenomena under consideration.

4.3.1.4 Value Freedom: Finding Out about Reality

Dovidio et al. present the experiment as having been designed to obtain evidence that would either support or disconfirm their proposed hypothesis. As it happens, they note that the findings that they obtained largely confirm their initial expectations (1997, p. 401). However, they suggest that it is on the basis of a (neutral) statistical analysis of the data, rather than on the basis of their own predilections, that their views on common ingroup identity have become confirmed through the experiment. The positivist tenet of value freedom suggests that any prejudice on their part in favor of the theoretical model of ingroup identity should not have affected the way they set up the experiment and interpreted its results. The theory itself should be supported only if it can be shown (in the context of this experiment, as well as in other contexts) that it provides an explanation, which is not unduly convoluted, of evaluative bias and other behaviors in ingroup/outgroup interaction. According to Dovidio et al., the theory can be shown to offer a parsimonious explanation of these human responses (1997, p. 416). From a positivist point of view, as long as Dovidio et al. are indeed able to show the links between their common ingroup identity model and the realities being explained, they can be argued to be acting without prejudice — and thus accountably as scientists.

4.3.1.5 Instrumental Knowledge: Offering Useful Knowledge

Dovidio et al. present the common ingroup identity model as providing an explanation of the realities of intergroup and interpersonal interactions to which the model refers. As a theory about such interactions, they propose that the model is useful in "informing practical interventions" (1997, p. 416). Once we know, for example, that "cognitive representations of group boundaries" make a difference to people's intergroup relations, this information clearly has "valuable practical ... benefits" (1997, p. 416). They suggest that this knowledge can be combined with (further) investigations of ways in which "the development of a common group identity creates a motivational foundation for constructive intergroup relations which can act as a catalyst for positive reciprocal interpersonal actions" (1997, p. 417). Together, the knowledge of intergroup and interpersonal relations can be used to good effect by those interested in "improv[ing] intergroup relations in lasting and meaningful ways" (1997, p. 417). The suggestions of Dovidio et al. concerning the way knowledge, produced via the scientific process, can later be used in the social world by actors wishing to create "practical interventions", are consistent with the positivist tenet of instrumental knowledge. In terms of this tenet, the accountabilities of social researchers to others in society consist in their trying to operate in accordance with the standards of professional inquiry in order to produce theorizing that can be used, if desired, by those acting as citizens in the social world.

4.3.2 A Non-foundationalist Assessment

4.3.2.1 Scientism Qualified: Attending to the Effects of Human Perceptions in the Social World

The non-foundationalist view of the relation between the natural and social sciences is that social researchers should operate in recognition of the similarity of the sciences, while also recognizing that in the human world additional factors might come into play due to the way humans perceive the world.

In the case of the experiment undertaken by Dovidio et al., it might be suggested that the manipulation of the single-group condition could have created certain outcomes as a result of the way participants perceived the experimental situation. The possibility — alluded to only briefly by Dovidio et al. and then ruled out — is that the participants in the manipulated single-group situation realized they were expected to regard themselves as belonging to one group and that this affected their subsequent evaluations and behaviors. (This might indeed have been more likely given the particular name assigned to them by the experimenters — the STARS.) Dovidio et al. argue that previous experiments have shown that cooperative behavior can be considered separately from the effect of recategorization manipulation, and that both independently affect people's group representations (1997, p. 413). But this still does not exclude the possibility that in this experiment (and in the others) participant responses to other people were affected by the ideas that they had about what was expected behavior. Even though the experimenters were at pains not to allow the participants to realize what the experiment was about (by giving them ostensible, rather than the actual reasons for their participation in the various tasks) the reactivity effect could have been operative. A non-foundationalist position might suggest that more attention be given to this issue, by, for instance, asking participants (later) how they had been responding to the experimental situation. By doing this, more information could be gleaned about the character of their responses during the study. This would mean that the results (conclusions from the study) would probably turn out to be more tentative than presented by Dovidio et al. But it would also mean that at least some indication of how the reactivity effect might have been operative, too could be presented.

It may also be noted that subsequent to the experimental process, and as a result of participants (and other audiences) having access to the information made available by Dovidio et al., people could decide to adjust their behavior. They could adjust it in line with what they now believe to be a "normal" response to group situations (in terms of Dovidio et al.'s statistical analysis of participant responses in the various situations). In other words, it is possible that people may decide to act in line with the results as stated by the researchers.

A non-foundationalist position would argue that Dovidio et al. could have shown more recognition of the fact that:

i) they are operating in a field of study where results can be influenced by the way people respond to scientists during the research process; and
ii) people may also respond to scientifically presented reporting of results.

However, a non-foundationalist position would accept that the scientific study of society should be based (as is the natural scientific approach) on the use of argument grounded in empirical evidence to support any statements that are made. Dovidio et al. have at least accepted the need to support their conclusions by pointing to the evidence afforded through their experiment.

4.3.2.2 Phenomenalism Reviewed: Improving Instruments for Observation

Dovidio et al. make reference to various phenomena that they believe they are discerning through their measuring instruments. From a non-foundationalist point of view, however, their techniques of data collection should be understood to result in more tentative statements about "the phenomena" than Dovidio et al. acknowledge. In all instances of their data collection, their processes of observation should be regarded with some tentativeness. Rather than assume that the data-collection process affords them the data that they hope to obtain, they should be aware that even what they may regard as basic observation statements (to use Popper's terminology), can still be subjected to doubt.

While a positivist position would require Dovidio et al. to provide a clear indication of how they are referring to the realities under consideration, a non-foundationalist position would emphasize that the process of attempting to validate measures is never fully dependable. Nevertheless, as Sapsford indicates, this does not prevent attempts being made to characterize the world in ways that can lead to a better understanding thereof (1999, p. 148). These attempts can be made along with an acknowledgment that "no evidence will be valid absolutely" (1999, p. 139). A non-foundationalist position would prefer that researchers operate in recognition of the tentativeness of their observation statements, coupled with an attempt to open these statements (as all statements) to more checks of self- and mutual criticism in the scientific community.

4.3.2.3 Empiricism Reviewed: Creating Theoretical Explanations Grounded in Observations

Dovidio et al. proceeded by developing a hypothesis derived (by deduction) from what they call the common ingroup identity model of intergroup behavior (1997, p. 416). The hypothesis was to be tested by "creating conditions that foster a stronger one-group representation" and examining whether this "significantly reduced intergroup bias in ... evaluations and behavior" (1997, p. 410). The observed results, as they interpreted them, were "generally consistent with the framework [the common ingroup identity model] that guided the present research" (1997, p. 411). According to a non-foundationalist position, a well-tested framework is one that has been repeatedly subjected to the test of possible falsification via the evidence. If results derived through repeated tests still do not disconfirm the framework, then it can be taken that it is (for the time being) tentatively corroborated. But it should not be taken as conclusively proven. Seen from a non-foundationalist position, Dovidio et al. should not, therefore, give the impression that the research process could be used to establish conclusive knowledge about the veracity (or not) of the theoretical framework employed to guide the process. The supporting results that they obtained may, however, be presented as evidence that the hypothesis, and the theory from which it has been derived, has withstood another test of falsification and has therefore become better tested.[28]

4.3.2.4 Value Freedom: Seeking Knowledge through Operating Self- and Collegial Criticism

Dovidio et al. suggest that although the theoretical framework of the common ingroup identity model has guided their research, the results of the experiment are a product of the evidence provided by reality. The conclusions that they have reached have not been pre-defined by the initial framework that they admit has guided the research. This framework, they contend, is indeed up for testing via the evidence. Dovidio et al. present the research as aimed at investigating the phenomena of intergroup relations. While their own values might have affected the topic that they have selected for study, they imply that once this selection has been put into effect, the investigation is no longer contingent on their (subjective) ideas/values that they have brought to it.

From a non-foundationalist point of view, their allegiance to the ideal of value freedom would be appreciated. However, it might be argued that they could have shown more convincingly that they are as intent on disconfirming, as on confirming, their theoretical model via the evidence. Part of their accountability as scientists is to harbor

---

[28]Cook indicates that the point of undertaking experimentation is to be able in some way to generalize from "specific manipulations to abstract causes" (2000, p. 3). But he also notes that we need to accept that complete causal knowledge (on the basis of inferences from observations to abstract causes) is not feasible. He sets out to offer a detailed account of how reasonable inferences might be made from experiments to draw conclusions about causes and effects (2000, p. 12). He calls his position a "modified falsificationist approach (2000, p. 19).

a suspicious (self-critical) attitude toward the framework guiding the research — in order to express their commitment to the principle of objectivity. By harboring this attitude they would be more likely to take on board possible criticisms that might be leveled by colleagues in regard to the way in which they organized the study.

4.3.2.5 Instrumental Knowledge: Providing Opportunities for Better-informed Action

Dovidio et al. remark that the knowledge that they have developed through this and other experiments can be put to good use in effecting more harmonious intergroup relationships. From a non-foundationalist perspective, it is fair to propose that knowledge generated through the scientific process has some utility for those wishing to create desired outcomes. What gives scientifically generated claims their instrumental value, according to a non-foundationalist position, is that they can be argued to have higher status than those which have not gone through the scientific effort of organizing rigorous testing of knowledge claims. From this point of view, though, it would also be emphasized that it is possible to "continually raise doubts about any claim to knowledge" (Hammersley, 1995, p. 43). Nonetheless, the better tested the knowledge claims, the more credibility can be placed in them as guides to informed action.

Thus, to act accountably as scientists, Dovidio et al. need to show that their model of group interactions is indeed being put up for rigorous testing through the scientific process.

4.3.3 A Scientific Realist Assessment

4.3.3.1 Scientism Qualified: Attending to the Reality of Generative Mechanisms, Including Societal Ones

The scientific realist argument is that when experiments are used in both natural and social science, they should not be organized in terms of experimental/control group logic. This logic is based on a conception of causality that fails to appreciate the generative causality of (hidden) mechanisms responsible for the triggering of observed effects. Remarking generally on the idea of setting up experimental and control groups and comparing them for significant differences in the behavior of their members, Pawson and Tilley ask (rhetorically):

> Do we understand the action of gravity on a falling body by observing the motion of a cannon ball dropped from a leaning tower and comparing it with the motion of one that remains atop? Do we understand the behavior of atoms smashed in a particle accelerator by comparing them with the actions of those not so treated? One does not need to be a historian of science to answer these questions. (1997, p. 57)

Dovidio et al.'s experiment is based on the logic of comparing (through statistical analysis) the responses of people from the manipulated single-group condition with those from the two-group condition. According to a scientific realist view, this way of organizing experiments is often a false start if one wishes to understand the behavior of that which is being studied — whether one is dealing with natural or social phenomena.

The scientific realist argument is based on the idea that "outcomes are explained by the action of particular mechanisms in particular contexts ... this explanatory structure is put in place over time by a combination of theory and experimental observation" (Pawson & Tilley, 1997, p. 59). As Bhaskar elucidates, experiments need to be designed so that the experimental scientist can "trigger the mechanism under study to ensure that it is active", and "prevent any interference with the operation of the mechanism" (Bhaskar, 1979, p. 53). This cannot be done by simply manipulating an independent variable and observing its effect on hypothesized dependent variables.

This means that from a scientific realist point of view, Dovidio et al. have not designed their experiment in such a way as to offer a sound explanation of the (social) realities involved. For instance, it might be proposed, in relation to Dovidio et al.'s study, that the structure of capitalist society acts as a mechanism that, inter alia, can dispose people to differentiate themselves into ingroups and outgroups on the basis of the idea that certain divisions in society are inevitable. The differentiation of people into "in" and "out" groups on the basis of certain defined criteria, might be hypothesized to be a product of the way capitalist society is itself organized. The entity of "capitalism", it may be argued, can be better understood once we realize that such group consciousness is a possible outcome that can be generated or triggered by it.

The appreciation of how people may come to develop and act upon group representations thus cannot be a matter of merely establishing causal paths in terms of sequential logic. Even if Dovidio et al. are able to set up various experiments that support the idea that group manipulation, through affecting group representations, can lead (sequentially) to types of evaluative and other behavior, this does not mean that they understand why this should be so. To develop this understanding, requires a different vision of causality.

4.3.3.2 Phenomenalism Reviewed: Developing Acceptable Observations

Dovidio et al. have set up a way of observing the phenomena in which they are interested. From a scientific realist perspective, the theory-ladenness of observation cannot be avoided in the process of doing research. But attempts should be made by researchers to enter into a dialogue with others holding different theoretical beliefs (guiding frameworks) in an effort to set up what might be taken to be "acceptable observations" (Keat & Urry, 1982, p. 53). A scientific realist stance applied to Dovidio et al.'s work, would require that they develop their observations in dialogue with other researchers who may be observing "similar" phenomena. Thus far, Dovidio et al. can be criticized for leaving little scope for opening up a dialogue around the way in which they have developed their observational evidence. They might counterargue, however, that in presenting the details of how they arrived at their observations, they implicitly do open the space for others to question them. But from a scientific realist point of view, it would have to be made clearer that the observations as they have dealt with them in this experiment, are a way of seeing that requires complementation with other ways in order to provide the possibility for a genuinely cumulative development of theory.

4.3.3.3 Empiricism Reviewed: Working at an Analytic Level to Understand Generative Mechanisms in Context

From a scientific realist perspective, Dovidio et al. would be considered as offering a theoretically impoverished approach to the evidence amassed in the experiment. From this viewpoint, the analysis proffered by Dovidio et al. in support of their common ingroup identity model, is insufficient for scientific theorizing. Their effort to link directly their theorizing to the realm of observational evidence is in itself ill-conceived, and excludes an investigation of the wider structural forces that may be at play in the situation. Dovidio et al.'s grounding of their theorizing in the evidence that they have collected, may be misguided — unless complemented by other analytic efforts based on a different logic of inquiry. Specifically, the logic of retroduction needs to be introduced into the process of inquiry, so that researchers interested in this topic can look at Dovidio et al.'s study anew, considering what relevance it might have for an analysis of mechanisms operative at the level of social structure. They may also have to design a number of different experiments — to learn more about intergroup and interpersonal living in relationship to social structural forces that are built into social settings, but not immediately visible.

4.3.3.4 Value Freedom: Considering Possible Reformist Allegiances and if Necessary Re-Organizing One's Focus

Dovidio et al. present their investigation as a neutral process of examining the facts in relation to their proposed hypothesis. A scientific realist argument concurs that experimentation in the social sciences should in principle be geared to finding out about social reality. But in this case, it might be argued, there seems to be a prejudice in favor of finding generalities that would hold across all societies. The proposal of Dovidio et al. is that there may exist some regular causal connection between single-group situations (and representations) and outcomes in terms of evaluations of others, self-disclosure, and helping. From a scientific realist point of view, this hypothesis already reflects a predilection in their approach in favor of the belief that human intergroup and interpersonal behavior is not different in, say, capitalist-oriented and socialist-oriented societies. Dovidio et al. do not mention the societal context in which the experiment was undertaken — referring only to the fact that 144 undergraduate students took part (1997, p. 406).

According to scientific realism, their theorizing therefore cannot be said to be sufficiently alert to the possible prejudices that may arise from their narrowness of focus; it can indeed be argued to contain implicit reformist allegiances. Their investigations could be improved upon if they were prepared to widen the scope of their analyses. Their accountability as scientists within the scientific community is to join (in dialogue) with others who have pointed attention to the need for this.

4.3.3.5 Instrumental Knowledge: Extending Information about (Inter Alia) Social Structures

Dovidio et al. propose that their theoretical framework can "inform practical interventions designed to reduce bias" (1997, p. 416). But, as indicated above, from a scientific realist point of view their reasoning in this regard involves offering "information" in a certain light. The idea that is perpetuated by their work is that interventions to foster harmonious intergroup relations can be accomplished within the basic institutional structure of the society — as this structure purportedly does not affect the outcomes that are produced. Insofar as people try to make use of the knowledge afforded by Dovidio et al.'s analyses, they may thus fall prey to the same narrowness of focus encouraged, implicitly, by Dovidio et al.'s approach.

A scientific realist argument suggests that to operate accountably in society, social scientists also need to offer people the intellectual resources to reflect more radically on what may be required to alter the patterns of human relationships in which they are enmeshed.

4.3.4 An Interpretivist Assessment

4.3.4.1 Scientism Opposed: Doing Social *Science*

From an interpretivist point of view, Dovidio et al.'s experiment would be criticized specifically for taking for granted that human behavior can be studied in the same way as objects of nature can — that is, by isolating certain phenomena for attention and locating their (possible) causal connections. From this viewpoint, what has been left unexplored in Dovidio et al.'s experiment are a number of issues, as follows.

Firstly, Dovidio et al.'s experiment was organized through a manipulation that was undertaken to create different group situations. This means that the situations in which the participants found themselves were experimentally (artificially) produced. Dovidio et al. assume, nevertheless, that the experimental procedure can tell us something significant about common ingroup identity and its effects in real social life. Expressing a more interpretive-oriented argument (against the scientism of positivism), Fielding contends that social researchers should not rely on trying to study human behavior in experimental settings. Humans need to be studied in what he calls "natural settings" in order that researchers can get sufficiently "close" to the real situations in which they interact (1993a, pp. 156-157). Any results obtained through an experimental manipulation are likely to be a product of the *meanings that actors attach to that experimental setting* — the results cannot be assumed to tell us about the everyday worlds in which people live.

Secondly, in terms of an interpretivist argument, the aim of social researchers is to try to gain an understanding of what Fielding calls the "symbolic world" in which people live. (By symbolic world he means "the meanings that people apply to their own experiences" — 1993a, p. 157.) From this perspective, Dovidio et al. could be accused of not utilizing the appropriate techniques to be able to obtain an understanding of the way people give meaning to their interactions with others. For example, Dovidio et al.'s use of a questionnaire asking people to decide using a 7-point scale whether "during the interaction it felt like one group" or "during the interaction it felt like two groups", would be regarded as an inept way of developing an understanding of how people were experiencing the group situation. Clearly, the different participants need not have understood the meaning of "group" in the same way as they answered the question whether "it felt like one group" or "it felt like two groups". Or again, the meaning of one-ness and two-ness might have differed across participants' conceptions. To get even a modicum of understanding of the way in which people attach meaning to their words, one needs to include some in-depth interviewing — which, as Fielding notes, is "more like a conversation" (1993a, p.157). The advantage of this is that instead of "imposing an outsiders' sense of what is going on", one can adapt the research to focus on what emerges as "available and interesting" to participants (1993a, p.157). Dovidio et al.'s own concern with the differences between single-group and two-group situations makes it difficult to determine whether and in what ways, if at all, the issues on which they were focusing are relevant to people.

Thirdly, also as a function of the fact that social research deals with subjects rather than objects of research, interpretivist-oriented researchers argue that social research requires some attempt to establish a relationship of mutual understanding between researchers and respondents. As Fielding indicates, "for interactionists, the data are valid when a deep mutual understanding has been achieved between interviewer and respondent" (1993b, p. 151). The problem with "prescheduled standardized interviews" (or questionnaires such as the ones employed by Dovidio et al., where pre-set, similar questions were asked to all participants) is that, as Fielding complains, the process rests on "fleeting encounters to which respondents are uncommitted, leading to possibilities of fabrication" (1993b, p. 151). In the case of Dovidio et al.'s study, the psychology students' "interaction" with the questionnaires that they were asked to complete, can be argued to be too fleeting and superficial for the researchers to be able legitimately to base conclusions about human behavior in group situations. The same holds for the judges' assignment of an intimacy rating to their interaction in pairs — without probing their experiences of the "intimacy". And likewise one can question the researchers' decision to regard as indicative of helping orientations, the number of locations that subjects indicated they would be prepared to put up posters — again without the researchers' probing their views.

For all these reasons, from an interpretivist point of view, Dovidio et al.'s experiment would be considered as failing to do sufficient justice to the essential differences between the subject matter with which natural and social scientists have to deal.

4.3.4.2 Phenomenalism Re-interpreted: Creating Meaningful Categories

The interpretivist approach to the study of social phenomena supports the building up of conceptual schemes that are grounded in human meaning-making. The creation of categories within social scientific investigation should lead to rich, complex, and dense theorizing, indicating the variegated character of human experience (Henwood & Pidgeon, 1993, p. 24).

Considering the example of this experiment, it would be noted that when organizing the group manipulation into a single-group condition, Dovidio et al. tried to induce participants to recognize certain features that they shared in common, such as their common student identity, a common color T-shirt (maroon), and a common name (the STARS). By doing so, they tried to create the impression of a group condition grounded in a sharing of certain characteristics. However, it could be argued that there are varied ways in which a sense of "we-ness" might be developed in social life, and that Dovidio et al. have failed to highlight this variation. From an interpretive position, an understanding of the phenomenon of "group identity" cannot be gained unless efforts are made to reveal the complex ways in which people might treat their relationship with others in different contexts.

Likewise, when considering people's propensity toward self-disclosure in the two-way interaction, Dovidio et al. allowed the two judges to define the amount and depth of information disclosed (via the tape). The judges did not at any point refer to the meaning of the "disclosures" for the participants involved. But for some people, the fact that another person is experienced as more or less a stranger, may affect the way in which they define their own "disclosures" — they may regard it as insignificant to have revealed their fears to someone whom they regard as, after all, not important to them. For others, the revelation of the "same" information (which the judges may regard as indeed same in terms of its amount and depth), may have a different import. A grounded theorizing approach would call for conceptual schemes to be developed in relation to such "data". Thus, for instance, different ways in which people define the meaning of their disclosures would be identified and created into "ideal types" that exaggerate, while still pointing to, the alternative meaning-constructions that might be developed by people in different contexts of interaction. The purpose of developing such "types" would be to explore the data corpus with a view to adding "variation and depth of understanding" around the issues being explored (Strauss & Corbin, 1990, p. 109).

4.3.4.3 Empiricism Re-interpreted: Developing Understanding of the Social World

The experiment organized by Dovidio et al. was based on the view that it is advisable to study the social world by concentrating on locating causal relationships between phenomena. But an interpretivist perspective questions this view, arguing that social researchers need to consider that there may be other ways of developing understanding of the social world, and that these other ways can be said, at least at times, to provide a "fairer reflection of the data" (Fielding, 1993b, p. 152). Fielding suggests that part of the accountabilities of social researchers is to work with the knowledge that their manner of organizing the process of research may be tied to "different theories of the social world" (1993b, p. 152). These differences, he argues, cannot simply be sidestepped as one organizes social inquiry. However, Dovidio et al. provide no indication of an awareness that their way of generating theoretical inferences is already shaped by a specific conception of the operation of causality in social life. Their empirical appeal to "the data" masks the fact that they have chosen to address it in a specific way.

4.3.4.4 Value Freedom: Paying Attention to Others' Worlds

Dovidio et al. express the purpose of their inquiry as being that of developing models that offer parsimonious explanations of intergroup and interpersonal behavior. In these terms, they pose their own model of common ingroup identity as offering an account of the realities to which the model refers.

From an interpretive point of view, their allegiance to the principle of using research to try to advance knowledge of social reality is laudable; but their manner of going about this is ill-suited to the task. In order to better appreciate the realities of intergroup and interpersonal life, they would need to consider the complexities of how people produce, and work with, meanings in different interaction contexts. In the case of an investigation of intergroup and interpersonal behavior, one would need to get much closer to "respondents" than can be realized through the techniques of data collection favored by Dovidio et al. While Dovidio et al. believe they are operating in a non-prejudicial way as social scientists, an interpretive argument would suggest that they are not operating with sufficient awareness of what may be involved in paying attention to the worlds of those they are studying.

4.3.4.5 Instrumental Knowledge: Creating Meaningful Information Relevant to Actors

Dovidio et al. offer a model that they present as of use to those wishing to generate practical interventions in society designed to reduce intergroup bias. From the point of view of an interpretive argument, however, social researchers should bring to the fore the way in which the patterning of social life is grounded in specific ways of creating meaning. So, for instance, one could highlight how people interact with others when they define "we-ness" as implying that others need to share certain characteristics with them, while at the same time highlighting how people interact with others when they assign a different meaning to "we-ness" (than that considered in Dovidio et al.'s definition). One could set out to examine the topic of "we-ness" with a view to what Weber calls "swimming against the stream" (1949, p. 47).

Considering Dovidio et al.'s research, one could attempt to swim against the stream of what may normally be meant (by both theorists and lay actors) when the term "group identity" is invoked. One could do this by trying to isolate what Strauss and Corbin call "negative cases" (1990, p. 109), which point to alternatives. Hereby one could offer some understanding of how different ways of creating meaning around the phenomenon of "we-ness" in social life may lead to different experiences of group life and different orientations of actors toward others.

A (theoretical) concern with the way in which actors' assignment of particular meanings to "group life" issues in particular ways of acting in relation to others, can help people in turn to review their actions in the light of the various "cases" as discussed by (social) scientists.

4.3.5 A Critical Theoretical Assessment

4.3.5.1 Scientism Criticized: Avoiding the Transference of an Engineering Approach to the Study of Social Reality

Critical theory criticizes research that is unreflective about its way of treating causality in social life. Habermas cautions that the scientific search for social regularities, can all too easily become tied to the belief that it is possible to engineer outcomes on the basis of knowledge of causal connections supposedly existing "in reality". He suggests that in order to avoid this way of seeing social reality, empirical regularities should be considered as being only quasi-causal. That is, they should be considered as existing only insofar as people continue to act them out (Habermas, 1976b, pp. 208-209). But Habermas warns in addition that the likelihood of people continuing to act them out is strengthened through the operation of a social science that treats them as "given". Thus when Dovidio et al. treat as given the fact that people's creation of a superordinate group categorization leads normally in the direction of being less biased toward members of the (superordinate) group, this statistical normality starts to become equated with a norm of conduct in social life. When Dovidio et al. "reveal" the fact that people's application of a superordinate group category renders their responses significantly different (in terms of a comparison of mean scores) from those not applying such a category, they simultaneously give this "fact" normative support.

Doing social science as if one is studying non-human objects, is detrimental seen from a moral point of view, as far as critical theory is concerned. The regular connections between phenomena as "uncovered" within such a social science become seen as the order of the day. The identification of them by scientists can operate to stifle moral thinking about alternative ways of creating "the facts" of human life. In the case of Dovidio et al.'s study, critical theory would prefer to concentrate on the possibility of people's experiencing forms of intergroup and interpersonal relations that express the vitalization of human communicative capacities. It would in any case call for a public discussion around the project of rescuing these capacities.

4.3.5.2 Phenomenalism Reconsidered: Creating Realities through the Research Process

From a critical theoretical point of view, the way in which Dovidio et al. have defined the terms in their hypothesis "objectifies" (makes into objects of study) the phenomena referred to. But what is observed through the scientific process (for example, through the observation of the characteristics of the single-group condition, through the coding of the various questionnaires, and through the content analysis of the pair discussions, etc.) is just that — a way of observing phenomena. According to critical theory, the definition of reality through these processes is a reflection of the manner in which the scientists involved have invoked rules for observing what is taken to be reality. The content of the empirical scientific information is "valid" only within the frame of reference that they have brought to bear in order to define how reality is to be approached (Habermas, 1976b, p. 209).

Dovidio et al. account for their observations by trying to demonstrate that they are drawing on acceptable practices for scientific observation. Insofar as they admit that future research might require a refinement or modification of (some of) their operational definitions of terms, it is implied that this process of redefinition can take place via a discussion within the scientific community. This can act as a check on the way in which observations are achieved. But what they do not subject to consideration is that the practices for observation agreed within (parts of) the scientific community, still rest on an intersubjective accomplishment of the scientists involved. Once this is understood, the way is open for approaching reality with a different agenda (or interest). For example, one could decide to "see" the realities of intergroup and interpersonal conduct by investigating people's potential to organize (genuinely) discursive encounters as a way of conducting their sociality. One could probe people's (participants') visions of what this discourse might entail; and, in the process of probing, one could introduce the critical theoretical conception of discursive encounter as an option for guiding their social interactions (within and across any groupings with which they might identify).

As far as critical theory is concerned, Dovidio et al.'s intention of rendering variables observable so that experimentation can be undertaken to test whether expected outcomes occur, is thus misdirected. If this is seen as the task of social researchers, it becomes difficult for them to make a contribution toward realizing alternative human experiences.[29]

4.3.5.3 Empiricism Reconsidered: "Discovering" Social Potentialities

Critical theory would criticize Dovidio et al.'s suggestion that they could test the truth of their hypothesis with reference to "the evidence" offered by social reality. Linked to this criticism would be a concern that the observed connections between variables "found" to be operative through the experiment, become taken by the researchers (and presumably by others) as providing information about social reality. The causal connections between, say, group representation (via superordinate categorization in which people (re)categorize themselves as belonging to one group) and evaluative and other biases, then are regarded as unbreakable — insofar as they are seen to refer to regular patterns similar to the regularities found in natural reality. As such, a discussion of whether it is possible to transcend this realm of causality that Dovidio et al. have located, becomes seen as nonsensical (irrational).

From a critical theoretical point of view, however, a discussion about the way in which (supposed) causal connections in social life should be treated, is to be welcomed. But such a discussion is in danger of being threatened by the narrow empiricist view of scientific rationality in terms of which the experiment (and others of its kind) has been organized. In other words, the discovery of potentiality in social reality becomes threatened by the very endeavor of seeking to locate social regularities in the way Dovidio et al. have attempted.

---

[29]See also Romm (1991, pp. 145-146) for a more detailed account of the Habermasian view of "experimentation" as a way of creating experiences/phenomena in social reality.

4.3.5.4 Value Freedom Undercut: Recognizing the Value of Human Discourse

Critical theory presents as a value guiding social research, the nurturing of intersubjective forms of social relationship. It is suggested that people can experiment with new forms of democratic existence — involving public argumentation around issues of concern as raised by participants.

Dovidio et al.'s way of organizing the experimental procedure (including their utilization of techniques of observation) and their development of their theoretical discourse around the explanatory power of the common ingroup identity model, does little to open the space for reviewing the (quasi-causal) laws relating to group life that they have identified. Their experiment is in this sense not value-neutral in its social consequences.

Critical theory would be cautious of endorsing a sense of group belonging that is determined (mainly) by people's application of some superordinate category defining group membership. Instead, critical theory tries to encourage people to reflect upon possibilities for developing social integration around what Delanty calls the "principle of discursivity" (1998, paragraph 6.5).

4.3.5.5 Instrumental Knowledge Undercut: Furthering Communicative Understanding

Dovidio et al. suggest that the common ingroup identity model of group relations provides information of relevance to those interested in generating more harmonious intergroup relations. From a critical theoretical point of view, however, their presentation of their knowledge as being (instrumentally) useful in this way, might already serve to stifle communication around the question of how "we-ness" in democracy can be differently created/pursued. Sharedness (as intersubjective communication) might be able to be cultivated in a different manner from that suggested by Dovidio et al. in their common ingroup identity model. Critical theory proposes that people should be encouraged to experiment with options for generating communicative capacity in a democracy. To participate in this experiment, people need to be enabled to take part in a discussion around the way in which any "scientific information" is created and presented in society. Otherwise, their communicative life and attendant interactions are likely to be unnecessarily constrained by what they take to be scientifically generated results.

4.3.6 An Anti-foundationalist Feminist Assessment

4.3.6.1 Scientism Challenged: Overturning Traditional Research Control

Anti-foundationalist feminists are wary of what Lincoln calls "the stranglehold of experimental models of behavioral research in the social sciences" (1995b, p. 53). Insofar as this mode of approach is utilized, it is considered crucial, in any case, to recognize that our ways of approaching our social inquiries say something about "who we are as individuals, who we are as professionals, who we are as people, and who we are as a culture" (Lincoln, 1995b, p. 53).

Dovidio et al. organized the experiment to extract from the participants what was useful for their own purposes — that is, the data needed for testing their hypothesis. Their (implicit) justification for this is that the experimental results can later be of benefit to the participants as well as to others who may want to make use of the knowledge gained via the study. In conducting their relationship with the students, Dovidio et al. draw on the idea that, as Denzin expresses it, "the participating citizen [can] contribute in the public good by being a scientific subject. Indeed the citizen has a responsibility to participate in public science. In this way, justice is served" (1997, p. 271). This idea of justice relates to the attendant view that the scientists organizing the studies are "competent researchers" whose research will "carry universal appeal to all parties" (Denzin, 1997, p. 271).

In the case of this example, Dovidio et al. did not see the need to discuss with participants how they might feel about participating in the project, or how they might be conceiving the (power) relation between the experimenters and themselves. The researchers did not call into doubt their own assumption of control during the project. They would indeed probably regard the students as having shown disrespect for their experiment if students' encounters with them and with their questionnaires and other forms of data collection appeared to call into question the mode of interaction being set up.

Anti-foundationalist feminists would suggest, however, that it should *not* go without saying that using students in the way Dovidio et al. have done, is acceptable practice in social research. This is because the organization of the study is at the same time an instance of a way of organizing human relations. To act accountably as social researchers they would need to take into consideration how their experimental mode of research as they have employed it is not only a way of knowing, but "has powerful implications for the way we live" (Palmer, 1987, as cited in Lincoln, 1995a, p. 281).

4.3.6.2 Phenomenalism Reconsidered: Working with Ambiguity

Anti-foundationalist feminists would question the way in which Dovidio et al. treat the terms used in their hypothesis as if they refer to phenomena in reality that can be discerned independently of the interjection of the researchers. Taking the example of their manipulation of a single-group condition, it would be argued that the manufacture of the experience of "belonging" via the researchers' production of common characteristics, already suggests (albeit indirectly) to participants that group consciousness can become constructed through a sharing of characteristics that are defined as "the same". Dovidio et al.'s way of manipulating the situation (by, for instance, giving people the same color T-shirt and offering them a common name — the STARS) *sets up this option for experiencing group identification.* This option is now given a certain credibility in people's consciousness.

Harrison and Hood-Williams note in regard to definitions of gender categories (and group membership on this basis), that what is at stake here is "the political *transformation* of experience by a process of re-signification" (1998, paragraph 3.5). As they state, when researchers act as if "membership of a group leads to shared experience which leads to shared views", this poses "a political problem" for anti-foundationalist feminism (1998, paragraph 3.3). The problem is that people designated in terms of some categorization as belonging to a group, now come to see themselves as having to show some identity therewith. But the meaning of identification — that is, its signification — is never subjected to re-examination to allow those involved to explore the specificity of their experiences. Indeed their experiences might become reconstituted so that they begin to feel that a particular way of understanding "sharedness" is the only legitimate way of expressing group membership.

Anti-foundationalist feminists would call on Dovidio et al. to appreciate that their supposedly scientific process of defining the phenomena under study, simultaneously might create an intervention in social life. Once this is recognized, anti-foundationalist feminism would suggest that the space is opened for seeing (and constructing) the realities of intergroup and interpersonal life in a variety of ways. For instance, the question of whether group "identification" need imply either a sharing of common (material) characteristics or a sharing of points of view, can be opened for consideration. In this way, the givenness of the phenomenon of, say, identity, could be displaced (along with the givenness of all social phenomena).

4.3.6.3 Empiricism Reconsidered: Appreciating the Situatedness of Research

Dovidio et al. suggest that they have organized a research design to discover information about the realities of intergroup relations. However, what would be of concern to anti-foundationalist feminists is that they fail to interrogate, or invite others to interrogate, the situatedness of the "data" generated through their study. The "immaculate perception" of the realist tale that they build up (Lincoln, 1995a, p. 280, citing Van Maanen, 1988), belies the partiality of their specific way of creating the evidence in support of their argument. They imply that they are building up their theory based on repeated experiments that they (and others) have undertaken. But the experiments are organized in such a way that participation in defining both its process and product becomes restricted to the experiences that they construct and utilize as part of their own research agenda. This, for anti-foundationalist feminism, is an unduly restrictive agenda.

4.3.6.4 Value Freedom Undercut: Appreciating the Value of Heterogeneity

Dovidio et al. present their analysis as offering a detached account of aspects of intergroup realities, based on the evidence afforded through their (and other) studies. Anti-foundationalist feminists would accuse them of failing to explore the preferences that might be underpinning their study. For instance, it could be argued that their "finding" that creating a superordinate category is a way to generate harmonious interactions between those (now) identifying with the superordinate group, expresses their own preference for ways of organizing people's identification with groups. The idea of producing superordinate categories as an avenue to harmonious group relations is an expression of a specific preference, which is obscured by its being presented as based on the facts of group existence.

Anti-foundationalist feminism thus uncovers the value position underlying the apparently factual account presented by Dovidio et al. By doing so, it calls at the same time for a multiplicity of ways of interpreting the import of the study — so that Dovidio et al.'s authoritative voice does not come to rule subsequent experiences of participants and other audiences along the direction dictated by the supposedly "scientific" story.

The feminist ethic involved in breaking the monopoly of scientifically created stories, is that of "making the space for the lifeways of others and creat[ing] relationships that are based ... on mutual respect, granting of dignity, and deep appreciation of the human condition" (Lincoln, 1995a, p. 284). Those calling themselves researchers are required to take seriously the possibility of respecting a variety of ways of seeing and of acting, and to express this respect in the manner in which they conduct and write up their studies.

4.3.6.5 Instrumental Knowledge Undercut: Questioning the Power to Define Realities

According to Dovidio et al., the practical benefit of their research springs from the fact that it can supply information for people to use in organizing action toward what are considered to be desired goals, such as the goal of, say, creating more harmonious group relations.

But anti-foundationalist feminists argue that in presenting their work as informative in this way, the instrument of knowledge becomes an instrument of power. In order to minimize the effects of power, it is suggested that researchers give attention to trying to develop minimalist texts as they present their "findings". As Denzin notes: "A minimalist text is saturated with theoretical understandings, but it does not announce or parade its theory" (1997, p. 40). Dovidio et al. would be accused of indeed parading their theory of common ingroup identity. Their presentation of their theory as offering a parsimonious explanation (of intergroup relations) that can serve to direct action, renders it the more difficult for what Denzin calls "the sound of the other's voice" to be heard in society (1997, p. 40).

4.3.7 A Trusting Constructivist Assessment

4.3.7.1 Scientism Revisited: Developing Trusting Relationships through the Research Process

Dovidio et al. assume that those assessing their experimental design will understand that social experimentation is an acceptable way of proceeding to uncover causal connections relating to intergroup life in society. All that remains for them, therefore, is to discuss the details of the design and to show how they used certain techniques of observation to measure the variables isolated for attention.

From a trusting constructivist point of view, Dovidio et al. have not given sufficient credence (in defense of their way of proceeding) to arguments and concerns springing from alternative views on what experiments can, and cannot achieve. Hence, when discussing suggestions for further research, it is noteworthy that they at no point mention that it might be advisable to perhaps consider an alternative way of approaching the issues that they have raised through their study. They state, for instance, that "the degree to which intergroup factors and interpersonal factors combine to facilitate self-disclosure may be a productive issue for further research" (1997, p. 415). In the same context they suggest that

> it is possible that creating a common group identity produces higher levels of interpersonal trust which, in turn, facilitates self-disclosure and other types of interpersonal cooperative behavior (Kramer & Brewer, 1984). (1997, p. 415)

They cite Kramer and Brewer's (1984) studies of simulated dilemmas as an example of how research into interpersonal relationships can be organized. But they do not confront possible queries concerning what can be gained via the experimental mode of inquiry employed by the researchers. They thus sidestep — without showing that they are taking seriously — the kinds of questions that have been raised in the various sections above (especially in Sections 4.3.4-4.3.6).

In considering their proposal regarding research on interpersonal trust, it might be argued that if one wishes to study the factors leading to what they call "high levels" of trust, it is important for the research process itself to be used as a forum to explore the experience of creating trust in human relationships. An experimental process in which researchers take it as read that the respondents should trust (by fiat) that they know best how to proceed to examine the issues, might not be an appropriate mechanism to explore the dynamics of human trust as an interactive event.

In the case of their own study, Dovidio et al. have not provided any indication that they have tried to come to terms with various contentions surrounding experimentation as a procedure in the social sciences. One way of taking these contentions seriously would have been to reflect upon some of the issues together with the participants at some stage. After all, these were students who might well have wished to spend time on discussions around their involvement in the study (as might other participants). Discussions in the case of this experiment could have been organized through individual and/or through group interviewing at some point. The latter could have been provided for in focus group forums, that is, forums focused around the topic of discussion — to wit, the experience of having been a participant in the experiment. Participants, in discussion with the experimenters, could, for instance, reflect upon the mode of interaction set up in the experiment in terms of the quality of "data" generated. Dovidio et al. dismiss concerns about subjects' possible responses to implicit researcher expectations, on the basis of previous experimentation undertaken by Gaertner et al. (see Section 4.2.2). But this issue could have been explored in this instance *with the participants themselves*. In doing so, Dovidio et al. could have treated as relevant to a discussion of the experiment, the manner in which participants deliberate on their experiences of it. And they could also then provide some account in their report of how they personally are responding to the discussions in terms of what they believe they need to take into consideration in future as they organize social inquiries (in order to develop more trust on the part of participants and other audiences in their work).

4.3.7.2 Phenomenalism Reconsidered: Evoking and Defending Observations

Dovidio et al. show how they utilized techniques of observation to measure the variables involved in the experiment. The data that appears (in the form of, say, answers to questionnaires, content analysis of (pair) conversations, reporting of the locations where posters are to be hung up, etc.) is henceforward taken by them as being an indication of the phenomena of interest.

A trusting constructivist position would suggest that Dovidio et al. need to conduct an argument showing how, in the light of differing possible interpretations of the observations that they have generated, they are choosing to see them in the way they do — namely, as indicators of the variables used in their hypothesis. It would be expected (as part of the trust earning process) that the researchers show some acknowledgement of the part they may be playing in evoking the phenomena that they are using as empirical evidence. They could try to account for their way of organizing observations, and their analysis thereof, while not giving the impression that we need be mesmerized by these. Audiences of the "findings" could then be invited to understand that the experiment cannot necessarily be treated as a way of uncovering causal connections that exist independently of the way in which researchers set out to observe them.

4.3.7.3 Empiricism Reconsidered: Exploring Alternative Ways of Seeing and Using Evidence

Dovidio et al. take it as a rational way of proceeding to use a theoretical framework to develop certain hypotheses, so that causality in social life can be examined through experimental/control logic. But a trusting constructivist position would complain that the organization of their study is not sufficiently discursively attuned to engaging with alternative ways of defining acceptable research practice. Colleagues in the "scientific" community who wish to enter into a discussion around the experiment might of course do so by referring back to the way Dovidio et al. measured their variables and analyzed their results. Yet they would seem (in view of the way in which Dovidio et al. have set the terms for debate) obliged to draw upon the rules for scientific inquiry that Dovidio et al. themselves invoke. Likewise, those in other communities (including participants in the project) are required to converse with the investigation in these terms (insofar as they converse with it at all).

From a trusting constructivist point of view, the respect for the experimental process that Dovidio et al. expect others to accept (on trust), would have to be extended to take into account a respect for some of the objections that have been raised against it. It is only in the light of an engagement with ongoing debates around the way in which experimental "evidence" is to be treated, that Dovidio et al. could develop an argument that might be read as more discursively accountable.

4.3.7.4 Value Freedom Undercut: Encouraging Discursive Accountability as a Value

A trusting constructivist position values the possibility of using the research process to cultivate discursively accountable practice both in research and other contexts of interaction in society. How might such possibilities have been explored in Dovidio et al.'s study?

As a starting point, the researchers could have discussed with participants any concerns that they may have in regard to participating in this kind of experiment. Although Dovidio et al. did ask students whether they wished to refuse participation at the point of discussing the moderately intimate topic (in the pair discussions), we do not know how they phrased their suggestion to students that if they wished they could refrain from participating herein. It is possible that their phrasing was prohibitive of withdrawal, and that students felt the pressure to "participate in public science" (see also Section 4.3.6.1 above). In any case, the students' participation in other parts of the experiment was assumed to be acceptable — although the procedures to be utilized were fully in control of the experimenters. Seen from a trusting constructivist point of view, Dovidio et al. could have taken more seriously their students' way of conceptualizing the research endeavor. Some of them might not have wished to participate in a project that they knew to be set up to deal with their reactions (to an arranged manipulation) in such a standardized fashion (using pre-set questions and rigid rules for interpreting responses). While Dovidio et al. believe that many of the tasks during the experiment had to remain ostensible (and thus could not be mooted with the participants), the overall idea of putting the participants in manipulated conditions and then using a pre-set format to collect their responses, could have been discussed with them. Out of the process of discussion around this, it is possible that the researchers and (some of) the participants might have chosen to proceed differently. From a trusting constructivist point of view, the organization of the research could and should thus include opportunities that allow both the researchers and the participants to account for their choices made in relation to a discourse around issues of participant participation.

Additionally, as mentioned briefly in Section 4.3.7.1, Dovidio et al. could have used the experiment as a device to open up a discussion (at some point) with participants around the achievements of this experiment, and other experiments. Consideration could have been given to the viability of organizing experiments such as this one as a way of establishing causal connections existing in social reality. Participants might find it interesting to recognize that they do not need to take on trust (without discussion) Dovidio et al.'s presented conception of the achievements of the experiment. And the researchers too could indeed have broadened their understanding hereof as they engaged with their students' varying conceptions of the experimental process and its creation of "results". They could take this into account for their future organization of, and deliberations on, their inquiry practices. Again, their accounting for their research could then include accounting for issues of concern raised during the various discussions.

As regards their presentation of their findings, Dovidio et al. have at this point presumed that these are sufficiently sound to be able to offer what they call "practical benefits" in the form of recommendations for realistic action. However, seen from a trusting constructivist point of view, they should rather have opened the space (in a way that could have been designed into the experiment, or into another form of inquiry) for considering varied experiences around the issue being studied (group categorization). Instead of expecting others to take on trust that their way of manipulating a single-group condition creates experiences that can aid their own scientific theorizing, they might have allowed other voices on the issue to be heard. Rather than expecting people to defer to their (scientifically developed) vision, they could try to develop an alternative way of relating to others and to their arguments and experiences. Accordingly, they could explore what it might mean to engage others' trust in their work through a discursive relationship.

4.3.7.5 Instrumental Knowledge Undercut: Tempering the Possible Self-fulfillment of Results

Dovidio et al. have utilized the experimental procedure to advance their theorizing about intergroup relations. Their theorizing is presented as offering instrumental knowledge — that is, it is useful to people because it is instrumental for taking action toward desired goals.

From a trusting constructivist point of view, the theorizing in which (professional) researchers engage should not be seen as having the status of instrumental knowledge in the way conceived by Dovidio et al. What Dovidio et al. *can* offer is an indication of how they have built up their story about superordinate categorization in recognition that there is a multiplicity of ways of conceptualizing "groups" and the "identification" of people therewith. Out of an engagement with alternatives, they could develop an account, which, for the time being, they express as their "results". But their suggestion that we need to pay heed to their results about superordinate categorization if we wish to consider developing more harmonious group relations, should at least be pitted against other ways of developing meaningful experiences (around the issues raised). In this way, they can temper the possibility that their presentation of "results" might act as a self-fulfilling directive (if people were to see and experience reality along the lines suggested as being "found"). The findings could rather be used to extend people's experiences of social life — in ways that do not close other options. In short, the experiment could support a type of discourse geared to cultivating people's taking responsibility for their choices of vision and of action (at points in time) while leaving space for others to do likewise.

4.3.8 Trusting Constructivism Summarized in Relation to Alternatives

Thus far in this chapter I have outlined Dovidio et al.'s experiment and commented on it from various perspectives — in order to draw out some of the debates surrounding experimentation as a procedure in the social sciences. I have shown how Dovidio et al. made efforts to render clear their mode of proceeding, and how they argued that their manipulation of the independent variable in the experiment allowed them to determine effects (on hypothesized dependent variables). I pointed to possible assessments of their approach as leveled from within positivist, non-foundationalist, scientific realist, interpretivist, critical theoretical, anti-foundationalist feminist, and trusting constructivist argumentation. Below is a summary of the alternative interpretations of the experiment as I explored them, with a brief indication of how a trusting constructivist position might develop an argument in relation to each one.

4.3.8.1 Positivism and Trusting Constructivism

From the point of view of a *positivist* position, Dovidio et al. can be said to have operated accountably as scientists in trying to ensure that their manipulation of the independent variable was such that its causal influence on the hypothesized dependent variables could be tested. The process of organizing an experimental manipulation and examining effects via statistical analysis of experimental and control situations, is acceptable research practice. What might perhaps be questioned is the way in which Dovidio et al. presented the validity of the constructs that they used to measure the phenomena in which they were interested. But apart from this possible line of questioning, it can be suggested that Dovidio et al.'s experiment accords with processes endorsed within a positivist position for proper scientific practice. The procedures that they followed in organizing the experiment, and in comparing it with results from other studies, can therefore be expected to lead to a product that represents an advancement in knowledge of social reality. These results can be of some instrumental benefit to those in society who wish to make use of them.

The *trusting constructivist* response to this interpretation of the experiment is that it assumes that the purpose of organizing social scientific study is to be able to get to grips with causal connections existing in social reality; and it assumes that experimentation is in principle a route to achieving this. It accepts as the audience for judging experimental processes those colleagues in the community who themselves operate in terms of similar methodological preferences. However, by setting the parameters for the assessment of the experiment in this way, the positivist position appears to rule out alternative languages for speaking about the research endeavor. It is for this reason that I have tried to enter opposing arguments and concerns into my own engagement with Dovidio et al.'s experiment — rather than concentrate on the technical concerns that might be raised from a positivist perspective.

4.3.8.2 Non-foundationalism and Trusting Constructivism

From the point of view of a *non-foundationalist* position, Dovidio et al. can be said to have overestimated the success of their attempts to ensure that their manipulation of the independent variable was such that its causal influence on the other variables could be determined. Although organizing an experimental manipulation and examining effects via statistical analysis of data from experimental and control situations is one way of arranging scientific practice, researchers need to be more cognizant of factors that can introduce error into the process of generating knowledge. For example, factors such as the reactivity effect in experimental designs (and indeed in other research situations) may need to be accorded more recognition than Dovidio et al. have done. The possibility of the reactivity effect affecting the results, as they affect participants' responses to the experimenters and to the experimental situation, should have been given more explicit attention by Dovidio et al. Also, they could have shown more explicitly that their attempts to validate the observation instruments used to observe the phenomena being investigated (and the consequent analyses of the data) are not to be regarded as error-free. Their experiment should therefore probably have been opened both to more self-criticism and to more collegial criticism in order to take into account some of these reservations. With this proviso understood, the product (findings) generated, can still be of benefit to people in society who wish to make use thereof — as long as they are aware that the results do not claim to offer certainty of knowledge about the realities investigated.

The *trusting constructivist* response to this (non-foundationalist) interpretation of the experiment is that it continues to regard as the purpose of social scientific study the quest to grasp the connections/patterns existing in social reality. Although it admits that the scientific processes geared toward this end are never infallible, it still places it in the hands of scientists to find a path that is likely to lead to an advancement of (some) knowledge. The non-foundationalist interpretation turns for the assessment of scientific work to colleagues in the scientific community who see the purpose of social science in society as that of advancing knowledge of patterns therein. From a trusting constructivist point of view, however, it is preferable for "scientists" to have to account for their work by engaging other perspectives on it. This engagement should embrace, inter alia, the perspectives of those who are to participate in the study (or have already done so) and those in other circles who have an input to make in the debate around, in this case, experimentation in society.

4.3.8.3 Scientific Realism and Trusting Constructivism

According to a *scientific realist* position, Dovidio et al. have begun their investigation with a narrow conception of what is involved in doing both natural and social science. Organizing an experimental manipulation and examining effects via a statistical analysis of data from "experimental" and "control" situations does not give us an indication of *why* whatever effects are observed, have been generated. Dovidio et al. may have contributed to advancing our knowledge about effects that can ensue when people apply superordinate categories in group situations (in certain instances). But according to scientific realism, it is possible that the generative mechanisms that trigger these effects in certain situations, can trigger other effects in others. To understand this, we need to organize experiments that allow us to explore the underlying mechanisms whose constitution explains the outcomes that may be generated. Seen in this light, Dovidio et al.'s experiment has limited value in offering an understanding of social reality. Furthermore, the practical danger of their experimental results is that by posing as offering information of benefit to people toward realistic action, they can endorse a reformist outlook. People are led to believe that a feasible avenue toward generating harmonious group relations is via the (re)application of categories (through, for example, recategorization and decategorization as Dovidio et al. understand it). People are thus discouraged from considering more radical restructuring of the basic institutions of the society. The instrumental knowledge that is proffered in terms of Dovidio et al.'s experiment therefore may have consequences in society that have not been sufficiently problematized in their discussion.

The *trusting constructivist* response to this (scientific realist) interpretation of the experiment is to question the underpinning hope that scientists can, if they proceed appropriately, find a path that is likely to lead to an advancement of knowledge of (social) causality. The scientific realist interpretation (still) accepts as its favored forum for organizing a debate around types of experimentation and their utility, the scientific community. And it accepts that the discourse in this community must be oriented toward the goal of seeking knowledge of (extra-linguistic) reality. From a trusting constructivist point of view, however, other issues need to be considered, such as the social relationship that is set up with participants in the process of experimentation, and the social relationship that is set up with other audiences in society. Scientific realism can be accused of confining its critique of Dovidio et al.'s (as others') experiments to those issues that it brings to the discussion — issues concerning the possibility of advancing knowledge of generative mechanisms in social reality.

4.3.8.4 Interpretivism and Trusting Constructivism

Dovidio et al.'s idea of manipulating an independent variable in order to examine effects on dependent measures, is borrowed from experimentation in the natural sciences. According to the interpretivist argument, Dovidio et al.'s study cannot be said to have contributed to advancing our understanding of the variety of meanings that might be created by people as they organize their interactions in group situations. To take account hereof, we need to reconsider whether and how, if at all, experiments can be used for this purpose. Perhaps if supplemented by other processes of inquiry, they can make a contribution to developing such understanding. But Dovidio et al. have not credibly shown that their experimental work has anything other than very limited scope in offering an understanding of the meaningful constitution of social reality. Furthermore, a practical effect of their work is that by presenting as "real" the responses created via their various techniques of observation, they may serve to preclude an understanding of the complex ways in which people make meaning.

The *trusting constructivist* response to this (interpretivist) interpretation of the experiment would be to agree that the definitions of terms developed by Dovidio et al. do not do justice to the different ways in which participants might attribute meaning to apparently "given" phenomena. However, the trusting constructivist position argues that the processes that interpretivism suggests need to be organized to correct this deficit, should be extended in terms of the possibility of evoking new forms of human relationship. The scientific process can be used to explore the possibility of, for instance, developing more trusting relationships with participants and other audiences touched by the research. However, in order to be able to make such a proposal, the critique of Dovidio et al.'s experiment has to be undertaken in terms of some suggested value-orientation. A value such as the intent to evoke trust in discursive interaction, is offered in the trusting constructivist position.

4.3.8.5 Critical Theory and Trusting Constructivism

What would concern *critical theory* in the case of Dovidio et al.'s experiment is that the specific practical advice that ensues from their study, namely, the recommendation to establish superordinate categories as a way of improving intergroup life, is seen by them as rooted in "the facts". But they do not invite public discussion around the way in which these facts became created in the scientific process; instead, they focus on their supposed reference to empirical reality.

Dovidio et al. have organized their research around the presupposition that it offers a way of addressing what are taken to be problems in society (for example, the problem of bias in intergroup relations). A *trusting constructivist* position accords with the critical theoretical concern that we should be careful of putting our faith in the scientific community to organize inquiries in order to "find out" about social reality. This assignment of faith may be misguided in the sense of threatening possibilities for public discourse regarding ways of identifying social issues to be addressed. A trusting constructivist position emphasizes that the understanding of what is "problematic" in society can be directed toward opening up vistas for people to consider anew the choices that they are making as they locate issues of concern. It is suggested that choices can be enriched as people engage discursively with different ways of seeing social realities.

4.3.8.6 Anti-foundationalist Feminism and Trusting Constructivism

Anti-foundationalist feminists would see Dovidio et al.'s defense of their experimental procedure as excluding certain kinds of considerations that should be incorporated within our discourse on the quality of social inquiry. As part of Dovidio et al.'s establishing of a politically sensitive relationship with participants and other audiences, it would be suggested that they should have made some provision for people's participation in defining the meaning of the study — and not solely in terms of the rules of the game of knowing that they have implicitly endorsed.

A *trusting constructivist* position appreciates the proposal of anti-foundationalist feminism that the organization of social inquiry can be seen as an opportunity for revisiting ways of knowing and living. What is added (or emphasized, rather) by a trusting constructivism is that when decisions are created by those involved in a study, they can account for these in a way that expresses their choicefulness (in the light of a consideration of ongoing controversies), while leaving scope for others' responsible choice-making too.

4.4 CONCLUSION

Dovidio et al.'s study was used in this chapter as a focal point to explore ways in which experimentation might be defended as a strategy of social inquiry. I explored a variety of assessments of the study, starting with a positivist one, using the tenets of positivism and debates around these to structure the discussion.

From a positivist perspective, social experimentation — if properly organized — can be used to set up conditions that allow researchers to make conclusions on the basis of evidence that they gain about experimental subjects' responses (by comparing data from experimental and control situations). Dovidio et al. can be said to have organized an experiment appropriately in line with this experimental/control logic.

From a non-foundationalist point of view, the idea of using experimentation to create advancements in knowledge about human behavior is accepted as an option for inquiry. This is provided that it is coupled with the understanding — which Dovidio et al. could have better highlighted — that results created are always (as in all scientific work) somewhat tentative. An important factor that can contribute to the tentativeness of experimental work in the social sciences, is the reactivity effect — that is, the effect that is created by the fact that subjects are responding to the research situation. Experimenters (as all social researchers) need to take this into account when considering how to set up experiments and interpret results therefrom. Also, from a non-foundationalist point of view, it is understood that people may subsequently take on board in their thinking the results that are presented as springing from experimental (and other) studies. It is therefore incumbent on researchers to indicate how "well-tested" the results are, so that others in society are aware of the way in which hypotheses have undergone tests via the scientific process. People can then recognize the ways in which scientific results differ from "ordinary" opinions about the topic under investigation. And they can hence act in a more informed way in the light hereof.

From a scientific realist perspective, a similar assessment of Dovidio et al.'s study would be made, but in addition it would be noted that it is important to set up experimentation with a view to making inferences about the structures generating the observed outcomes that are triggered in the research process. Dovidio et al.'s study falls short on this score and is therefore ill-equipped to provide knowledge of the generative mechanisms that could explain the responses on the part of those participating in the experiment.

From an interpretivist point of view, a specific concern that would be raised in regard to this experiment (as others conducted in similar fashion) is that it does not afford the opportunity to get sufficiently close to the natural settings of human conduct so as to offer depth of understanding of human life. Its superficiality cannot be corrected by doing more research along similar lines. Dovidio et al.'s failure to indicate how in this or in other research designs researchers might attempt other ways of addressing human subjects, is indeed cause for concern.

Critical theory focuses on the effects that might be created in society when experimentation is regarded (by researchers and others in society) as a forum for gaining knowledge about causal regularities existing "in reality". From this point of view, the fact that people might see their social world as a realm of causality similar to that operative in natural reality, is something that should be re-appraised. But unless researchers are careful with the way in which they organize experiments and report on their results, they may well contribute to perpetuating the operation of the quasi-causal patterns that they "find" in society. Dovidio et al. have not been sufficiently alert to these considerations.

From an anti-foundationalist feminist perspective, prime attention would be given to the process by which those conducting experiments interact with participants being investigated. Insofar as experimentation as a research strategy is chosen for some study, efforts should be made to render the relationship a less controlling one than is currently accepted practice in experimental (and much other) research. Instead of accepting that subjects are duty-bound to contribute to the creation of public knowledge (through research designs devised by researchers), the interaction that is being set up between "researchers" and "subjects" could become a point of discussion as part of the experiment. Only then may it be said that the research can be humanly accounted for. However, Dovidio et al. gloss over this issue (on the grounds that they are justified in doing the research because of its potential contribution to public knowledge).

From a trusting constructivist position, Dovidio et al. would be taken to task for not being sufficiently oriented to developing their decisions about how to interact with participants and others concerned in view of some of the issues that have been raised in debates on the proper conduct of social inquiry. Had they shown more sensitivity to these debates, then this in itself would leave more space for both themselves and other social inquirers, to consider what alternative strategies and techniques can offer in terms of ways of approaching the topic that they have isolated. By showing a propensity to engage with these as potentially viable options (as they organize the research process and as they write it up) they could help both themselves and others to enrich their understanding of possibilities that can be tried for setting up social inquiry processes.

As far as their account of their findings is concerned, a trusting constructivist position would advise that they present their work in a way that allows audiences to reconsider how they have collected (or rather, constructed) and utilized the data. They could, for instance, point to areas of contention around the manner in which they created the single-group condition and measured the effects of the manipulation. They could also show what issues they consider to have been excluded through their study, and how this might have affected the outcome thereof. In this way, they could earn the trust of readers differently from trying to defend the "parsimonious explanations" that they present based on their observation of the evidence (to date). Instead of trying to account for their task as an effort to arrive at such explanations, they can account for it as an effort to enter into debate about what status the product of the research should be accorded. It is in this way that they can act responsibly to temper the possible self-fulfilling effect that scientifically reported results otherwise may have in society.

5
Exploring Survey Research

5.1 INTRODUCTION

In this chapter, I organize a discussion with reference to an example of what I characterize as survey research. I myself was involved in this project — as the named principal researcher. I outline the example and then show how it can be conceptualized from various perspectives. I treat the study as an example of survey research in line with De Vaus's suggestion that surveys amount to "collecting information about the same variables or characteristics from at least two (normally far more) cases and ending up with a data matrix" (1996, p. 3). De Vaus compares survey research with doing experiments by noting that in each type of method (or strategy, in terms of the definitions in this book), data are collected in the variable by case matrix form. But in experiments, researchers set out to manipulate a situation in order to create a variation in the independent variable (as we saw in the example given in Chapter 4). A survey approach differs from experimentation in that researchers do not create variation by organizing an experimental manipulation, but seek "naturally occurring variation" in the data that is examined (De Vaus, 1996, p. 6).

De Vaus notes that in survey research, as in experiments, a range of data-collection techniques may be employed. Questionnaires (which consist largely of closed questions, but which sometimes also include some open-ended ones) are normally associated with the doing of surveys. Through questionnaires, questions are asked in the same order to all respondents (that is, the questions are scheduled); and all questions are worded in the same way for all respondents (that is, they are standardized). A closed response format means that respondents are given set options from which to choose their answers to the questions, while an open-ended response format means it is left open for them to construct their own answers. Besides the use of questionnaires, other techniques can also be used to gather the data needed on the variables being examined. These techniques can include: structured interviewing (with pre-set questions), in-depth interviewing (semi-structured or unstructured), (direct) observation, and content analysis (De Vaus, 1996, p. 6). As De Vaus explains,

> the technique by which we generate the data need not be highly structured so long as we obtain each case's attribute on each variable [in which we are interested]. Because questionnaires are the easiest way of ensuring the structured data matrix they are the most common technique used in survey research. But there is no *necessary* connection. (1996, p. 5)

Following De Vaus, Seale and Filmer also comment that social surveys do not "always involve rigidly fixed questioning devices" (1998, p. 128). They note that "qualitative data" as gathered through, for example, interviewing processes, can later be quantified by putting them into a form that allows us to measure differences — if it is decided that this fits in with the aim of the project being undertaken (1998, p. 128). In De Vaus's view, the data would need at some stage to be expressed in terms of variables, so that each case's attributes on the different variables can be compared with others. (This is in order for the research to meet the requirements for it to be classified as a survey.)

Issues on which I concentrate in this chapter revolve around considerations of the way in which different techniques for collecting/evoking data through surveys may be conceptualized, and the way in which any statistically established likelihood of association (or correlation) between variables studied might be treated. The question of the status to be given to these associations is addressed. The practical relevance of the information generated is also considered.

To conduct the comparison between different arguments, I present first a positivist understanding of the example (in terms of positivist tenets). This is followed by other understandings. Each heading used in Section 5.3 expresses the manner in which the positivist argument is being treated at each point — whether accepted, revised or reworked — as a way of creating criteria for judging research examples. In the light of these expressions of acceptable research practice, the example given here is examined in depth.

The example is of a research project aimed at evaluating an Adult Basic Education and Training (ABET) program run by the ABET Institute at the University of South Africa (UNISA). The evaluation (done from 1996-1998) was funded by the United Kingdom Department for International Development (DFID). As principal researcher on the project I was supported by Robert Flood and Peter Adman at various stages.[30] The final report (presented both to DFID and to ABET staff, who in turn distributed to those interested) was written largely by me, with support from co-workers. I refer to it as Romm et al. (1998). (As in the final report, I take responsibility here for my way of making sense of the project.)

---

[30]Robert Flood took part in three (out of eight) of the arranged research visits to South Africa for the evaluation, helping to envisage data sources, helping to lend (additional) structure to the questionnaires, and helping to develop an outline of issues that could be taken to one of the tutor workshops for discussion. He also held discussions with the Vice-Principal of UNISA on several occasions, in order to gain a sense of how management was seeing the ABET Institute. Peter Adman visited UNISA in the last stage of the project. His contribution was as follows. Firstly, he aided the analysis of the (qualitative) data collected through the interview guide. He helped to consider how these data could be organized, and to consider especially the relevance (for tutors and students) of the question asked on how students experienced their family. Secondly, he held a number of discussions mainly with the administrative staff in order to gain a sense of the administration of the system and to help consider ideas that may be relevant thereto.

5.2 THE EVALUATION OF ABET AT UNISA (ROMM ET AL., 1998)

ABET at UNISA was set up toward the end of 1994, in line with the Minister of Education's comment on the need to

> launch, as rapidly as possible, a national initiative in Adult Basic Education and Training aimed at transforming both the learning opportunities and earning potential of the millions who have been educationally disenfranchised in the past (Education Minister, Bhengu, 1994, in his first speech to the National Assembly).

The director (and founder) of ABET at UNISA (Veronica McKay), saw the potential for UNISA, being a correspondence university, to contribute to the process referred to by Bhengu — by catering for large numbers of students, at a relatively cheap cost. Student intakes for a certificate course began in 1995 (and intakes for a diploma course began in 1998). The evaluation conducted by Romm et al. was aimed at assessing the delivery, relevance for students, and impact on students' lives of the work done by the ABET Institute in relation to the certificate course. The administrative operations undertaken by the Institute and by the ABET Administrative Registry at UNISA were also examined.

The research took place over the period from August 1996 to December 1998. At the same time as we were organizing our evaluation, Elijah Sekgobela from the ABET Institute and a researcher based at another university (Humphrey Glass) arranged for a way of collecting data that would allow the students themselves to set the questions that they thought would be relevant to ask in order to undertake an evaluation of the program. Also, other internal evaluations, based on responses given by students to questionnaires sent out from the ABET Institute (regarding their experience of aspects of the program) were undertaken (under the auspices of Mike Sarakinsky).

DFID was (and still is) funding some of the activities of the ABET Institute. For example, it contributes to the funding of the tutorial system — which involves more than a hundred tutors across the country, who offer monthly classes for students. The person responsible for project funding at DFID at the time requested an evaluation to be done (by myself and co-workers) that focused on the way students were experiencing their course, including the tutorial system. He was also interested in the way tutors and regional co-ordinators (who co-ordinate tutors in a region) likewise were seeing their involvement in ABET.

Figure 5 (overleaf) offers an indication of the manner in which the research project was organized in view of this "remit".[31] It should be noted, however, that the organization of the various processes as expressed in Figure 5 was not all pre-planned, and indeed much of the planning emerged in the course of the project. As processes were set in motion, so new ideas developed as to how we might proceed further in the light of (what we saw as) currently emerging options.

---

[31]As with other figures, this figure was drawn by Peter Adman. In this case, Peter had some acquaintance with the research through his involvement in it.

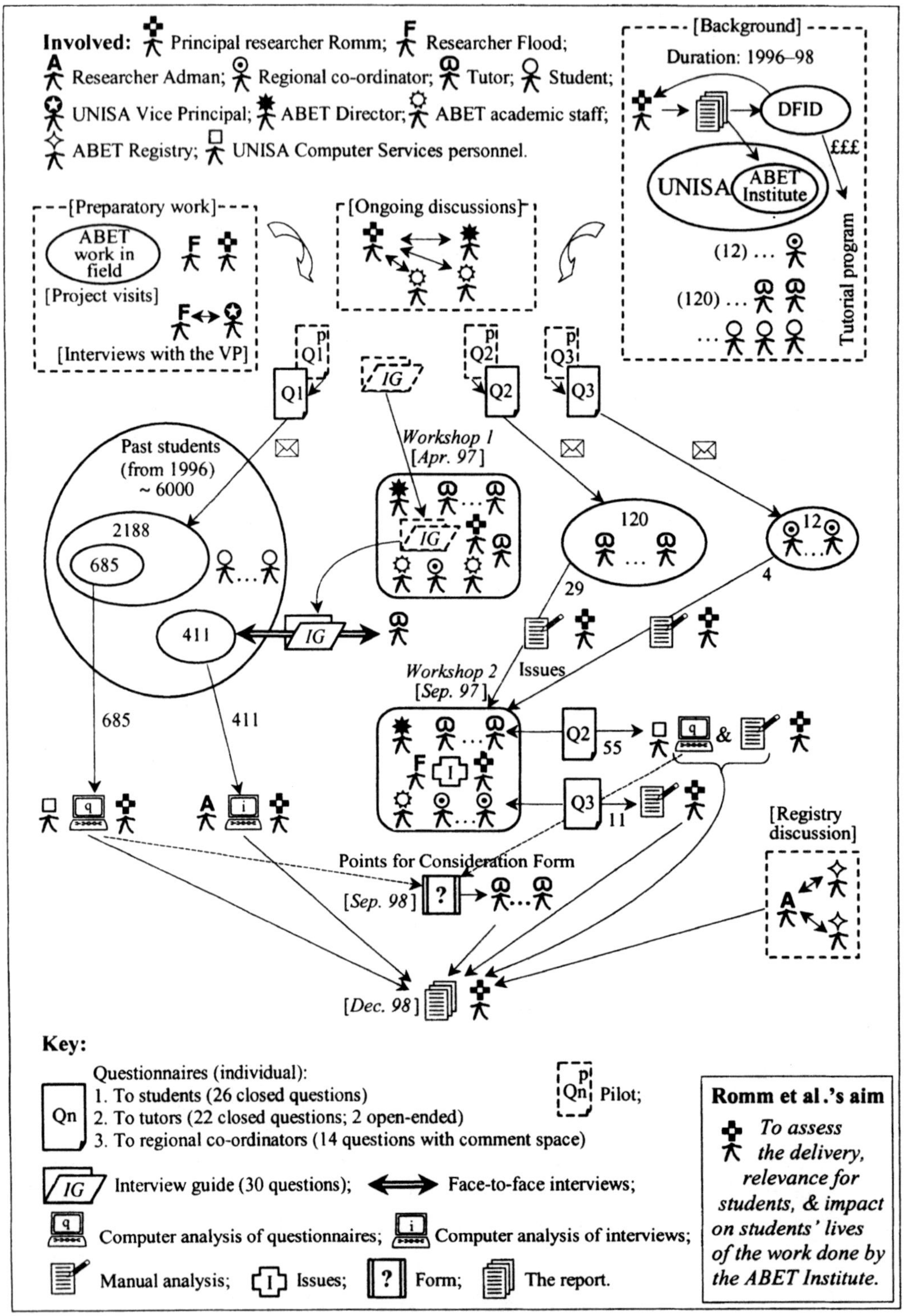

Figure 5: Romm et al.'s evaluation.

5.2.1 The Organization of the Research Process

Much attention was given in the research process to the development and administration of different questionnaires that were sent out, and to the formulation of an interview guide that was created. One questionnaire was developed to be sent out to a sample of past students (Q1 in Figure 5). One questionnaire was developed to be sent out to all tutors (Q2 in Figure 5), and one to be sent out to regional co-ordinators (Q3 in Figure 5). All questionnaires were first piloted with the various types of respondent and then mailed out in April 1997. (In the piloting stage, students, tutors, and regional co-ordinators offered some suggestions not only for clarification of certain questions, but also for modification of the questionnaires in terms of their expressions of areas of interest that they felt should be included.) Tutors were invited to participate in the development of the interview guide to be used in interviews with past students. A draft guide was first created (and piloted with some students). Time was then set aside in a tutor workshop held in April 1997 to discuss it (Workshop 1 in Figure 5).[32] I facilitated a discussion around the interview guide, and ensuing from this, the final guide was created (IG in Figure 5). Tutors were involved further by being employed — on a voluntary basis in terms of their expression of interest — to administer the guide to past students (through face-to-face interviewing).

As part of the research process, we also tried to organize some group discussion with tutors and regional co-ordinators along the way, by bringing some themes to one of their workshops. These were themes arising out of our analysis of their responses to the first set of questionnaires sent out to them in April 1997. The issues were discussed in time set aside in a workshop held in September 1997 (Workshop 2 in Figure 5). At the same time, we handed out a repeat of the questionnaires that had been sent to them in April. This was done in order to increase the response rate on these questionnaires.

And much later in the project (September 1998) — just before the write-up of the report — a "Points for Consideration Form" was issued to tutors to allow them to comment on what appeared to be some discrepancies (on certain questions) between tutor and student responses. The individualized comments of (some) tutors were returned through the post to us. The detail on all of these processes is discussed in Section 5.2.1.1. But aside from these ways of focusing our inquiry, we involved ourselves in the following:

- To begin with, some project visits to sites where past students were administering projects, were undertaken (by Bob Flood and myself). The students were interviewed about the relevance of the course to them for their work in the project. From these interviews, and observations of activities being undertaken by those on site, we tried to develop a sense of the "feel" of ABET UNISA in the community. (These are classed as "Project visits" in Figure 5.)

[32]This was one of the twice-yearly workshops that are attended by ABET academic staff and by tutors and regional co-ordinators, where any matters that tutors wish to raise for attention are discussed and where ABET staff arrange events such as lectures and participative discussions around them.

- As a way of exploring the managerial context outside of the ABET Institute, it was decided that Bob Flood should undertake a number of interviews with the Vice-Principal of UNISA. This was in order for us to gain some appreciation of the latter's views on the operation of ABET at UNISA. (These views were later discussed with the ABET staff, and the final report includes an indication of some differences of opinion in regard to the "placing" of ABET at UNISA — that is, differences of opinion concerning the degree to which it can and should be independent from larger institutional arrangements.)
- Informal discussions with the academic wing of the ABET Institute were ongoing throughout the evaluation. These discussions were primarily with Veronica McKay (the founder of the Institute), but they also took place with Elijah Sekgobela and Mike Sarakinsky. This was the full contingent of academic staff in the Institute, as the contact with the students was run largely through the tutorial system, that is, via the tutors and regional co-ordinators. These discussions, which I facilitated in all cases, were largely on a one-to-one basis. Discussions in the beginning helped to gear us toward considering what questions to include in the questionnaires with the various types of respondent (before they were piloted). Discussions also involved considerations of how we could organize the design of the research process so as to involve tutors appropriately therein. And, as findings emerged, these were shared and discussed with staff also from time to time.
- Discussions with (and demonstrations by) the head of the ABET Administrative Registry were arranged — followed up by discussions with, and demonstrations by, other staff. These discussions occurred toward the end of the project and were facilitated by Peter Adman. In the process of their discussion, Peter and those involved in the Registry, discussed options for addressing issues that were being experienced as problematic in some way and as creating dilemmas — in regard to the timing of registration of students and its process. (These issues had also appeared in the workshop held in September 1997.)

5.2.1.1 Gaining Data from Past Students

As indicated above, we concentrated in the evaluation on finding a way of examining past student responses from 1996. We chose to do this via a postal questionnaire (in order to reach the large numbers that we hoped to reach) and also via an interview guide.

The postal questionnaire (Q1) was sent out in April 1997 to 2,188 past students from different geographical regions — 685 of which were returned. (In that academic year, the number of students who had passed the certificate course was nearly 6000.) The top of the questionnaire indicated that as researchers we were conducting some research on students' experiences of the course that they had studied. 26 questions were then asked. Information was firstly requested from students (using a closed response format) in regard to their gender, place of residence, workplace, type of employment, level of education, and age. Other questions targeted the students' experiences of classes and of their tutors, the desired frequency of the classes, ease of access to the venues, ease of contacting their tutor, the relevance of the study material (and its level of detail), and the UNISA administration. These questions were all in the format of asking a question and then offering three response options, which allowed for some "rating" to take place on the student's part. For instance, a question asked: "How valuable did you find the classes?" Options offered were: i) without value, ii) some value, and iii) very valuable. In regard to the utility of the tutorial classes in helping students to learn, a question asked: "Did the class discussions help you to learn new ways of learning?" Options offered were: i) in no new ways, ii) in some new ways, and iii) in many new ways. As another example, a question (regarding tutors) asked: "Did the tutor show willingness to listen to your concerns and ideas?" Options offered were: i) unwilling, ii) some willingness, and iii) very willing. And as an example of their experience of the study material, it was asked: "How easy was it to read the study guides?" Options offered were: i) very difficult, ii) quite easy, and iii) very easy.

The responses given through the questionnaire were summarized via a computer-based analysis organized by the Computer Services Department at UNISA (and summarized again by me for the report, as I compared some of the responses from students and from tutors in regard to certain issues relating to similar questions that had been asked to students, tutors, and regional co-ordinators). In regard to the responses from past students, the Computer Services Department was asked to provide general descriptive statistics as well as to provide information on a number of potentially interesting cross-tabulations (aimed at seeking correlations between some of the variables).[33] For instance, it was asked to check whether there was any (statistically significant) association between the type of residence (small town, big town, and rural area) and ease of getting to the tutorial venue, as well as whether there was an association between type of residence and ease of contacting the tutor (outside of the tutorial classes). It was also asked to check for a possible association between type of employment (such as, ABET practitioner, nurse, development worker, self-employed, unemployed, etc.) and ease of reading the study material, as well as between level of education of the student and ease of reading it.

At the same time as we were formulating the questionnaires to be sent out to past students, we also formulated the draft interview guide. In the workshop in April 1997, tutors helped us to consider possibilities for modifying/reformulating some questions and for generating others. At the end of the workshop, tutors were asked to write to UNISA indicating whether they wished to be involved in administering the guide. Due to the large expression of interest in this, we had to allocate a small number of interviews to each tutor (in different geographical regions). Tutors were asked to choose their allocated number of students from within that region. Altogether, 411 past students were interviewed in this way. Tutors were advised to indicate to students that they were interested in their experience of the course, and then to proceed with asking the (pre-set) questions. Although the questions were as given in the guide, the tutors recognized that they needed to ask them in a way that they felt the students would understand. The tutors in turn interpreted and wrote down (in summary form) the students' answers. There were 30 questions in the interview guide. Questions focused on: students'

---

[33]This was done using chi-square converted into an expression of correlation. (See also De Vaus, 1996, p. 166; and Sapsford, 1999, p. 165.)

impression of the course, whether it suited their purposes (and what their purposes were), their experience of setting up and managing projects, and whether they felt that there were issues that still needed to be addressed — in the certificate or advanced course. The responses given in this guide were analyzed using largely qualitative analysis — with the aid of the Nud*ist computer package (Peter Adman and I organized this analysis). (However, one question that asked past students what their purposes were, was also analyzed through a quantitative package, and cross-tabulations with students' gender were made through this package.)

5.2.1.2 Gaining Data from Tutors and Regional Co-ordinators

While sending out the postal questionnaire to a sample of past students and arranging for the process of administering the interview guide to begin, we also sent out (in April 1997) questionnaires to all tutors and regional co-ordinators. (There was a contingent of 120 tutors and a contingent of 12 co-ordinators.) The postal questionnaire to tutors focused on tutors' experience of the classes in terms of issues such as their frequency, students' participation, the process of tutoring itself, how they perceived that students experienced the study material, and their own relationship with UNISA. There were 24 questions in this questionnaire. Most of them were in similar format to those in the questionnaire sent to students — that is, questions were set with three options from which to choose answers. There were also two open-ended questions at the end, which invited comments on their conditions of employment and on any other issues of interest to them.

Because we received only 29 responses to this questionnaire, we decided to hand out a repeat of it in one of the later workshops (September 1997). (This was the same workshop where we discussed with tutors and regional co-ordinators some of the tutors', as well as regional co-ordinators', responses to the first questionnaire — Workshop 2 in Figure 5.) The tutor responses to the two questionnaires were analyzed by the Computer Services Department at UNISA (although beforehand, for the purposes of the workshop, we were able to do a manual analysis on the first set of responses).

Of the questionnaires that were sent out to the regional co-ordinators, we received only four responses to the first send-out, and therefore, as with the tutors, organized a repeat by handing this to them at the September workshop. To the repeat we received 11 responses. The questionnaire for regional co-ordinators asked similar questions as those asked to tutors — this time focusing on the regional co-ordinators' perceptions of their own relationship to students, tutors, and UNISA. A similar format to other postal questionnaires was used. But spaces were also provided in this questionnaire for respondents to explain all their answers. There were 14 questions altogether here. Because of the small numbers of regional co-ordinators (and hence of responses), the responses could all be analyzed manually.

In the workshop held in September 1997, we arranged for discussion around some of the issues that we isolated as possibly relevant to consider. For instance, we introduced the issue concerning tutors' differing views on students' ease of reading the study material. 17 of them had reported (in the questionnaire) that students had some difficulty in reading them, while 12 had suggested that the students found them very easy to read. Through the discussion, the view was expressed that the guides could be regarded as pitched at the right level in the sense that they seem to allow the students to be "stretched" — although this may not apply to the better-qualified students.

Another issue that was discussed at the workshop was the felt utility of the system of regional co-ordination. Tutors had differing views on how the regional co-ordinators were helping them and their students. These differences were explored (in conjunction with the views of some of the co-ordinators, and with the views of the person at ABET responsible for the arrangement of the tutorial system — Elijah Sekgobela). Ideas on the evolving role of the co-ordinators were expressed. It was suggested that it was important that they are seen as acting as a support to tutors, rather than being interpreted as having a policing function. A discussion ensued about ways through which the relationship between tutors and regional co-ordinators could be developed in keeping with this understanding.

A further point of discussion in the workshop related to the question of the starting date of the tutorials after the process of registration of students had been completed. Problematic issues revolved around the late registration of many students, which affected the starting date of tutorial classes; however, the ABET Registry were trying not to discourage students from joining the course, and needed also to take into account their financial situation (which sometimes led to late registration). Some options for addressing issues connected with registration as well as other administrative issues were later discussed between those involved in the ABET Registry and Peter Adman (September 1998).

Finally, before regarding the evaluation as completed for purposes of the final report, a "Points for Consideration Form" (see Figure 5) was handed to tutors in September 1998 (at a workshop venue). About 20 tutors returned this form (through the post). Tutors were asked via the form to consider possible explanations for some apparent discrepancies between students' and their own responses to certain issues. For instance, one ongoing issue related to the frequency of tutorial classes held. From the returns to our questionnaire sent to past students, 23.6% found them not enough, 67.4% found them enough, while 9% found them too many. Tutors' responses in this regard (judging from the 55 returned on the second send-out of the questionnaire) suggested that eight tutors found them not enough, 30 found them about right, and 17 found them too much. The "Points for Consideration Form" given to the tutors asked them to interpret these answers. In their responses, tutors indicated that they felt that the desired frequency of classes depended on different areas, and on different situations. Some explained that students sometimes found it difficult to get time off (which would perhaps provide some explanation of the 9% of students who did find them too many — see above). Others noted that students in fact requested more classes from them (this would explain the 23.6% of the students who found them "not enough"). Taking all this into consideration, it was suggested by us in the final report that a way of arranging tutorials between tutors and students to suit the particular classes and cases (as is often done) is what seems to be needed.

Another issue that we raised in the "Points for Consideration Form" concerned ways of interpreting students' as well as tutors' views relating to the students' ease of reading the study material — in the light of all the responses that we had received and analyzed at that point. Some tutors suggested that the ease of reading the material as expressed by many of the students, could be interpreted as implying that the students had already received some guidance on this from the tutorial classes. Others indicated that it may not be so relevant if tutors' and students' views (considered as groups) differed in regard to the ease of reading the study material — as long as the tutors were committed to aiding students to come to terms with it where this was needed.

5.2.2 Some Findings Mentioned in the Report

After elucidating the processes undertaken in the project (as explained in Section 5.2.1), the final report presented what were called findings attained through the process. But before the discussion of findings, a section was devoted in the report to explaining that the data we obtained from all data sources — whether questionnaires, interviews, informal discussions, workshop discussions, etc. — should be treated as being a product of the way those involved were possibly responding to the contexts of interaction that were being set up. It was explained that, from our point of view, no research should be considered as unsituated. We therefore indicated that we preferred to make visible the contextual character of the data that was evoked. That said, the following were some of the findings mentioned in the report.[34]

As far as the tutorial support system is concerned, the results of the evaluation were interpreted in the report as suggesting that it is by and large fulfilling the purpose of guiding students in their learning throughout the course. The vast majority of the students responded (in the questionnaire sent to them) in very positive terms regarding the value of the classes and their aid to their learning, and also regarding the willingness of tutors to help. To the question aimed at assessing if the students found the classes of value, only 2.2% answered in the negative, saying that they were "without value". 22.4% found them of "some value". 75.4% suggested that they were "very valuable". Furthermore, only 6.1% of these students (who returned the postal questionnaire) said that the tutor aided their learning "very little". 32.5% replied that they aided them "to some degree", and 61.4% said "very much". Interestingly, only 1.1% suggested that the tutors were "unwilling", while 84.4% said they were "very willing". (14.5% said that the tutors had "some willingness".)

The postal questionnaire responses indicated that 20% of the students were finding it very difficult to reach the venues for the tutorial classes, and 25% were finding it very difficult to contact their tutors. However, in cross-tabulating the responses regarding ease of access to venue with the type of residence of students (small town, big town, rural area), as well as in cross-tabulating ease of contacting the tutor with type of residence, it was shown that there did not seem to be an association between these factors. (Later, based on a number of considerations, the ABET Institute set up a special

[34]Later (in late 1999) one of the people from DFID in the United Kingdom who read the report (a person other than the one who had initially sponsored it) indicated to the director of ABET that she regarded the study as methodologically complicated but academically sound. The person who had initially sponsored the evaluation expressed a similar opinion.

"hotline" as a way of fielding additional calls from any students needing to discuss issues relating to the course.)

It was also noted when cross-tabulations were organized between type of employment and ease of reading the study guides, that there was no statistically significant association. In other words, it seems that whether people were practicing as ABET practitioners, nurses, development workers, etc., or were self-employed or unemployed, did not have an influence on their ease of reading the study material. There was, however, not surprisingly, some association between level of education and ease. Those already possessing a degree were far more likely to find it very easy to read the guides. Nonetheless, very few students did indeed possess a degree. Most possessed matriculation, that is, a school leaving certificate, while less than a fifth possessed "standard eight" (having left school early), and less than a fifth had some kind of diploma. (This information was combined with interpretations expressed by tutors in the Points for Consideration Form for purposes of the report.)

Findings in regard to the perceived impact of the ABET course on students' lives — as developed from an analysis of responses to the interview guide — suggested that it had made a substantial difference to the lives of nearly all students and to some of those around them, in the workplace/community and in their families. Many of the students indicated that it was through the course that they initiated projects in their community and/or that they learned how to manage better existing projects in which they were involved. They also reported improvements in their ways of approaching people (relating with them, communicating with them, regarding them with respect, etc.) in the community/workplace/family. (Contrary to what we initially thought, cross-tabulations between different purposes for studying the course and gender of the student, showed that there were no significant differences in relation to any of the purposes.) Students indicated generally that they believed they had become more enabled to address issues that they faced in their lives and had become more motivated to act to deal with life's challenges than prior to the course. This was often related to their reported improvement in self-worth. Most students also reported differences to their family relationships (and especially to their relationship with their children). The course had also apparently helped many of them to aid their children to learn new study skills, as well as to encourage a positive attitude in their children toward "meaningful" (rather than rote) learning. (It was suggested in the final report that some of the ways in which past students had reflected on their ways of relating to their partners and to their children could be used by tutors in future to enable other students to consider these as possibilities.)

When asked in the interview situation about aspects of the course that could be covered in more detail, many students ventured to offer ideas in this regard.[35] Answers given with some frequency included: How to motivate learners because of dropouts;

---

[35]Interestingly, a similar question that was asked to students in the questionnaire in regard to level of detail of the material led to only 4.6% of the respondents saying that there was too little detail, with 43.2% saying the detail was about right, and 52.2% indicating that it was too much. However, it seems that the way in which the question was posed in the interview (when students were asked for aspects that could have been covered in more detail) encouraged some students to volunteer ideas in this regard. (About half of them offered certain suggestions.)

more consideration for rural women (issues of empowering women featured in this context); the kinds of problems the youth might have; and more detail on managing projects. Proposals for the advanced course involved suggestions about training for more skills (including skills of business financial management or what was called "commercial subjects", and computer skills). Students also suggested using the advanced course for "more practical work". And suggestions were made regarding marketability of what they were being taught.

Some students did express disappointments after completing the course. These were related to their expectations (and/or those of their family) that after doing the course they would find paid employment. When this did not materialize for them, they were not sure whether all their efforts had been worthwhile. Some of these students felt that their failure to find employment was due to the course not being sufficiently accredited by government bodies, and felt that steps needed to be taken to grant the course more credibility. Steps were taken (later) by the Institute in this regard, by gaining accreditation by bodies in South Africa that were at that point being set up for the task. The Institute was also (at the time of writing) moving in the direction of trying to raise awareness in various other government departments (not only the Department of Education) of the need for ABET practitioners within a multitude of sectors.

No specific credit was claimed in the evaluation (during the process of conducting it or in the write-up of the report) for having brought up particular issues for notice — as if they would not have been noticed otherwise. It was indeed remarked explicitly in the report that it is quite possible (indeed, probable) that they would have been — and solutions discussed and developed. (This was a question that had to be handled sensitively, as there were many concurrent ways in which issues could be brought to the attention of ABET staff.)

5.3 EXPLORING ALTERNATIVE ASSESSMENTS OF THE EXAMPLE

5.3.1 A Positivist Assessment

5.3.1.1 Scientism: Doing Science

The positivist tenet of scientism suggests that scientists should try to organize research designs geared to locating causality in the world. The location of causal connections in the social world can possibly be organized via (well-designed) survey research. It can, for instance, be attempted via the use of longitudinal surveys, that is, surveys in which respondents' responses are examined at different points in time and changes in the data noted. In longitudinal surveys either the same or similar respondents are used in the different time periods, so that comparisons can be made between earlier and later responses. Because of the time element involved in such surveys, they may be employed with a view to making inferences about ways in which independent variables are affecting dependent ones. However, cross-sectional surveys — where a repeat of a data collection instrument might be sent out (following closely on the first one) only for purposes of increasing the response rate of participants — do not involve this time element. Nevertheless, statistical reasoning can be used to determine whether variables are associated with one another in the sense that they vary concomitantly, and how strong the association (or correlation) is. For this reason, as Bryman (1992, p. 31) notes, cross-sectional surveys are still regarded within positivism as a possible way of coming to grips with causal relationships between variables studied. If correlations are found between the variables examined in the study, it is deemed possible sometimes to infer the operation of causality. This is as long as researchers are theoretically sensitive to seeing whether causality can be argued to exist in view of the correlations that are appearing. As Bryman remarks:

> The old maxim — correlation cannot imply cause — ostensibly implies that the social scientist's ability to establish causality from social survey research is severely limited. However, survey researchers have by no means been deterred and have developed a variety of procedures for the elucidation of causality by means of ... reconstruction of the "logic of causal order" ... . (1992, p. 31)

In the case of Romm et al.'s survey, it would be noted that some attempts were made to check for correlations between certain variables. For example, it was considered of interest to check to see if there was a correlation between, say, gender and types of purposes for which students registered for the certificate course. It was also checked to see whether there was a correlation between, say, level of education and ease of reading the study material prepared by the Institute. A relationship between gender and type of purpose, or between level of education and ease of reading study material, could (if found) be interpreted as implying that in each case the first variable was functioning as an independent one, influencing the second one. So in this sense the study could conceivably aid the investigation into causal influence operative in social reality.

From a positivist point of view, nonetheless, not sufficient emphasis was placed by Romm et al. on locating areas of possible connection between the variables examined. Scientific activity is an activity that must be directed at the cumulative development of theory. And thus far, there is scant indication of how the project helped toward this end. Nor do Romm et al. provide sufficient pointers to how additional research might pick up on the study and take it further (either through additional survey work or through other research designs).

5.3.1.2 Phenomenalism: Organizing Measurement Validity

The positivist tenet of phenomenalism suggests that the process of observation should achieve the purpose of getting to grips with the data provided by reality. In the case of survey research, much attention is given to the way in which questions are asked to respondents so that it can be argued that their responses tell us something about the realities that the researchers are setting out to reflect. The positivist view is that, as Bryman puts it, researchers have an "avowed obligation to specify the meaning of particular concepts precisely and to develop sound measuring procedures which will stand for them" (1992, p. 23).

For instance, considering Romm et al.'s study, if one wants to find out whether tutorial classes are perceived by students as helping them to learn, one needs to set a question (or questions) in such a way that students' responses will give a good indication of this. One of the questions on learning was set in the questionnaire by asking students whether the class discussions helped them to "learn new ways of learning". Response options were: i) in no new ways, ii) in some new ways, and iii) in many new ways. From a positivist point of view, though, it might be remarked that Romm et al. did not make much effort to show (to those assessing the study) why they regarded this question as being a valid measure of the experience of learning. They rely on its apparent face validity. That is, the measure seems, on the face of it, as if it should provide an indication of the experience. Romm et al. may be able to defend its validity in this way by referring to the fact that practitioners in the field (such as those working for the ABET Institute) seem to regard it as valid. Seale and Filmer (1998, p. 34) point out that this is one way of proposing face validity of definitions. But in terms of positivist discourse, Romm et al. could have made some attempt to check the results created via the use of this definition, with established indicators of it as have been utilized in other studies around the topic; and/or they could have tried to determine to what extent their use of this definition conforms with expectations derived from prior theorizing in the scientific community. Because they did not organize any of these ways of checking the validity of their way of measuring the experience of learning, it is difficult for us to know the likelihood of its (real) validity. The same could be said of all of the variables studied by Romm et al.

In terms of the positivist tenet of phenomenalism, Romm et al. could have better accounted for this possible criticism of the way in which they set up their questionnaires (and other instruments of observation).

5.3.1.3 Empiricism: Seeking Causal Connections through Data Analysis

Romm et al.'s study developed descriptive statistics in regard to variables that they isolated — with a small amount of analytic work done to check for certain correlations. It also offered some "qualitative" impression of the way in which the course was experienced by the various participants (students and others). Impressions were gained mainly by analyzing the answers given in the administration of the interview guide; by looking at (and summarizing) answers to the open-ended questions that were set in the questionnaires sent to tutors and regional co-ordinators; and by noting issues raised in the (second) workshop and in ongoing discussions held with staff of the ABET Institute.

As an attempt to collect data on a range of variables from a large number of cases, Romm et al.'s study was probably illuminative in reflecting (some of) the realities. (As noted in the previous section, exactly what realities were being tapped might be questioned from a positivist point of view.) The study was likely to have high external validity in terms of its ability to generalize findings (in this case, largely descriptive ones) from the respondents studied, to the various populations of interest — that is, the past ABET students at UNISA, and the tutors, regional co-ordinators, and other staff involved in the program. Romm et al.'s sampling can be appreciated in terms of its likelihood of being representative of the populations of interest. (See also Gill & Johnson, 1991, p. 78, for a discussion of so-called population validity.) As far as the population of past students was concerned, the questionnaires were sent to a sample consisting of more than a third of the past students (from the year of study in which the researchers were interested), and the response rate was also relatively high (in relation to what is often considered as acceptable to work with statistically — see Weaver, 1997, p. 41). Furthermore, the characteristics of the respondents compared well with the known characteristics of students studying ABET at UNISA. (Records are held of this, and so checking was easily possible.) As far as the populations of tutors and regional co-ordinators were concerned, the questionnaires were sent to all those involved. The response rate of tutors and regional co-ordinators could again be considered as high (especially taking into account the responses to the repeat send-out and taking into account that opinions could also be offered in the form of workshop discussions).

Nevertheless, to give the study theoretical substance, Romm et al. could have ventured to propose certain hypotheses expressing connections of correlation and/or of causation that could be tested with respect to data from past students of ABET at UNISA, and with respect to data from tutors of the course. And they could have offered leads toward the generation of hypotheses pertaining to adult education more generally in South Africa (or any other context). (For any results to be generated in respect to the larger populations, these would of course have to be checked against samples thereof.)

5.3.1.4 Value Freedom: Finding Out about Reality

Romm et al. developed questions and organized discussions in order to be able to comment on the way in which ABET at UNISA was operating — as far as those involved in organizing the course were concerned and as far as its students were concerned. But Romm et al.'s initial expectations that ABET was indeed being experienced as fruitful for students and that it was worthwhile to continue to operate the tutorial system as an aid to their learning, might have affected their approach to the study. Romm et al. indicated in their report that their account of the operation of ABET was to be treated as a product of the specific research contexts that were set up in the research process between the various respondents and the researchers. Nevertheless, because it was couched as a study of ABET itself, those funding the report (and others reading it) might not understand the import of the qualification that Romm et al. put in the report in regard to the way they built up their findings. This is especially because the use of surveys is normally associated (by lay people) as offering data untarnished by subjective influence. Romm et al. could have made more effort to reflect upon the extent to which their own hopes and expectations might have affected the way they organized the study and interpreted its results.

5.3.1.5 Instrumental Knowledge: Offering Useful Knowledge

Romm et al.'s research was aimed at assessing the delivery, relevance for students, and impact on their lives of the ABET course at UNISA. Information in this regard was of particular use to certain clients. However, it is possible that because the study was directed in view of the interests of these clients, it was not sufficiently directed toward creating an unbiased account of the realities that were being studied. Those involved in delivering the course, and those funding aspects of it, probably preferred to find out that it was being experienced as beneficial on a variety of scores. But from a positivist perspective, the specific interests of those funding a study or of others affected thereby, should ideally not mar a scientific study. Its utility in offering information consists in the fact that the results can be taken as a product of the attempt to strive for value freedom, rather than as a reflection of some political agenda. The real utility of Romm et al.'s study thus lies in the extent to which it offers information free of political interests.

5.3.2 A Non-foundationalist Assessment

5.3.2.1 Scientism Qualified: Attending to the Effects of Human Perceptions in the Social World

The non-foundationalist argument is that social researchers should operate by giving due attention to the possible effects of people's perceptions in the social world. These perceptions might make a difference to how people respond to the researcher's presence (and to his or her observation instruments), and they also might be operative at a later stage when results are reported. At that stage, how people regard the findings can affect the way they continue to act in the social world.

In the case of the research undertaken by Romm et al., a non-foundationalist argument might suggest that there is a chance that the "reactivity effect" was operative in the application of all of the instruments of observation that Romm et al. drew upon in the course of the investigation. It is likely that, for instance, when the researchers undertook project visits, the students were responding to questions about the relevance of their ABET course to their work in the light of their perception that the researchers had some involvement with ABET staff. Or when students were sent the questionnaire stating that the researchers were interested in their experience of the course, there is a chance that they thought the researchers were hoping that positive responses would be given. Or again, when tutors administered the interview guide, the students might have given their responses in view of their knowledge of what they took to be the tutors' expectations. And when tutors and co-ordinators were given questionnaires to fill in, and were asked to participate in workshop discussions, they too are likely to have been influenced by their understanding of the purpose of the research. A non-foundationalist argument would suggest that with respect to all these contexts of interaction, the likelihood of reactivity being operative in the various contexts, should have been examined and reported upon as part of the study.

A non-foundationalist position might also suggest that as Romm et al. began to share the results with others (e.g. with those involved as participants in the study and with wider audiences) these people's ideas about the issues studied might have been altered. This means that people could come to interact differently with one another, in consideration of the information developed in the study. Romm et al. need not be too concerned about this, as long as the information can be taken as being, indeed, informative to the best of their (scientific) ability at that point in time. However, they would need to show more evidence of the scientific status of the findings in terms of their being a product of a well-organized scientific procedure.

5.3.2.2 Phenomenalism Reviewed: Improving Instruments for Observation

From a non-foundationalist point of view, all ways of referring "evidence" from the real world are subject to doubt. But we must still "aim for as much proof as we can obtain" (Sapsford, 1999, p. 139). Romm et al.'s attempt to characterize the various phenomena in the way that they did, may be a starting point. But their constructs "can [perhaps] be superseded by other ways of framing the world which offer better understanding" (as Sapsford, 1999, p. 148, comments when he speaks about the development of constructs). So, for example, in the case of Romm et al.'s study, in order to understand whether tutorial classes aid the process of learning, we might need to develop better indicators of learning than those utilized in the study. When Romm et al. posed their question(s) on learning they probably had some theoretical vision that guided them. But this could have been made more explicit. It is possible that if Romm et al. had opened themselves up to the test of self- and collegial criticism, they would find that they could do better with their measuring instruments. How this might be achieved, could be a subject for rational discussion within the scientific community. (De Vaus argues in this regard that "if all else fails we ... [can] give the measure to other people (referred to as a panel of judges) to see what they think" (1996, p. 57).)

5.3.2.3 Empiricism Reviewed: Creating Theoretical Explanations Grounded in Observations

Romm et al. did not appear to be overly concerned with offering a theoretical account of relationships of causal connection that could be inferred from the survey. De Vaus notes, nevertheless, that even when such theoretical work is not the purpose of doing a survey, one can still make "*ex post facto* explanations ... which are inductively developed after making observations" (1996, p. 302). De Vaus considers that the ex post facto approach has (theoretical) value insofar as it can provide scope for "researchers to discover quite new patterns in data and develop some quite innovative ideas" (1996, p. 303). However, as De Vaus notes, when such inductive theorizing is attempted one must be aware that there may always be competing theoretical explanations, and that inductive theorizing cannot be used to decide on their relative merits. As he puts it, "the evidence which could disconfirm the explanation [offered] is often unavailable or not examined so we are in no position to evaluate the explanation" (1996, p. 302). Seen from this point of view, some of the work that Romm et al. did may be relevant to theorizing by offering material for creating some conjectures about possible causal connections. Though this empirical material on its own cannot logically be linked to scientific theorizing about causal connections existing in reality, as soon as it becomes the basis of testable hypothesis, it is relevant to such theorizing. A variety of tests can then be set up (some of them involving further surveys, others perhaps using other research designs) to see if the data support or disconfirm any theoretical statements that might be proffered.

De Vaus indicates that it is possible to use surveys in a more concerted fashion to (immediately) test hypotheses that have been deductively derived from some theory. In such cases data are collected with the intention of testing those hypotheses, along with the theory (De Vaus, 1996, p. 302). The advantage of the hypothesis testing approach, as far as De Vaus (following Popperian argumentation) is concerned, is that "evidence which could potentially disconfirm the hypothesis is specifically collected" (1996, p. 302). In the case of Romm et al.'s research, no explicit hypothesis testing was undertaken. Nevertheless, some prior theorizing could be argued to have influenced their suggestions as to which variables should be checked for association with others. For instance, they probably had some theoretical vision that because of the way gender differences are conceptualized by many people in South African society, women would tend to see their purposes for studying differently from that of men. When they checked for an association between gender and purpose of study, however, this was not found. This means that Romm et al. would need to rethink their initial ideas about the perceptions of gender, taking into account the evidence gathered. Or at least others theorizing around gender in South African society, might wish to take this evidence into account. The same goes for any of the other associations (or failure to find associations) that emerged through the research.

5.3.2.4 Value Freedom: Seeking Knowledge through Operating Self- and Collegial Criticism

Romm et al. did not profess an allegiance to the ideal of value freedom. But this might have meant that they were not sufficiently attuned to considering the influence of their own values on their conduct of the research. For example, it is possible that they hoped that students would experience as fruitful some contact with tutors. This might have affected the way they constructed questions and interpreted results relating to the tutorials. Self-criticism on the part of Romm et al., as well as the application of collegial criticism, could be used to locate possible biases in their approach. Romm et al.'s question-construction, and their interpretation of what the answers indicated, could then be checked in the light of the problem of error or bias. It is important that such checking takes place so that accountability (of those engaged in the practice of science) to others in society can be enhanced.

5.3.2.5 Instrumental Knowledge: Providing Opportunities for Better-informed Action

Sapsford indicates that surveys "come out of a history of counting, measuring and [hopefully] putting to good use" the studies (1999, p. 161). He cautions, however, that when scientific information is presented, it should be couched as being an abstraction that represents some simplification of "the world" (Sapsford, 1999, p. 148). It should be displayed as being a result of people (scientists in this case) deciding to "latch on to features that [they] think important [and] fit them into overall models of what the real world is like" (1999, p. 148).

Romm et al. operated their survey by responding partly to what they thought might be important factors for those who would be using the results to take into consideration. Those wanting to make use of the findings of the survey could consider to what extent the abstractions offered of the complex realities in which they find themselves, are of utility. A non-foundationalist position would aver that insofar as the study was not unduly driven by prior political agendas, its abstractions, understood as such, could afford some basis for better-informed action on the part of those wishing to use them.

5.3.3 A Scientific Realist Assessment

5.3.3.1 Scientism Qualified: Attending to the Reality of Generative Mechanisms, Including Societal Ones

Pawson and Tilley give some examples of designing what they consider to be realist evaluations of social programs introduced by policy-makers. Their examples show why they would be wary of the kind of evaluation organized by Romm et al. in terms both of its theoretical significance and its practical implications. One of their examples is of a study aimed at evaluating the success of the engagement of prisoners with some course experience (in the form of educational rehabilitation) in lowering rates of recidivism (1997, p. 113). With respect to this example, they note that in one context, prisoners had little or no previous education and had a growing string of convictions. In this context, a new mechanism was introduced that consisted in their engaging "modestly" with a course aimed at triggering a habilitation in which they could experience self-realization and social acceptability for the first time. The outcome that could be observed for this group of inmates was a low level of reconviction as compared with the statistical norm for such inmates. In another context, prisoners were serving a majority sentence in maximum-security penitentiaries. In this context, the new mechanism introduced consisted in their "high level" engagement with a course aimed at triggering a habilitation in which they could experience changes in powers of self-reflection and reasoning. The outcome that was observed was again low levels of reconviction compared with the statistical norm for such inmates (1997, p. 113).

According to Pawson and Tilley, it was important to design the evaluation so that it could be understood why the introduced policies were having the effects that they appeared to have. They suggest that "moderate engagement" with a course can trigger rehabilitation in a specific context because of its workings in that context, and likewise with "high engagement". The reason why, say, the moderate engagement works (for some prisoners with a string of convictions), is because of its specific way of creating an experience of self-realization and social acceptability. Likewise, the reason why the high engagement mechanism works (for some prisoners serving majority sentences) has to do with the specific way in which it creates an experience of self-reflection and reasoning. And these experiences in turn can be linked to the wider social contexts that explain how the mechanisms have been rendered operative. These social contexts also offer us a window to provide explanations for the cases of failure of the programs in question (1997, pp. 113-114).

Pawson and Tilley would argue that Romm et al. appear to have recognized that the variables they employed in their survey were constructed as people gave meanings to the questions asked (whether in structured or less structured format). But for Pawson and Tilley, following realist argumentation, this is not sufficient to create an understanding of *why* the course was being experienced in the way that it was by different participants, and what outcomes were created by what mechanisms. Romm et al. also gave no indication that, or how, one could use the investigations that they did undertake, in order to consider the setting up of realist evaluations (in future).

5.3.3.2 Phenomenalism Reviewed: Developing Acceptable Observations

From a scientific realist point of view, theoretical preconceptions always affect the way observations are undertaken. It is therefore necessary to attempt to set up some dialogue within the research community around different ways in which the observations of phenomena should be generated and treated. For instance, Romm et al. could set up a dialogue in relation to theoretical literature on what may be involved in the process of learning. They could examine to what extent their own view of learning links up with other definitions of learning. And springing out of this dialogue, they could try to confront the controversies in order to construct a definition of learning that seems to be acceptable to various contenders. In any case, in future, it would be better to proceed in this way in order to be able to defend better the observations about learning (and about other phenomena) that they arrived at through their survey.

5.3.3.3 Empiricism Reviewed: Working at an Analytic Level to Understand Generative Mechanisms in Context

Romm et al.'s organization of the evaluation was not geared toward aiding the development of ideas about the workings of the generative mechanisms that might account for the observations reported. Hence we still do not know in what contexts, and due to what mechanisms, different approaches to, say, the organization of tutorial classes work better, and why. From the survey as it stands, it is very difficult to make much (in terms of theory development) of the data offered. Or, as another example, in the case of Romm et al.'s suggestion that adult learning could be considered (judging from students' self-reports) as contributing to their children's attitudes and skills in regard to learning, this exploratory observation needs to be tied better to the development of theorizing around the conditions that might account for this.

Romm et al. need to offer some pathway for thinking at the appropriate analytic level about the mechanisms that might render explicable the observations that they have generated through their survey. They could also suggest different ways of organizing studies in future that are designed to develop theorizing at a different analytic level (in order to complement the work that they have done).

5.3.3.4 Value Freedom: Considering Possible Reformist Allegiances and if Necessary Re-organizing One's Focus

Romm et al. organized their study according to their conception of what questions they might ask so as to be able to gain some idea of past students' experiences of the course, as well as others' experiences (including tutors and co-ordinators). They chose certain issues on which to focus (although they also sought advice from funders of the evaluation, staff at ABET, past students, and tutors in regard to this). Their choice of focus could be considered as acceptable from a scientific realist viewpoint only if it can be argued not to display a bias in favor of a particular way of seeing the realities being studied. But in this case, it might be argued, there seemed to be a bias in favor of concentrating on the range of concerns of those funding the study, and on the areas of interest of staff in the Institute, as well as of others whom the researchers consulted. The danger is that these allegiances to what they took to be their clients, might have led the researchers to concentrate on seeing adult learning outcomes out of the context of the wider social structures in which they are embedded. The research could have been improved if Romm et al. were prepared to widen the scope of their analyses so as to recognize how an inclusion of these wider structures might make a difference to the achievements reported.

5.3.3.5 Instrumental Knowledge: Extending Information about (Inter Alia) Social Structures

As it stands, Romm et al.'s chosen points of focus can detract attention away from larger issues that according to scientific realism would require investigation when considering adult education in South Africa. The idea that might be perpetuated by their study is that those involved in hoping to spread adult literacy in South Africa, can effectively transform the life conditions of those who have historically been educationally disenfranchised in South Africa. But it is quite possible that there are social structural mechanisms, which have not been excavated by Romm et al., which make it difficult to achieve anything approaching an overhaul of the patterns of disadvantage in the society. It is possible that even though adult educators are trained in the thousands (by, inter alia, ABET at UNISA), and can make some positive differences to the communities and workplaces of which they are part (as experienced by themselves and others around them), their contribution in the society will be underplayed. This could arguably be a result of the basic capitalist orientation written into the society in favor of what is deemed to be "productive work" (even though this orientation is not immediately visible).

Romm et al. have not shown how their study could be followed up with a view to extending inferences about the operation of forces such as these, including the constraints that might be exercised on those seeking to alter patterns of disadvantage in the society. Failing such additional studies (or at least an additional element introduced into Romm et al.'s one), the chance is that the results of the research could become misleading to actors, rather than (sufficiently) informative.

5.3.4 An Interpretivist Assessment

5.3.4.1 Scientism Opposed: Doing Social *Science*

From an interpretivist point of view, Romm et al.'s study would be regarded as having worked only insofar as it took due account of the fact that observations made by the researchers refer to meanings created in the process of human interaction. According to this argument, when respondents answer researchers' questions they give meaning to the content of the questions as well as to the contexts of interaction set up. Hence it is quite probable that in Romm et al.'s research the students did not all give the same meaning to the question about, say, ease of reading the study material that appeared in the questionnaire sent to them. The meaning assigned would have depended, at least partly, on what they saw as the point of asking the question from the researchers' perspective. As far as the "relationship" established with students via this question is concerned, it is possible that many of the students decided not to tick the response category suggesting that it was "very easy" to read the guides because this would mean that they would be sending the message that the material was not sufficiently challenging. Also, it is possible that they did not want to send a message suggesting that they found the tutorial classes redundant in helping them to appreciate the value of the material. For reasons such as these, they might have chosen to tick the "appropriate" response category in terms of how they thought those interpreting their answers would read their response. Once we understand something of this context, we can better understand the responses. Indeed without understanding this context, we are likely to be misled by the responses. This also means that it is impossible to make any sense of the numbers that are eventually given (in terms of adding up the various responses).

This does not imply that what Romm et al. "found" through their analysis of the questionnaires and interview guide responses should be discarded. The real experience of the interaction of respondents with researchers is itself a meaningful experience that, as such, could be the subject of study. Romm et al. appear to have appreciated this when they referred to the significance of these experiences in deciding how to interpret the various responses. Nonetheless, when taking into account the kinds of interaction developed between researchers and respondents via questionnaires (and even semi-structured interviews) we need to remember that these techniques are not an ideal way of appreciating the real social contexts in which subjects live their lives. This is because responses are still restricted to the range of variables focused upon (even when some new variables are established by creating new codes out of the answers to open-ended questions). More importantly, as Bryman indicates, according to interpretivism, we need also to take account of the flow of events in which "variables" are located (1992, p. 101).

Romm et al. did indicate their hope that "results" they created could become starting points to open up further discussion (in forums such as tutor workshops in which they or ABET staff could act as facilitators) on the meanings that different respondents might have attached to the questions, considering how this relates to the lifeworlds of the participants. This was Romm et al.'s way of expressing their recognition of the essential differences between the subject matter with which natural and social scientists have to deal. But Romm et al. could have been clearer about what investigations in future might take place in order to further capitalize on the possibility of generating data through less fleeting encounters between researchers and subjects in the research process.

5.3.4.2 Phenomenalism Re-interpreted: Creating Meaningful Categories

The interpretivist approach would emphasize that it is doubtful whether respondents utilizing the "same" response categories to offer their answers to questions implies that they are using the words with the same meaning. This renders it difficult to determine what phenomena are being discerned via the respondents' answers. Thus, for example, in the case of Romm et al.'s study, unless the researchers understood that "ease" of reading might have different meanings for different students, they could be misled by the answers given. They would be comparing the responses, as if they meant the same to the different respondents. They would be unaware that they were actually not comparing like with like through this coding method.

Romm et al. did attempt to correct this (potential) shortcoming of the research by trying to tap into meanings via some open-ended questions (for example, as given in all the questions in the interview guide and as given in some of the questions in the questionnaires to tutors and regional co-ordinators). And they also arranged for additional interaction with tutors and co-ordinators to explore further some of their answers (via the second workshop) and to consider interpretations of some student responses compared with some of their own (via the Points for Consideration Form).

These additional ways of accessing the meaning of participant responses can be regarded as lending more credibility to the story presented in Romm et al.'s report (than if only questionnaires with a primarily closed response format were relied upon).

5.3.4.3 Empiricism Re-interpreted: Developing Understanding of the Social World

Some attempts were made by Romm et al. to comment on patterns in the data that were located by using quantitative and qualitative computer packages as aids in the analysis. The quantitative package was used to organize an analysis around the variables dealt with in the questionnaires; and the qualitative analysis located themes that had not been pre-determined by the closed questions in the questionnaires (and that sprung mainly from responses to the interview guide). From an interpretivist perspective, we should not be misled into thinking that variables (including ones that could be created out of an analysis of the qualitative data converted into variables) exist as things that can be isolated by researchers' coding of the responses of participants. The problem with research focusing on variables is how to justify the contextual validity of results in terms of an appreciation of its relationship to the "internal logic" of participants' answers. (See Gill & Johnson, 1991, p. 35.)

The interpretivist position is that prime attention needs to be given to gaining an in-depth appreciation of the lifeworld of those answering questions. This can occur more readily arising out of different instruments of observation from the questionnaires and interview guides on which Romm et al. seemed to base the bulk of their analyses. According to interpretivism, the aim of future studies should therefore be to investigate in (far) more depth the world of the adult learners being studied, as well as of those with whom they are relating in various aspects of their lives. (The worlds of learners who are not adults could also become a topic for investigation at some point, and comparisons could be made in an attempt to add insight into the way learning is experienced by people.) Romm et al.'s study could be argued to have left open the space for other researchers to regard the findings as providing exploratory material in the sense supported by an interpretivist position; the findings can be seen as offering material that can support the additional exploration of meanings in various social arenas.

5.3.4.4 Value Freedom: Paying Attention to Others' Worlds

Romm et al. defined the aim of their study as providing an evaluation of the operation of the ABET Institute at UNISA. The way it was being experienced in relation to issues that students, tutors, regional co-ordinators, staff of the Institute, and others concerned seemed to consider as worthy of investigation, was focused upon. From an interpretivist point of view, as long as close attention was paid to issues of concern to those being studied, and to the way in which they were living with the issues, Romm et al.'s focus can be defended. However, there are points at which the results of the study could be misleading, unless due attention was given to the nature of the "comparisons" organized through both the quantitative and qualitative computer packages. Words or phrases in questions set by researchers might not have the same meaning to all respondents; and answers that could appear (to the researchers) to have a certain meaning, might have been seen differently by different respondents. A distortion of respondents' reported experiences — seen in relation to their experiences in their everyday settings — could easily result from the way the analyses were undertaken. Romm et al. could have given a clearer indication of how they were paying attention to the lifeworlds investigated in such a way that they could argue that they retained a fidelity to the data as meaningful for those involved. Only if they can defend their study in this way, can they be argued to have operated accountably as social scientists/researchers.

5.3.4.5 Instrumental Knowledge: Creating Meaningful Information Relevant to Actors

Romm et al. organized the information arising from the study, in such a way that it could supposedly inform the actions of those who wished to make use of it. They showed readers/audiences how information generated from different (observation) sources could be interpreted by referring back to participants' meaning-making. So, for instance, they suggested that although an association was found between level of education and ease of reading the study material, this "information" needs to go hand in hand with a recognition that students' responses to the question might have been tailored by their desire to offer particular messages to the researchers. (In this regard Romm et al. drew also on tutors' interpretations of the students' answers to the question concerning ease of reading the material.) Those wishing to take into account the information would have to consider it in terms of the meanings that it might have for the various respondents, and bearing this in mind, consider what to make of it. As long as Romm et al.'s report was able to alert readers to the importance of taking into account the relevant meanings, it can serve as a guide to action that is likewise considerate of the meaningful character of social life.

5.3.5 A Critical Theoretical Assessment

5.3.5.1 Scientism Criticized: Avoiding the Transference of an Engineering Approach to the Study of Social Reality

De Vaus comments that survey research is seen by Frankfurt Marxists (also labeled as critical theorists in terms of the categories used in this book) as "scientistic" and "technistic" (1996, p. 9). In terms of De Vaus's conception of critical theory, it regards surveys as "intrinsically manipulative" in the sense that they "give power to those in control" (1996, p. 9). This is especially if the survey research is underpinned by a vision of social reality as comprised of patterns of causality to be uncovered. When surveys are geared to uncovering patterns of correlation (and ideally causation) in "the data", they can readily become manipulative, according to critical theory (as De Vaus remarks).

According to critical theory, if we wish to use surveys in a different (less engineering-oriented) fashion, much theoretical work needs to be done so that their utility can be understood in an alternative light. Habermas indeed speaks favorably of surveys used to offer some indication of whether supporters of, say, ecological ("green") societal movements, or anti-nuclear ones, seem to express resistance to being organized into parties (1979, p. 164). He suggests that the results of these surveys can be said to point to a radical "protest potential" in the social contexts in question. But this is only insofar as subjects themselves come to accept the proffered interpretation of the (radical) meaning of their resistance as pointing to such a type of protest in society. In this sense the results are "revealing" of the potentiality. (See also Romm, 1991, p. 146.)

In the case of the study organized by Romm et al., critical theorists would hope that it could become an opportunity for the researchers, in conjunction with some of those being studied (and other audiences), to consider what it might mean to create information about the experience of ABET. Critical theory would encourage the information to be treated as creating a springboard for argumentation around its meaning through social discourse.

5.3.5.2 Phenomenalism Reconsidered: Creating Realities through the Research Process

Romm et al. paid tribute in their report to the notion that the discerned data that they evoked, arose out of a particular way in which questions were posed and interactions set up between the researchers and participants in the study. They recognized that they were, for instance, trying to encourage people to focus more on their *experience of learning* than on the content of what was "delivered". This directedness toward a concentration on learning *processes* was implicit in most of the questions in the questionnaire as well as in the interview guides; and it also informed the way in which Romm et al. facilitated workshop discussions with tutors and co-ordinators.

From a critical theoretical point of view, the use of questions to encourage respondents to regard their own experience in a new light (especially one favorable to communicative interaction) is acceptable practice. According to critical theory, one can never in any case simply report on others' "given" experiences (Fay, 1996, pp. 213-215). Research is necessarily at the same time an intervention in social life that sets up options for experiencing social reality.[36] Romm et al.'s focus on learning processes might indeed have aided those involved with ABET to (re)envisage their experience of learning. From a critical point of view, the study could thus be interpreted as contributing to creating social phenomena in this fashion. That is, the reported "phenomena" relating to learning experiences, could be seen as linked to the way in which participants came to reflect upon these, in engagement with others (including engagement with their questionnaire items and interview questions).

---

[36]This is whether or not the researcher is placed in an authoritative position in relation to others in the situation (see also Midgley, 2000, p. 119).

5.3.5.3 Empiricism Reconsidered: "Discovering" Social Potentialities

Critical theoretical inquiry is not inspired by an attempt to find out about (factual) social realities, but by an attempt to discover potentials that can increasingly be activated through the research process. So, rather than focusing on whether the students thought that tutorials had improved their chances of, say, performing well in the assignments or the examination, Romm et al.'s focus (as noted above) was on the *ways of learning* that they might have developed through the tutorials. Or again, rather than focusing on what impact the course might have had in terms of whether, say, they got promotions in their job or whether they were able to find formal employment (if they were unemployed), Romm et al. organized the questions so that the students could consider not only these, but also other "purposes" (such as, say, their ability to relate to others). This was done by Romm et al. through the response options given in the questionnaires (regarding possible purposes of studying the course); and it was also done through the way questions were worded in the interview guide. Via the guide, of course, students interviewed could still raise issues that were of particular concern to them — which Romm et al. might not have highlighted at that point. And their concerns could also be raised indirectly via the tutors. Nevertheless, in the research process considered as a whole, Romm et al. hoped that the different participants could reflect further upon the purpose of the course and its possible value. In this sense the research seemed to be geared to creating an arena in which people could come to see themselves anew (in relation to what they might have learned through their ABET experience).

5.3.5.4 Value Freedom Undercut: Recognizing the Value of Human Discourse

Romm provided an indication that the results as laid out in their report were admittedly developed through the specific data-collection processes utilized. In the presentation of the report Romm et al. kept open the opportunity for audiences to decide how to treat the information that was presented. This was by indicating the processes by which the information had been derived, and by indicating how it had been interpreted by different people invited to engage with it at the time. (For example, through the workshop discussions and the "Points for Consideration Form" some provision was made for this engagement.) The value endorsed by Romm et al. appeared to be that of developing a discourse in relation to results being created. Critical theorists would appreciate Romm et al.'s efforts to embrace this value as part of their research activity, but they might remark that the quality of the discourse still needs to be assessed.

5.3.5.5 Instrumental Knowledge Undercut: Furthering Communicative Understanding

Romm et al. made some effort to ensure that information arising from the study would not become detached from communicative endeavors on their own or others' parts. So, for instance, while the study might have shown a correlation between different levels of education of students participating in tutorials and group dynamics in the class, Romm et al. indicated that this correlation could be interpreted in a variety of ways. To create the "outcome" of favorable group dynamics, one need not argue that students with similar levels of education should be grouped into classes accordingly. One could rather engage discussion around how the correlation between the factors is to be understood in the first place. One way of understanding it would be to suggest that those with different levels of education might be encouraged to develop the facility to learn from one another. Hence the "fact" that differing levels of education of students in the tutorials appeared to be related to an experience of difficult group dynamics, could serve as an instigation for people to consider options for perhaps working past this pattern. A possibility for shattering the (quasi-causal) connection was for the ABET Institute to, say, organize for tutors to develop their facilitation skills. (Tutors at one point had requested a workshop on this; so the action option did not seem to be anathema to their own concerns.) In any case, in whatever way the results were treated, critical theory would draw attention to the need for people to recognize how, in social life, they can always be re-engaged. "Found" correlations and statements of probable causal connections between variables, do not mean that we need to accept them as fact. So, in this case, "varied education levels" and "group dynamics", and their association, became "facts" and a "connection" between them that, because imbued with meaning, could be made meaningful in a different way.

5.3.6 An Anti-foundationalist Feminist Assessment

5.3.6.1 Scientism Challenged: Overturning Traditional Research Control

Romm et al. organized the study in such a way that some voice could be given to the large numbers of students who had been involved in the ABET course at UNISA, as well as to others concerned with the running of it. The questionnaire sent to more than 2000 past students could be conceived as an invitation to them to comment on their experience of the course. Also, tutors and regional co-ordinators, and others staffing the ABET Institute and Registry, were given the opportunity (through various forums) to add their views. It might be argued nevertheless by anti-foundationalist feminists that Romm et al. could have tried to devise a research design that allowed students to discuss their points of difference with one another as well as with others (such as, say, tutors) holding alternative views. Romm et al.'s design of the research permitted students to enter the "discussion" only through the questionnaires and through their speaking to tutors during their administering of the interview guide. The questionnaire format that was used, as well as the interview guide, might have been seen by many of the past students as a way for Romm et al. to extract information from them that could be passed on to others. Romm et al. did not indicate how, if at all, they were able to develop a relationship with these participants that differed from the traditional relationship where researchers control the research process.

Notwithstanding the above comments, the study can be seen as an interesting effort to use surveys with an intended purpose differing from that normally associated with the use thereof (within scientific and other communities). More detail is needed on how Romm et al. built up their relationships with the various participants and with different audiences. But in any case, Romm et al. did seem to be bent on forwarding the study in a way that invited those concerned to reflect upon the traditional epistemological status of survey "results". Hence they tried as far as possible to engender discussion around these, and to show their contentious character. But as Alldred points out, organizing any research that is grant dependent "does not lend itself to radical critique of the research enterprise" (1998, p. 159). Romm et al. were placed in (or saw themselves as placed in) a "research role" by virtue of being funded to evaluate the ABET course. Also, the various participants too were aware of this role that Romm et al. appeared to have assumed. This might have made it specifically challenging for Romm et al. to operate in their investigation differently from that which is normally expected of researchers in society.

5.3.6.2 Phenomenalism Reconsidered: Working with Ambiguity

Anti-foundationalist feminists would find commendable Romm et al.'s admission that data was created through the interactions between the researchers and participants in the different contexts in which responses were elicited. So, for instance, Romm et al. were aware that their question on "ease of reading" the study guides elicited responses based on how different students might consider the relevance of the question. To the extent that they saw the question as asking whether they felt the need for tutorial support, this would influence what "ease" meant to them. Or again, to the extent that they wondered about whether ease of reading was being opposed to, say, stretching their capacities, or to some other opposition, this would affect their responses. Romm et al. did not believe that further piloting of the various questions could have ironed out their so-called "ambiguity". Their ambiguous character springs from their necessary contextualization by people in differing ways.

Anti-foundationalist feminists would furthermore appreciate Romm et al.'s recognition (as reported by them) that information relating to variables and their correlations that were found in their study, needed to be seen in the light of the ambiguity both of the questions and of the respondents' answers to them (which appeared as "data"). The constructions that Romm et al. created as they worked with the "data" could then be considered as opportunities (by those concerned) to debate their meaning.

Nevertheless, Romm et al. could perhaps have shown more sensitivity to what Alldred calls the dilemma of the power of language (1998, p. 153). As soon as some category is used (by, for instance, those presenting themselves as researchers), it already starts to structure the ensuing interaction around how reality is to cognized. So, for instance, already by the researchers' asking about "ease" of reading the study guides, participants become directed to think around this issue. This is a dilemma that anti-foundationalist feminists would say is not solvable by trying to efface the voice of researchers. Romm et al.'s solution would have to be considered in the light of the dilemma faced by all social researchers: the dilemma that they themselves do have some involvement in constructing phenomena, but also wish to make provision for various entries into the social construction of "the phenomena".

5.3.6.3 Empiricism Reconsidered: Appreciating the Situatedess of Research

Romm et al. tried to develop an approach to the research that allowed them and others to appreciate the processes involved in generating the information emerging through the study. Anti-foundationalist feminists would accept that there is no recipe for deciding how to proceed as researchers in the face of conventional expectations that they should produce reports with findings that supposedly refer to realities existing outside the knowing process. Although Romm et al. did present some findings (labeled as such) in their report, they also did at least remark on their situatedness in the contexts in which they were built up. They indicated in the report that findings should be treated as being a product of the interactions undertaken in the research process. By indicating this, but not dwelling on it, they tried to cater for a variety of audiences and to a variety of epistemological positions that might have been at play in judging the research project.

5.3.6.4 Value Freedom Undercut: Appreciating the Value of Heterogeneity

Romm et al. did not present their analysis as offering a clear-cut account of the operation of ABET at UNISA, seen from a would-be neutral "observer" perspective. Anti-foundationalist feminists would welcome Romm et al.'s effort to create an investigation that, while giving voice to various participants, did not pretend that the researchers' way of presenting the ensuing information was the only possible (or best available) presentation. The study can be viewed with favor in terms of its intention to produce some categories that served as an invitation to further continued conversations around (and beyond) them.

Anti-foundationalist feminism would side with Romm et al.'s expression of commitment to work with the recognition of their personal involvement in the process of "understanding". They would see this as a way of making provision for continued heterogeneity of understanding.

5.3.6.5 Instrumental Knowledge Undercut: Questioning the Power to Define Realities

Romm et al. seem to have given pointers in their report to aid people's awareness that any statements made by the researchers were a product of the way they were generating and addressing "the data". They gave an indication that any interventions that might be undertaken by actors in social reality should take account of the way in which the study had been done. From an anti-foundationalist point of view, if actors can appreciate the personhood of researchers behind their display of "results", then researchers accordingly do not need to submit to the pressure to produce results that can inform action on the basis of some "scientific authority". Romm et al. could perhaps have given more of an account in their report of their quest to discourage an instrumental vision of their research contribution.

5.3.7 A Trusting Constructivist Assessment

5.3.7.1 Scientism Revisited: Developing Trusting Relationships through the Research Process

Romm et al. wished to defend their research by showing that it took into account, and operated in view of, debates about the character of social scientific inquiry. It is not clear to what extent various participants and audiences were aware that Romm et al. were trying to act as researchers while at the same time engaging in debates about the function of social research in society. But it seems that Romm et al. were attempting to establish trust with people through a different mechanism than presenting themselves as finding out about realities as if these could be accessed independently of the knowing process.

Romm et al. dealt with the possible influence of the so-called reactivity effect by making visible to people interpreting the "results", that responses obtained during the research process would have depended on how participants were seeing the purpose of the research. This also meant that when treating "the data" (and any found correlations), people were implicitly encouraged by Romm et al. to consider their constructed character. A trusting constructivist position suggests that as people come to terms with any presented information, they need to take into account the requirement to defend their way of addressing it.

5.3.7.2 Phenomenalism Reconsidered: Evoking and Defending Observations

In utilizing a variety of sources of observation, Romm et al. recognized that it is possible that some of the sources would be accorded more credibility than others (by audiences) in terms of the quality of "data" produced. For example, it is possible that the data produced through the questionnaires (especially the closed questions therein) would be given more credibility — because it was generated and analyzed seemingly without "subjective" input on the part of the researchers. Romm et al. nevertheless tried to show in the report that none of the data sources that they utilized could be seen as more faithful to the phenomena than others. The different observation instruments were shown to have generated data through different contexts of interaction between the researchers and others.

Romm et al. tried to account for observations in such a way that others could enter into further discussion around their meaning, while (hopefully) appreciating the researchers' input too. While we may accord some credibility to Romm et al.'s mode of defending their observations, we do not thereby need to consider that the discussion around them (in further trusting interactions) should be constrained by their presentation.

5.3.7.3 Empiricism Reconsidered: Exploring Alternative Ways of Seeing and Using Evidence

Romm et al. began their preparatory work for the investigation by arranging visits to projects of past students (and interviewing these students) and by discussing with staff of the ABET Institute possible ways of proceeding with the evaluation. Ensuing from this, they generated initial questions for all the questionnaires and for the interview guide and they piloted these, with a view also to modifying these observation instruments. (The interview guide was also reworked with the help of the tutors.) Romm et al. discussed — with those who offered suggestions — what questions (and what wording thereof) should be considered as priority; and they took these discussions into account as they proceeded.

Apart from analyzing the data created via the closed questions, Romm et al. drew out themes from the answers to the open-ended ones in the questionnaires, and tried to connect these with themes arising from the (qualitative) analysis of the interview guide responses from past students. Romm et al. also engaged tutors and regional co-ordinators in some discussion around their various responses, and they engaged them further in considering some apparent discrepancies between their own and student responses on certain issues (the latter being through the "Points for Consideration Form"). Having taken all this into account, they offered a discussion in the report around the way they had developed the findings. In line with trusting constructivist argumentation, their accountability is to be assessed by considering the way in which they tried to come to terms with differing views on "the evidence" as these became presented.

5.3.7.4 Value Freedom Undercut: Encouraging Discursive Accountability as a Value

From a trusting constructivist point of view, Romm et al. should ideally have done more in the way of organizing for discussions with various participants around the way the project was being organized and findings generated. This would have allowed them better to account for their research by expressing their manner of dealing with discussions entered into. It is of course possible that many of the participants might have been less interested in the research than were the researchers; they might not have wished to share in more detailed discussions on this. But Romm et al. could have tried to develop the relationship with participants further, if only to the point where their desired involvement could have been voiced.

As far as reaching the students (the largest body of participants) was concerned, if Romm et al. felt it impractical to discuss results with (a sample of) students, they could, at a minimum, have provided a space in the questionnaire asking them to comment on the overall relevance of the questionnaire to them. They could have asked them to comment on which questions they found more important than others in terms of their interests, and which questions might have been introduced that were not in the questionnaire. (The same could have been done in the administration of the interview guide.) This would perhaps be a way of gaining some feedback from students in some form. And in their report Romm et al. could then have shown that they had entered into a discourse with the students in this manner, and taken into account their concerns/suggestions for future studies.

Researchers (including those organizing surveys) should in any case bear in mind the importance of trying to explore avenues for building up trust with participants and other audiences touched by their projects. This is not a matter of researchers having to concede to (felt) expectations that they strive for value freedom in their work. It is rather a matter of participants/audiences becoming party in some way to a discussion around the generation of (admittedly value laden) results. The accountability of researchers consists in their making provision for visions/results to be recognized as expressions of concerns that need to be defended in engagement with alternative concerns (and visions).

5.3.7.5 Instrumental Knowledge Undercut: Tempering the Possible Self-fulfillment of "Results"

Romm et al. regarded it as part of their responsibility to consider the possible impact that their report might have on those who had been involved in some way in the study (including those funding it) and other audiences. They were specifically wary of presenting a vision of realities (or even a simplification of these realities) that invited a use of it by actors in an instrumental fashion — with possible self-fulfilling effects. For instance, considering the example from Section 5.3.5.5, if people had read the results concerning the found correlation between varied levels of education and group dynamics in the tutorials as a given "fact", this might have geared them in the direction of accepting this as, indeed, fact. Then, to find an efficient way of creating better group dynamics in classroom situations, it might have been seen as best to divide group tutorials according to levels of education, and to organize for tutors to work at the level appropriate. However, Romm et al. showed that people instead could reconsider what "education level" means and could decide that the experience brought by those with lower (formal) education levels might be drawn out by a practiced facilitator. And "group dynamics" could be (re)interpreted to mean that debates around people's differing life experiences could become envisaged as fruitful. By allowing the correlation between the apparently stable categories (variables) to be seen as a *construction*, Romm et al. indicated (by implication) that people could organize their actions differently than if they assumed an attitude toward this as "given" by reality. But Romm et al. could have further elucidated the way in which they saw the constructions that they were producing as indeed inviting reconstruction (by themselves and others) — through discursive activity in which people are still called upon to defend their constructions.

5.3.8 Trusting Constructivism Summarized in Relation to Alternatives

I provide below a summary of the alternative interpretations of the example that I have discussed in this chapter, with a brief indication of how a trusting constructivist position might develop an argument in relation to each one.

5.3.8.1 Positivism and Trusting Constructivism

From the point of view of a *positivist* position, it is prudent of Romm et al. to admit that their study was aimed primarily at generating descriptions rather than explanations of the phenomena studied. They generated some data about aspects of the course on which they chose to concentrate. Using closed questions in various questionnaires, they were able to generate information about variables in which they were interested, by analyzing responses given. And they were able to use the more qualitative data (from open-ended questions in questionnaires, and from the interview guide and other sources such as group discussion) in developing further ideas about the experience of the ABET course. But their study was weak seen from the perspective of its contribution to theory building. Whether or not they could have organized their study to make a theoretical contribution by considering possible causal connections between the different variables, is a matter that Romm et al. had to decide. These decisions would admittedly have to be made in the light of practical considerations such as available funding. These considerations might legitimately have affected Romm et al.'s choice of topic and their choice of procedure for pursuing the investigation. Be that as it may, in its current design, the study clearly did not aid theory development in any significant way. Nevertheless, the data that did emerge through the study can still be regarded as of some scientific value. The main value is in pointing to areas for (future) research into possible causal connections that might hold between certain of the variables isolated.

The trusting constructivist response to this interpretation of the evaluation is that it assumes that its scientific value can be accounted for only in terms of conventional expectations given by (some parts of) the scientific community in regard to what is properly involved in doing science. But positivist-inspired expectations on this score do not do justice to debates around alternative visions of the purpose of social science in society. Romm et al.'s study could be regarded as defensible in terms of an assessment that introduces different conceptions of the proper practice of social research (than implied in a positivist position). From a trusting constructivist point of view, it is preferable to give leeway to researchers to account for decisions as to how to apply their skills in specific projects by appreciating that their "doing science" does not necessarily have to be equated with striving to locate causal connections existing in reality.

5.3.8.2 Non-foundationalism and Trusting Constructivism

From the point of view of a *non-foundationalist* position, Romm et al.'s manner of proceeding with the investigation could have been more opened up to the test of self- and collegial criticism within the scientific community. For example, Romm et al.'s way of taking into account the reactivity effect as they organized their relationship with the participants (through the various data collection instruments) could have been more systematic. Although granting that we can never know quite how this effect may be operative, it is important to try to arrange the data-collection process with due regard for it. Romm et al. could have considered more carefully how their phrasings of questions, their manner of arranging for the administration of the interview guide, and their manner of facilitating discussion with tutors and others, might have been tailored differently to try to minimize potential reactivity on the part of participants. Had the researchers monitored the reactivity effect, then in this study or in future ones better ways of collecting the evidence could be tried. Nevertheless, although Romm et al.'s collection and interpretation of data may be subjected to some doubt, they can be regarded as having furnished certain information (which might be improved via further studies). Romm et al.'s evaluation can thus be seen as offering some information of interest to actors who might wish to make use thereof. Romm et al.'s study did also provide interesting leads to further studies that could be aimed at investigating patterns in reality through the creation of some hypotheses to be put to the test of the evidence.

The *trusting constructivist* response to this (non-foundationalist) interpretation of the research is that it assumes that doing social science entails endeavors to get to grips with patterns existing in social reality: scientists are considered as duty bound to try to find a path to generate results of this sort (or at least to offer leads in this direction). The trusting constructivist position challenges the idea that research must be directed in terms of this understanding. This idea was indeed challenged by Romm et al. in the process of their organizing the research and in the way they presented the status of findings in their research report. They worked in terms of the understanding that they might be able to earn the trust of others by trying to shift this expectation. In other words, they tried to operate an alternative mode of trust building — in keeping with the values of trusting constructivism. But Romm et al. could have offered to readers of the report more detail on ways in which their own practices might be conceived as still accountable. This would allow readers to be placed in a better position to consider their trustworthiness.

5.3.8.3 Scientific Realism and Trusting Constructivism

According to a *scientific realist* position, Romm et al. did not make sufficient effort to develop an appreciation of the underlying causal mechanisms that might be operative in the realities they were studying. Nor did they make efforts to connect their study up with other studies concerned with developing such understanding. Because they failed to operate at the level of excavating social structures, there is a real danger that their research can be used to advance a reformist political agenda. Romm et al. seem to have been too naive in regard to the consequences of their way of proceeding with the study (and presenting findings from it).

The *trusting constructivist* response to this (scientific realist) interpretation of the evaluation is that it concentrates attention on the requirement for researchers to develop strategies for advancing knowledge of (structures of) reality. From a trusting constructivist point of view, the social relationship that is set up with participants in the process of doing social research and the social relationship set up with other audiences, have to be considered as crucial to the investigation. Romm et al.'s evaluation rightfully accorded such considerations an important role in the study. Romm et al. did not wish to take it upon themselves to develop statements about social structural forces that they believed could be inferred to exist "in reality". A defense of any statements made about posited structures would require creating an argument around how these statements have been discursively developed in the first place. There is nothing to stop the making of such statements (in future studies, for instance) — with the proviso added that they are to be treated as aiding continued discursive encounter around their content.

5.3.8.4 Interpretivism and Trusting Constructivism

Romm et al.'s style of survey research did not seem to be directed first and foremost toward uncovering variables that became defined as given by researchers' categorization of them. Romm et al. tried to use all the data (from all the instruments of observation) to consider different ways in which categories might be interpreted in the experience of those studied. They thus appear to have attempted to use survey research not simply in order to uncover facts about specific variables and their possible connections, but in order to display meanings in the data. An *interpretivist position* would appreciate that they made some efforts in this regard.

The *trusting constructivist* response to this (interpretivist) interpretation of the study would be to agree that Romm et al.'s style of survey (and self-understanding thereof), is preferable to an approach that assumes that common meanings can be attached to "same" questions made by researchers, and to "same" answers given by respondents. However, the trusting constructivist position argues that an important contribution of Romm et al.'s approach is that it opened up spaces for participants and others to explore different ways of relating to the found "research material". Romm et al.'s attempts to avoid posing as an authority in making sense of the material, was significant to the extent that it helped people to learn that credible sense-making is not the preserve of scientists. In this way Romm et al. can be argued to have helped install a different relationship between themselves and participants/audiences from that which is conventionally associated with doing survey research.

5.3.8.5 Critical Theory and Trusting Constructivism

According to *critical theory,* Romm et al.'s study can be justified as an attempt to use surveys in a manner that lays bare the researchers' involvement in creating a way of addressing "the phenomena". Romm et al. conceptualized the data that was generated as produced through meaningful encounters between the researchers and those being studied. And they also at the same time showed (with reference to some examples in their report) that considerations around the way the information had been produced, could affect people's conception of any courses of action that might be envisaged in relation to it.

A *trusting constructivist* position accords with the critical theoretical suggestion that one of the contributions of Romm et al.'s study lies in its refusal to display information as if it could be used as a guide to proffering realistically informed recommendations for action. For example, Romm et al. chose not to treat the correlation between varied levels of education and group dynamics in the tutorials as an indication that to reach better dynamics one would need to re-arrange the classes by levels. Romm et al. showed that this apparently efficient "means" toward the supposed "end" (better group dynamics), itself contains a value that had to be made explicit and discussed. Depending on the way in which people assigned meaning to the "found" association, this would affect options for action considered. As does critical theory, a trusting constructivist position appreciates that Romm et al. did not try to use the authority of science to point to efficient solutions that appear to be value free. A trusting constructivist position would emphasize that Romm et al.'s study needs to be judged by reflecting upon its manner of engendering trust between those involved with the research (in whatever capacity) in a different fashion.

5.3.8.6 Anti-foundationalist Feminism and Trusting Constructivism

Anti-foundationalist feminism would see Romm et al.'s organization of the survey as a possible way of giving voice to large numbers of people. However, to give voice appropriately, one cannot simply define categories (as variables) that researchers then use to control the research process and its results. The conventional way of conducting surveys to extract answers around issues that reflect researchers' pre-defined agendas, needs to be reviewed. Romm et al. seem to have attempted some alternative way of going about their survey, while capitalizing on opportunities to reach a large number of participants in the research process.

Anti-foundationalist feminism recognizes that the practicalities of the situation (as viewed by those involved) might make it difficult/impossible to satisfy everybody's potential demands. For example, in the case of this study, the funding body (DFID) probably favored, at least initially, a large-scale survey with data that could be arranged numerically so that a picture of what was going on could be obtained in this way. However, Romm et al. indicate that they tried not to accede to an assigned role of simply "finding out" (whether in numerical terms or otherwise), but tried to operate their research role differently (more in keeping with the epistemological commitments of anti-foundationalist feminism).

A *trusting constructivist* position would comment that Romm et al. developed a way of working with different ideas concerning their research role, as they tried to create an arena of trust between themselves and others in regard to the research purposes. Romm et al.'s defense in their report of the way they proceeded in this case, can be considered as an entry into debates around the practice of accountable social research.

5.4 CONCLUSION

A number of issues were explored in this chapter in regard to survey research, using Romm et al.'s study to co-ordinate the discussion. One of the issues that arise in the use of surveys is what status should be given to any associations (between variables) that can be created via a statistical analysis of results. According to positivism and non-foundationalism, survey data can conceivably be used to develop statements about causality operative in social life. One survey alone is, of course, not regarded as sufficient to create statements about causal patterns. But as long as further checking of these statements takes place in future studies (which may include surveys or other strategies), it is seen as possible to use survey data to make a contribution toward advancing our theorizing around the patterning of social life.

Scientific realism introduces a different issue into the debate about survey research by questioning the successionist view of causality that it sees as implied in the way survey data are normally analyzed. The suggestion of scientific realism is that efforts to find out about the (underlying) structures of reality need to be introduced explicitly at some point, in order that a theoretical contribution can be made and in order that a reformist political agenda does not become buttressed by the research.

From an interpretivist perspective, yet another angle from which to assess survey research is offered. Here, the focus is on the character of the data created via survey research — seen in the light of the fact that in social life people assign meaning in different ways to questions that are posed by researchers and to the interaction contexts in which they are posed. In order to provide for better quality data, interpretivism suggests that some other means of data collection must be drawn upon, other than by researchers posing (primarily) pre-set questions to respondents and giving them largely closed response formats for their responses. Open-ended response formats are regarded as affording some space for researchers when interpreting results to get a better idea as to how the various respondents are interpreting the questions asked. But forms of less structured interviewing (including group interviewing at times) are considered as preferable in order to create a more plausible account of the way actors are assigning meaning to their world. Nonetheless, interpretivism queries the attempt of survey researchers — even when some open-ended response formats are used by them in their questionnaires and/or less structured interviewing — to compare the responses given so that "similar" ones can be categorized on the basis of researchers' conceptions of similarity. Even though Romm et al. attempted — through their style of survey — to give extra meaning to the variables examined, it is preferable, according to interpretivism, to supplement this still with a more detailed examination of actors' assignment of meaning to their world.

Critical theorists and anti-foundationalist feminists for their part are concerned that in survey research it can easily become forgotten that variables and found associations between them are a product of a particular means of approach to the study. From a trusting constructivist point of view, similarly, it is considered important to render visible the processes by which variables have been constructed, so that it can be recognized that ways of creating constructions need to be defended (as part of an ongoing conversation) through discursive encounter.

It was also shown in the chapter that the question of how researchers should relate to funders of their research is treated differently depending on the perspective adopted. Survey research is sometimes considered to be relatively expensive because of the large numbers of "cases" involved, relative to other forms of research. This is in order to defend the likelihood of the samples utilized being representative of the population or populations studied (cf. De Vaus, 1996, pp. 4-5; Weaver, 1997, pp. 112-113). Admittedly, one can also see other styles of research as expensive because of their being time-consuming and demanding for researchers (cf. Schofield, 1993, p. 205; Fielding, 1993a, p. 168). From positivist, non-foundationalist, scientific realist, and interpretivist points of view, if researchers manage to gain the required funding for research investigations, they might rightfully feel obligated to present results in such a way that audiences can understand their possible relevance in terms of their practical (instrumental) use. From these perspectives it might be argued that Romm et al.'s insistence on qualifying the cognitive status of the findings in the way they did, could have made the funders (and others) wonder to what practical use they could indeed be put.

Nevertheless, non-foundationalism cautions that those with access to the results ought to be made aware of their somewhat tentative character. They should be advised to treat any proffered findings (arising from survey research as well as other forms of research) as providing less than certain knowledge.

Interpretivism cautions in addition that the results developed via survey research must be understood in terms of their intrinsically meaningful character. Otherwise, the information supplied may be fundamentally misleading.

From a scientific realist point of view, the caution would be that survey research — if too narrowly focused — can all too easily slip into supporting a reformist political agenda. Its "practicality" thus may serve certain, rather than other, clients. Romm et al.'s study would be seen as a case in point where more attention to the scope of the study was (arguably) needed. Without this, the value of the research as aiding an appreciation of the kinds of changes that might be needed to correct the disadvantages associated with educational disenfranchisement in South Africa is questionable.

From a critical theoretical point of view, Romm et al. were, notably, wary of supporting a realist conception of the worth of their study (realism here being understood in broad terms). According to this argument, had they been more realist-oriented, communication around findings being generated through the study might have embodied an unnecessary power relation; the researchers' authority would have been likely to rule the terms of the debate.

From an anti-foundationalist feminist point of view, it would (likewise) be disapproved had Romm et al. addressed themselves to participants and audiences by posing as having the skills to find out about the realities as if these could be known independently of the knowing process. Anti-foundationalist feminism would welcome Romm et al.'s attempt to capitalize on the ambiguity in regard to the point of their doing the research.

From a trusting constructivist point of view, the way Romm et al. created a research role was a matter (though not an easy one) of their trying to earn the trust of others in their skills to organize a research process that could be considered as defensible, while not pretending that results could then be regarded as (more or less) authoritative.

6
Exploring the Ethnographic Study of Lives

6.1 INTRODUCTION

In this chapter, I organize a discussion with reference to an example of research that I label as "ethnographic" in terms of Fielding's definition of ethnography as "a form of qualitative research which combines several methods, including interviewing and observation" (1993a, p. 154). Fielding indicates that the "qualitative" character of ethnographic research derives from the "emphasis put on 'depth', 'intensity', 'richness', and so on" (1993a, p. 155). He also indicates that some effort needs to be made "to 'think' oneself into the perspective of the members" (1993a, p. 157). Bryman similarly indicates that the focus in ethnographic research as a style of research is on attempting to develop findings that "reflect [or resonate in some way with] what subjects deem to be important about their lives" (Bryman, 1992, p. 9).

The gathering of detailed material can be experienced as very demanding for those organizing ethnographic research. Hence sample sizes are normally much smaller than the samples used in some other kinds of research — such as, for example, surveys (Fielding, 1993a, p. 155). Nevertheless, the definition of ethnographic research does not rest on the size of the samples utilized, but on the way in which the material is approached. It is approached with a view to exploring the way in which life is experienced in the "natural" setting (Bryman, 1992, p. 59; Fielding, 1993a, p. 157). The idea is to build up an "illuminating description of and perspective on a situation that is based on and consistent with detailed study of that situation" (Schofield, 1993, p. 202). It is suggested that building up a plausible account of people's differing experiences of life in the setting can be organized with relatively small sample sizes (Fielding, 1993a, pp. 155-156; Arber, 1993, pp. 72-73).

It is sometimes suggested that the organization of ethnographic research need not detract from researchers' attempting to develop insights that can be recognized as being of analytic value toward theory building. This requires seeing ethnographic research as operating in terms of a different logic of inference from survey research — where inferences are made through statistical probabilities from the sample to the population of interest. Yin notes that in forms of "case study" research, cases can never be argued to be representative of the population of interest in the statistical sense of the term (Yin,

1994, p. 37).[37] He suggests that the analytic power (theoretical value) of the concepts generated through case study research is based on our comparing the situations studied with other ones in which we are interested. However, Yin suggests that if ethnography (as a contender for the label of case study research) is to be seen as a vehicle for developing theory, then this intent should be built into the design from the start. Yin himself wonders whether ethnography is "typically" thus oriented (1994, p. 27). But what is important for Yin is that the "generalizability" or "transferability" of results (of case studies) to other situations requires that plausible comparisons can be set up with other settings (for example, through organizing other case studies or through referring to those done by others). This process Yin calls "analytic generalization" (1994, p. 36, 2000, p. 241).

Whether or not Yin's particular view of theory development is adhered to, his argument is relevant in helping to draw attention to the ways in which ethnographic studies may be defended as (potentially) having wide-ranging "analytic power". Just because the research is located in such a way as to study people through getting close to their "natural" environment (the setting of everyday life experience), this may increase the chances of results being transferable to (or relevant in some way in understanding) different social settings (Gill & Johnson, 1991, p. 122; Henwood & Pidgeon, 1993, p. 27; Maxwell, 1996, pp. 96-98; Page, 2000, pp. 30-31).

The reason I have placed the research project detailed in this chapter under the banner of ethnography is because it seems to be oriented to exploring, in Bryman's words, "what subjects deem to be important" about the (broad) topic of interest located by the researchers. The remit of the researchers was to conduct an investigation around inheritance practice and law in a (small) country — Swaziland. The team conducting the research consisted of seven women who were exploring inheritance issues in Swaziland as part of a larger research program in Southern Africa on Women and the Law in Southern Africa (with national teams organizing research in different countries). The overall program was funded through the Women and the Law in Southern Africa Research Trust (W&LSART).[38]

My own involvement in the study (which forms the background to my way of outlining it here) was as a (so-named) consultant to the researchers in Swaziland at a certain stage of the research process.[39] I was invited by the co-ordinator of the team to adopt this role, which I shared with Nina Romm. The consultancy took the form of a

---

[37]Hammersley and Atkinson make a similar point when discussing ethnographic research, which they note does not rely on using statistical probabilities to organize generalizations from samples to populations (1995, pp. 41-43). Henwood and Pidgeon, for their part, draw out what they consider to be the theoretical implications of our appreciating that in "qualitative research" sampling decisions are not made on statistical grounds. Their argument is based on defending the need for "rich and dense grounded theory" (1993, p. 27).

[38]Contributors to the Trust were organizations such as: The Ford Foundation, SAREC, DANIDA, NORAD, SIDA, CIDA, and UNICEF.

[39]At the time I was Dean of the Faculty of Social Science at the University of Swaziland and was thus known to most of the researchers on the team.

workshop held over three days in Swaziland with all of the researchers to discuss the project — with a focus on the co-ordination of the researchers' various efforts.

The emphasis in this chapter is on considering how the researchers chose to involve themselves in the lives of the participants as they proceeded with the research. The discussion is set in the context of considering different views on the (proper) role(s) of ethnographic researchers. To conduct the comparison between arguments, I organize the discussion in similar fashion to Chapters 4, 5, and 7. The positivist understanding of the example is presented first (in terms of positivist tenets). This is followed by other understandings, discussed by exploring alternative modes of assessment. Each heading in Section 6.3 indicates how the positivist argument is being treated at each point as a way of creating criteria for judging research examples. As in other chapters, the example given in this chapter is discussed in terms of the judgments provided by the headings.

6.2 INHERITANCE PRACTICE AND LAW IN SWAZILAND (APHANE ET AL., 1993)

6.2.1 An Outline of the Project

Figure 6 (overleaf) offers an indication of how the team of researchers in Swaziland organized their investigations.[40] Regular meetings were held amongst the team, to consider ways of co-ordinating their work. (These meetings are labeled as "Ongoing discussions" in Figure 6.) Aphane took responsibility for the overall co-ordination task, including the task of co-ordinating the final report (for the W&LSART).

Some of the researchers in the team were professional lawyers, some professional sociologists, and some were nonprofessionals.[41] (All of the researchers were women, although in other national teams — in other countries in Southern Africa — the research teams included men.) Aphane et al. indicate that the research project could be considered as activist in that

> it aims not only at bringing change at a personal level [in aiding people to envisage options for action in relation to inheritance issues], but at the formal level as well, e.g., on policy, law and administration. Both the researchers and the researched benefited in that [some of] the various requests of the interviewees were responded to. Some were on the spot, e.g., giving of advice; others later, e.g., producing a pamphlet on inheritance. (1993, p. 7)

---

[40]Peter Adman created this figure after reading my account of the project as reported in this book.

[41]Romm suggests that the interdisciplinarity of the inquiries in the case of this research consisted in the researchers being "sensitive ongoingly to inputs provided from other perspectives (that is, other than some initial starting point or 'source discipline')" (1998a, p. 65). The quality of reflexive consciousness on the part of researchers engaged in teamwork is what can be said to distinguish *inter*disciplinary from *multi*disciplinary inquiries. The interplay in consciousness between alternative ways of seeing issues of concern is specifically focused upon in the former.

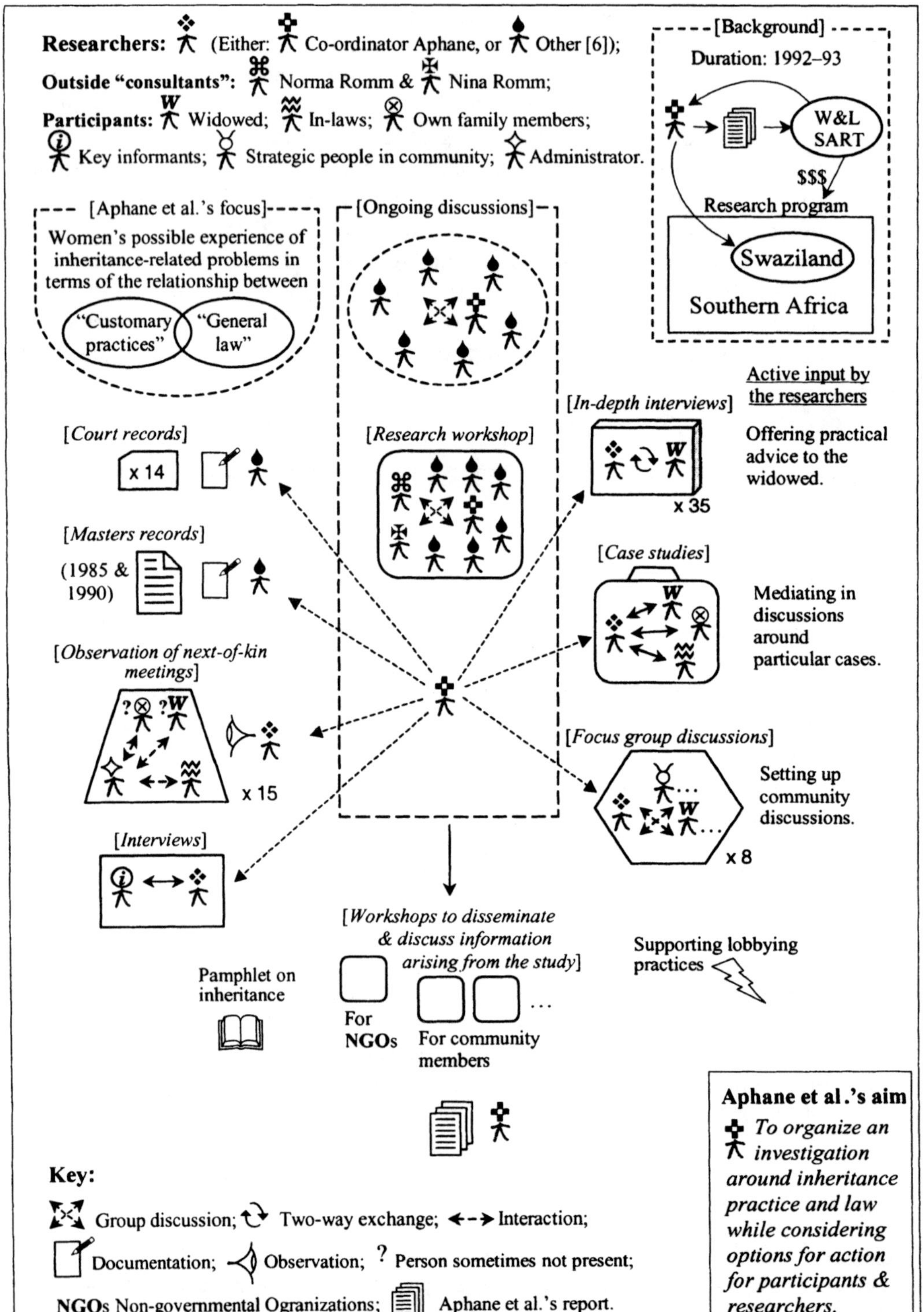

Figure 6: Aphane et al.'s ethnographic study.

The concern of the researchers was to explore, with participants, the relationship between customary practices relating to inheritance, and the so-called general law. Reference to the latter included formal legal proceedings to execute estates, and court case proceedings to address cases of dispute in relation to inheritances. One perspective that the researchers brought to the study was that, as they put it,

> if the laws which regulate inheritance and other spheres of the legal status of women could sufficiently provide for women, they would have less inheritance problems. However, women have to be encouraged to utilize such laws. (1993, p. 6)

The researchers drew on what they took to be their own life experiences (1993, p. 6) to suggest that the laws on inheritance are likely to have a differential effect on men and women. Laws relating to inheritance, which may have been meant to be gender neutral, can turn out to be gender specific. This is because

> they [women] cannot access the deceased estate, except through proper legal channels. This is as a result of another law, which prohibits women married in community of property from registering property in their own names. (1993, p. 6)

Another perspective that the researchers indicate they brought to the study (based on their life experiences) was that "women are not a homogeneous group and therefore are affected differently by the various laws and customs which have a bearing on inheritance". This meant, as they note, that "strategies faced [as options] for women with inheritance-related problems" could differ (1993, p. 6).

With these prior ideas in mind, the researchers began their investigations, making use of the following investigative techniques (1993, pp. 8-13).[42] (The different researchers used different mixes of the techniques as they divided the research task amongst themselves.)

- *Documenting Court Records* (of High Court proceedings). All court cases that the researchers could identify (since the introduction of a system of general law in Swaziland in 1968) were perused. 14 such cases were identified. The records were selected from the civi! cases register.

- *Documenting Masters Records* (of administrative procedures relating to dissolution of estates under general law). The researchers decided to collect records from two years — 1985 and 1990. These years were chosen because it enabled the researchers to explore the effects of the law on Death Duty and of the law on Succession in terms of estates that had been dissolved. The records were selected randomly by choosing every third record from the list of records.

---

[42]The researchers chose not to conduct surveys as part of their research — although some of the other national teams were doing so. As Armstrong indicates (in commenting on an earlier project concerned with maintenance for women in Southern Africa), the research choices in various countries were based partly on how researchers perceived that (potential) audiences might assess the research process (Armstrong, 1992, pp. 17-22).

- *Undertaking Observations of Next-of-kin Meetings* (where executors are chosen under general law). 15 such meetings were observed. The observations were directed in terms of the researchers' intention to "observe the interaction between the surviving spouse and their relations (e.g., in-laws), the conduct of the surviving spouse at the meeting, and the role and behavior of the men and women at this meeting" (1993, p. 10). Also observed was "the manner in which Administrators conduct themselves during the hearing, and the interaction with the surviving spouse and others" (1993, p. 10).

- *Interviewing Key Informants*. This involved creating contact with persons regarded as specifically "knowledgeable" in the field. For example, princes, princesses, and chiefs were interviewed as the researchers assumed that if any persons knew the customary practices, they did. The aim was, however, also to look at "divergent views, as Swazi law and custom [as opposed to the general law also operating in the country] were not codified" (1993, p. 12). The researchers thus expected to come across different interpretations of (informal) customs. Officials from courts and other institutions that "administer inheritance issues" were also approached as key informants. In interacting with the various key informants, the researchers raised specific questions for them to answer (1993, p. 13).

- *Undertaking In-depth Interviewing of Widows and Widowers*. Interviewees were chosen here through "strategic people" in the community who helped to locate the target group (1993, p. 11). (35 such interviews were undertaken.) In some cases the chief suggested that the leader of the community's women's group aided the selection of interviewees. At other times, selection took place according to the fact that potential interviewees were known to the researchers, or that they were still wearing mourning garments (and thus were easily identifiable), or by asking others to assist in identifying widows and widowers (1993, p. 11). The researchers carried an interview guide with them, but referred to it only "where the interviewee did not address a certain aspect that the researcher was interested in" (1993, p. 11). Aphane et al. indicate that they considered their method of in-depth interviewing to be very important to their research as it "has the lived experience of the interviewees. It gives the researcher the opportunity to talk to the interviewee within the environment which she/he lives in" (1993, p. 12). Interviewees were encouraged to "relate their stories" (1993, p. 11). However, Aphane et al. also note that "the information we got could be what the interviewee wanted us to hear". They suggest that this occurrence could have been "influenced by the fact that we were [sometimes] introduced by a Chief's representative to the interviewees" (1993, p. 12).

- *Organizing Case Studies*. This amounted to considering various players' roles in a case of a single inheritance. In following up some of the in-depth interviews with widows and widowers, the researchers interviewed others who had been identified (for example, the in-laws or other family members) as relevant to their case. In this way, the researchers hoped to "gauge different opinions on the same issue". They were interested in exploring the different perceptions of the issue (and at times they tried to mediate between them). They took both "harmonious and problematic cases" (of dealing with inheritances) — so as not to bias the discussion only around the problematic ones (1993, p. 12).

- *Arranging Focus Group Discussions.* This involved bringing together groups of people to focus on specific issues for discussion. Eight such groups were organized. Three groups consisted of participants from company towns (towns established by companies); three groups consisted of participants from urban areas; and two consisted of participants from rural areas. Discussants were identified "through strategic people in the community as they were [taken to be] conversant with the community" (1993, p. 11). The style of focus group interviewing was organized (facilitated) to allow group members to consider one another's differing views (including those that the researcher sometimes injected into the discussion).[43]

6.2.2 The Details of the Research Process

The researchers met at regular intervals in the course of the project. The research path organized by them (in terms of a consideration of techniques to be drawn upon and ways of drawing upon them) emerged along the way, as they met together to discuss ways of proceeding. In the (three-day) workshop in which I was involved (called "Research workshop" in Figure 6) we concentrated on exploring the potential of the various research instruments to carry information/views across the various contexts in which the research was being undertaken, and we also considered ways in which researchers could operate their "activist" aims.

One suggestion was to try to carry information across the observation of next of kin meetings and the interviewing of certain key informants. In some cases it had been observed (inferred) that women felt that there was pressure on them not to be assertive (and indeed rather to be submissive) in the next-of-kin meetings. It was suggested in the workshop that this issue could be mentioned to the administrators of the meetings, who, as key informants, could be interviewed with an eye to discussing this matter. The raising of the issue by researchers could draw attention to it as a potential issue that needed some form of addressal. It also could become an issue that — drawing on the responses of administrators and taking this into account — could become the basis for a wider "education" program. The church could, for instance, be one forum for introducing awareness of the (potential) purpose of next-of-kin meetings. (The research process revealed that many women were unaware of the meeting date when the

[43] This is more or less in accord with Stewart and Shamdasani's definition of what they call "the contemporary focus group" as consisting of individuals who "discuss a particular topic under the direction of a moderator who promotes interaction and assures that the discussion remains on the topic of interest" (1990, p. 10). However, the "moderators" in the case of Aphane et al.'s focus groups assumed a role of additional participant as they facilitated the interaction between members of the group by themselves sometimes adding input into the discussion. Gregory and Romm (1996, 2001) offer an account of different styles of facilitation that might be adopted by researchers, and a defense of the practice of their adding content into focus group discussions (rather than their just being concerned with group processes).

executor of the estate was to be chosen, and hence failed to attend — even though the decisions made at the meeting affected them profoundly.)

In the course of the project, various ways of redefining possible courses of action around issues of concern raised by participants in in-depth interviews were explored. The viability of some of the possibilities was explored in practice. For example, in one case a widow (a teacher) had been barred from her teaching activities during the mourning period (two years) — due to the custom surrounding the wearing of mourning garments. A mediation with the relevant chief was set up and a solution proposed whereby the widow could wear the garments only at night — after her working day (Aphane et al., 1993, pp. 73-78).

In the interviewing process, as in all their involvements, the researchers tried to be sensitive to the points of view of the various interviewees — accepting challenges to their own preconceptions, while also opening up new avenues for discussion, with a view to considering possible options for action. An interesting area of "expansion" (of consciousness) between researchers and widows revolved around the possible engagement of the general law as a route to addressing inheritance concerns. Aphane et al. note their own concern that

> the widow may be disadvantaged [in terms of Swazi custom] because it is another family, rather than her own family, which makes decisions after the death of her husband. This is the family which she joined by inkhonto [a customary expression of allegiance to her mother-in-law]. This differs from the situation of a widower, as when his wife dies it is his own family which makes decisions. (1993, p. 48)

Nevertheless, the researchers' views on these disadvantages had to be tempered by considerations of how others were seeing the potential use of the general law (in its juxtaposition to customary practices); and they also had to consider how others were seeing the possible working out of the relations between widows (and widowers) and in-laws in the community. Aphane et al. admit the uncertainty felt specifically by widows due, for instance, to their not knowing how precedents will be interpreted in the courts (operating in terms of the general law, which makes use of precedents).[44] Also, tensions felt with in-laws during the time at which widows are trying to make use of the general law, had to be recognized. As Aphane et al. note, "sometimes widows felt that they were too dependent on, or obligated to, their in-laws to threaten them via a court case" (1993, p. 112).

The researchers still sometimes indicated, in offering "advice" to widows, that even those who had been married under customary law have some legal recourse to the general law. But the researchers also had to recognize that widows might not have the

[44]Aphane et al. indicate (1993, p. 69) in regard to the operation of the general law (as opposed to customary law or tradition) that in a particular instance of contestation by a widow, the court ordered that the Master could administer the estate because the customary marriage was legally valid (in terms of the general law) in Swaziland. This decision has meant that the Master can handle estates of parties married by customary law, which might otherwise (in terms of Section 68 of the Administration Act) have been seen to be excluded (1993, p. 69). This in effect means that the general law can in principle operate as a force allowing for contestation over customary practices.

same view of the law and its utility as they may have. In this way the researchers tried to prevent a uni-directional focusing of the discussion with participants during their in-depth interviewing. While exploring the potential opportunities provided by the *law*, they also tempered this exploration with a sensitivity to the way in which *tradition and obligation* might add layers of complexity to the women's (and others') conceptions of options for seeing their situation.

In line with their intention to organize community discussions, the researchers considered focus groups to be an important forum for generating debate around issues of concern. Researchers conceived their role in such groups as creating spaces where people could reconsider customs in the light of a moral concern (supposedly written into the customs) to offer protection to women and children. The researchers' facilitation of the group discussions was practiced accordingly.

Aphane et al. indicate that issues concerning the effect of the money economy on the lives of the various participants were brought up in some focus groups. In one such group the issue arose as to how it could be ensured that those appointed to look after the deceased male's property did not abuse their position (Aphane et al., 1993, pp. 50-52). The issue of the "guarantee" of care for women and children was raised in terms of its connection to the possible effects of the money economy — specifically in terms of people's involvement in earning money in "the city". Sometimes the person customarily required to be "keepers" of the property (whether during the mourning period or thereafter) may not have the time, inclination, or sense of moral pressure, to account fairly for the property. As the research report indicates, the customary keepers during and after the mourning period (umphatseli and inkhosana respectively) "may be too busy in the city to have the time for the traditional duties" (Aphane et al., 1993, p. 50).

In the focus group discussions, some participants indicated that it was important that the traditional custom be altered so that people may bequeath their property to "responsible children rather than the customary inkhosana" (that is, the keeper appointed by custom) (Aphane et al., 1993, p. 61). This was seen as one solution to the problem that otherwise the property might be left in the hands of the less considerate or less responsible. Some participants — including some men — suggested that it would be appropriate if "the inkhosana was the most considerate and responsible child who had contributed during the lifetime of the deceased irrespective of sex, rather than simply the eldest son" (Aphane et al., 1993, p. 51). The question was therefore discussed as to whether the needs of widows and their children "today" can be met within the traditional interpretation of the customs — and whether alternatives in the light of the commitment to protect women and children, needed to be given attention.

An issue related to customs surrounding property "keeping" on behalf of women and children, was the contention concerning women's ownership of property. Focus group discussions revealed various viewpoints pertaining to women's right to property under customary law. Many people in the groups indicated that what women acquire through their own industry belongs to them — as long as the amounts do not add up to "substantial commercial value"; at this point people in focus groups began to contend the idea. For example, sometimes it was argued that the wealth is still the husband's, because it is he who allowed her to work and earn the money (1993, p. 44).

The researchers for their part considered it important to allow for the entry of the viewpoints of women more openly into the discussion in the community (than had they not initiated and facilitated the focus groups). The researchers were concerned that traditionally women's voices have tended to be somewhat muted in discussions of "fair" practice. And springing from their involvement in the project, the researchers could make up their own minds as to activities that they may consider supportable for activist lobbying. For example, they still considered it relevant to continue to generate awareness of the possibility of employing the general law as a way of tackling what they felt to be the undue dependence of widows on their in-laws. And they still considered it relevant to lobby against an administrative system that failed to notify women when meetings that concerned them were to take place.

This does not mean that through the researchers' lobbying activities they closed options for those who might have wished to act otherwise. Indeed the research pointed to the need to open up arenas for other ways of addressing issues in the community. In short, the researchers' own ideas for possible activities had to be tempered in the light of their engagement with alternative practical and moral considerations raised by others in the course of the research project.

In order to disseminate and discuss the information arising from the study, Aphane et al. organized a number of workshops. They were invited by non-governmental organizations (NGOs) to disseminate and discuss their findings in a workshop; and sometimes they organized their own workshops for community members (1993, p. 13). They also presented their final report on the study to the W&LSART.

6.3 EXPLORING ALTERNATIVE ASSESSMENTS OF THE EXAMPLE

6.3.1 A Positivist Assessment

6.3.1.1 Scientism: Doing Science

The positivist tenet of scientism suggests that ethnographic exploration can form a backdrop to scientific studies into the connection between variables in social life. Out of the process of exploring issues of concern to people in everyday life, variables referring to aspects of social reality can be isolated. These variables can then be defined clearly (in standardized fashion, so that we are aware of what is being referred to in each case) before considering if there is any connection (of correlation and/or causality) between them. For instance, in following up Aphane et al.'s exploration, researchers might decide to investigate whether widows are less likely to be able to attain access to inheritances arising from insurance policies taken out by companies for which their (deceased) spouses worked, than are widowers. There may exist a correlation between the variable of "gender" and that of "access to company inheritance" in the situation under study.

To organize such an investigation, records, questionnaire items, or interviews, could be utilized to locate the gender of people. And "access to company inheritance" would have to be defined clearly so that people using the term can recognize what is being referred to. It might be used to refer to the fact that the company actually hands over the insurance money to the surviving spouse (through the latter having a bank account of their own, for instance). Or alternatively, it could mean that the company makes the money available, leaving it to the person in question to negotiate with the family/in-laws as to where it will be placed and how it will be handled. Once we know what is meant by "access" on the part of the widow or widower, we can use the term in a scientific hypothesis.

Following this, whether or not a correlation exists between the variable of "gender" and of "access to company inheritance", could be studied by taking a sample of widows and widowers whose spouses worked for companies. The sample would have to be chosen in a way that allows us to make reasonable inferences in relation to widows and widowers in the area being studied (for example, Swaziland or some part thereof), with a view to testing the hypothesis.

From a positivist point of view, then, some potentially interesting avenues of investigation could be pursued springing from the initial exploration undertaken by Aphane et al. However, the study, as it stands, needs to be drawn into the arena of scientific study by locating some variables whose connection can be tested.

6.3.1.2 Phenomenalism: Organizing Measurement Validity

The positivist tenet of phenomenalism suggests that measurement validity should be designed into the process of using observation. Scientific observation cannot be equated with the way in which terms are used in everyday life, where they can be used with differing meanings in different contexts of use (by the same speaker) and can be understood differently by different people. The process of science requires that terms be used with more rigor. Because Aphane et al. tried to stick closely to the language use of participants, they did not concentrate sufficiently on organizing the measurement validity of their concepts. For instance, if one wants to find out whether women and men have asymmetrical access to company inheritances (as explained in the section above), one needs to know what the term "access" refers to in this case. It is unhelpful to use terms in a loose way if one wishes to organize a scientific study.

The non-scientific use of language is one of the things that mark the domain of everyday conversation. If scientists are to fulfil their research remit as scientists, they cannot merely mirror the way in which language is utilized in everyday life. They cannot use terms in a way that reflects convenience of use in a specific context of use (as is often the case in everyday life). The more rigor that is injected into the use of terms in the discourse of actors in social life, the more such discourse can be conducted in a rational way. This is one of the contributions that science can make in society.

6.3.1.3 Empiricism: Seeking Causal Connections through Data Analysis

Aphane et al.'s study was aimed at exploring issues connected with inheritance practice and law in Swaziland. In their report, Aphane et al. allude to certain patterns relating to differential inheritance outcomes for men and women. But we do not know whether their reported "results" should be taken as being a result of their initial preconceptions that they admit they brought to the project, or a result of the observations that they have undertaken. It appears that their fieldwork was directed primarily in terms of their prior belief that gender has an influence on the way in which inheritance-related issues are worked out in the community. So, for instance, in their observations of next-of-kin meetings the researchers observed that women tended to be non-assertive, and that this made a difference to the way in which the inheritances were administered. But this "observation" on the part of the researchers could have been a result of their initial idea that women tend to submit to male authority in Swazi society. Aphane et al. did not give sufficient indication of how their initial preconceptions that they brought to the study could be tested and revised in the course of the study. It is difficult, if not impossible, for us as audiences to determine the extent to which the researchers' preconceptions influenced their manner of observing (in all cases of their use of instruments of observation), and whether these were ever likely to be revised via the observations. The link between their data gathering and their testing of theory has not been well accounted for by the researchers.

6.3.1.4 Value Freedom: Finding Out about Reality

Aphane et al. acknowledge their predilection toward encouraging women's use of the law if need be (at times) in opposition to certain customary practices. Although Aphane et al. state that they wanted to use the research process to allow the participants to raise issues that were of concern to them, they also indicate that they themselves wished to use it to "bring change" at a personal and formal level (1993, p. 7). But we have no way of knowing how the researchers' values regarding the direction of desirable social change affected their way of conducting themselves in their exchanges with participants. They might, for instance, have conducted their in-depth interviews in such a way as to draw out views in keeping with these values. We do not know the extent to which subjects were influenced in their responses by what they felt to be the values of the researchers; and we also do not know to what extent the interpretations given by the researchers of the results of the various interviews are a good representation thereof.

As far as the "case studies" are concerned, the researchers indicate that they chose both harmonious and problematic cases to follow up — purportedly so as not to influence unduly their findings. But their way of seeing "harmony" and "problematic cases" could have still reflected their own initial biases. "Harmonious cases" could have meant to them ones where in-laws effectively allowed the women to keep control of the inheritance, while "problematic" ones may have been defined in terms of the researchers' sense that too much dependence was operative. The researchers' particular views of both harmonious and problematic outcomes, might have affected their way of directing their interactions with interviewees as they conducted the case studies, as well as their way of reporting the results.

Again, the same could be said for their involvement in focus groups. They admit that they raised issues sometimes in these groups so as to highlight areas of contention in the community. But we do not know from their write-up of the results the exact nature of their involvement in the focus group discussions. It is impossible to separate out their own views from those of the members of the focus groups. But this means that we do not know how far the information presented by the researchers (later in their dissemination workshops and in their final report) is a reflection of their own value-laden way of conducting, and recording the results of, the various exchanges.

6.3.1.5 Instrumental Knowledge: Offering Useful Knowledge

Aphane et al.'s research was avowedly "activist" in orientation (1993, p. 7). They consider it a strength of their study that they were, for instance, able at times to offer advice to interviewees during the interview processes in order to respond to their requests; that they were able sometimes to conduct a mediating role in their involvement in case studies; and that they took an active role in facilitating group discussions in the community. They indicate that in these ways they fulfilled one of the remits of their study — as they interpreted its purpose. They were acting in line with what they took to be the spirit of the Research Trust that was funding the project exploring women's relationship to the law in Southern Africa.

However, from a positivist point of view, their practical agenda overpowered the research agenda. Too much attention was given in the study to helping people to envisage options for change as they met with the various people during the study. As a result, it is difficult to know what was being found out about reality through the project. To fulfill the task of advancing knowledge that can function as a guide to informed action, it would have been better if the researchers had separated out their role as researchers from their perceived role as facilitators of change. They could then have given due attention to the research role and concentrated on collecting the information relevant for actors in their continuing action.

6.3.2 A Non-foundationalist Assessment

6.3.2.1 Scientism Qualified: Attending to the Effects of Human Perceptions in the Social World

Hammersley and Atkinson explore the implications of adopting a non-foundationalist position in relation to the doing of ethnographic research. They note that one of the issues that affects the scientific status of ethnographic, as well as other forms of research, is that "there is a sense in which all social research ... involves participating in the social world, in whatever role, and reflecting on the product of that participation" (1995, p. 17). But according to Hammersley and Atkinson, there is no need to draw the conclusion from this that the commitment to realism should be abandoned. This commitment can be expressed through the process of subjecting commonsense notions (in everyday life) to "systematic inquiry" (1995, p. 17). Ethnographic research, as one form of inquiry into the social world, can proceed by exploring the way in which the behavior and attitudes of members of society are influenced by contexts of social interaction. The fact that "people respond [also] to the presence of the researcher", can itself be exploited by ethnographers as an opportunity to explore human reactions (1995, p. 18).

Thus, in the case of Aphane et al.'s project, Hammersley and Atkinson might argue that if the respondents responded in a particular way to the fact that, say, the researchers were introduced to them via the Chief (or a representative) in the community, this tells us something about the social setting under study. The reactivity effect as it operated there can be used to understand something about the influence of the Chief in the context of the community — in terms of people's perceptions of his or her (normally *his*) status. Likewise, say, the fact that women might have become vocal in focus group meetings where a researcher was facilitating the discussion (against their traditional inclination to submission) says something about the way in which people respond to a sense of being "supported" in some way. It says something about the way in which participants' perception that they would be given some support (by a facilitator) in expressing views, might have affected their interactions with others. So, although it is possible that the views of the members of groups as mentioned in Aphane et al.'s report were offered partly in response to the known and felt presence of the researcher, this is not necessarily detrimental to the research effort to find out something about the operation of human life in the setting studied (and indeed in other settings with which it can be compared).

Hammersley and Atkinson also note that in the human world specific social dynamics might come into play by the way ethnographic (as other) research is reported. For this reason they remark that

> the contemporary ethnographer ... cannot remain innocent about the conventions of ethnographic reporting. .... *How* we write about the social world is of fundamental importance to our own and others' interpretations of it. (1995, p. 255)

Referring to Lofland's (1974) advice concerning the writing up of ethnographic research, they suggest that:

> It is ... important that the successful text demonstrate how existing ideas are being developed, tested, modified, or extended. Equally, the reader looks for how the evidence cited in the ethnography bears on such conceptual development. .... The analytic claims need to be "grounded" or anchored in the particularities of observed social life. (1995, p. 257)

Hammersley and Atkinson propose that ethnographers need to develop the skill of striking a balance between "the concrete and the analytic, the empirical and the theoretical" (1995, p. 257). Readers can "evaluate the adequacy of the textual representation" in these terms (1995, p. 257). However, Hammersley and Atkinson note that audiences may differ in the "expectations they bring to the ethnographic text". It is never possible to tailor ethnographic writing to "match the interests of all potential audiences simultaneously" (1995, p. 259). But ethnographers should at least "take account of the fact that our monographs, papers, and more popular texts may be read by our hosts or informants themselves" (as one audience of readers) (1995, p. 261).

Applying these ideas to Aphane et al.'s undertaking, it might be suggested that as long as Aphane et al. made efforts to develop an understanding moving between the concrete and the analytic, their insights might help people in developing their own conceptions. This understanding may then feed back into the way in which people practice their continuing relationships in their own life settings. As far as researchers' accountability to more academic audiences is concerned, Hammersley and Atkinson suggest that the expectations of more "academically" inclined audiences can be catered for by, for instance, using particular studies to create theoretical discussion around issues debated in the literature. As long as the write-up can be argued to offer insights into social life in the reported setting, as well as offer clues to their relevance for understanding other settings, it can be justified in scientific terms.

6.3.2.2 Phenomenalism Reviewed: Improving Instruments for Observation

Hammersley and Atkinson point out that many ethnographers see ethnographic research as being geared to attaining some fidelity to the phenomena under study by trying to get close to the natural settings of ordinary human interaction. Hammersley and Atkinson indicate that in such views of ethnography, fidelity to the phenomena sometimes implies that researchers try to

> interpret the world in the same way as they [the people being studied] do, and thereby learn to understand their behavior in a different way to that in which natural scientists set about understanding the behavior of physical phenomena. (1995, p. 8)

Hammersley and Atkinson comment that a non-foundationalist position (such as their own) does not make the assumption that the task of social research is to represent social phenomena in some "literal fashion" (1995, p. 11). For them, the aim of ethnographic, as all, social research, is to produce some account of the phenomena that can be justified in scientific terms as researchers develop concepts that move between the empirical and the abstract (see Section 6.3.2.1). Ethnographic research should be geared toward making a contribution through the "development of a set of analytic categories that capture the relevant aspects of the data, and the assignment of particular items of data to those categories" (1995, pp. 208-209). As Hammersley and Atkinson note, data in ethnographic research are not "already structured in terms of a finite set of analytic categories ... in the way that most survey research data are" (1995, p. 208).

Aphane et al. might be criticized in these terms from a non-foundationalist point of view for not developing sufficiently clear concepts to structure the data. For example, Aphane et al.'s suggestion that women are disadvantaged in inheritance issues, could have been clarified by their offering empirical instances of the concept of disadvantage and by showing in what sense the empirical material was being used to find patterns in the data. Whether the proposed analytic categories are identified by members studied, or are "observer-identified" (that is, identified by the researcher), what is crucial is that they become "developed and tested" with reference to the data (1995, p. 209).

6.3.2.3 Empiricism Reviewed: Creating Theoretical Explanations Grounded in Observations

Hammersley and Atkinson suggest that theoretical explanations can be developed through ethnographic research by the researchers'

> look[ing] to see whether any interesting patterns can be identified; whether anything stands out as surprising or puzzling; how the data relate to what one might have expected on the basis of common-sense knowledge, official accounts, or previous theory ... . (1995, p. 210)

Aphane et al. have provided some indication of the pre-understanding (derived from their own life experiences) that they brought to the study. For instance, they believe that while supposedly being gender neutral, the laws relating to inheritance are disadvantageous to widows in particular; and they believe that they affect some women more adversely than others. But what is not clear from Aphane et al.'s account is the extent to which their "commonsense" ideas and prior understandings about inheritance law and practice in Swazi society, were meant to be tested in the study. We also do not know what stood out for the researchers as specifically surprising — calling on them to review their initial ideas.

From a non-foundationalist point of view, some of the work that Aphane et al. did could perhaps be converted into creating conjectures about patterns in the discerned data. It is possible that the rich empirical material generated by Aphane et al. can be used to develop and test some theoretical ideas. But Aphane et al. did not themselves appear to proceed in terms of the scientific requirement to create and test theoretical statements in the light of the evidence gathered.

6.3.2.4 Value Freedom: Seeking Knowledge through Operating Self- and Collegial Criticism

Aphane et al. indicate that they operated in terms of the "perspective" that women should be encouraged to utilize the general law to aid them in addressing their "inheritance problems" (1993, p. 6). On their own admission, this perspective incorporated a value-orientation that they brought to the study. From a non-foundationalist point of view, as long as this affected only their *choice of topic* (that is, a study into the experience of inheritance in Swazi society) this would be acceptable. But the researchers would have to take special care not to allow their own value-orientation to affect features of the research such as: the kinds of samples they drew, the way they conducted their various observations, the way they probed interviewees when the latter offered responses that did not appear to fit the researchers' preconceptions, and the way they facilitated the focus groups. Otherwise it is possible that the understanding that they developed ostensibly through the study, simply reflects their own values that they brought to it in the first place.

It would have been easier for us as readers to judge their objectivity if they had included some checks on their interpretations. For instance, they could have asked those being "observed" in the next-of-kin meetings — including the women, their family, in-laws, and the administrators — to comment on the researchers' interpretations of what had been going on. Or again, they could have set aside time to check with individuals and groups interviewed, whether the researchers' proposed interpretations of the interactions/exchanges seem credible. As noted by Walsh (1998, pp. 231-232), "Hammersley advocates a subtle form of realism which comes close to the Popperian argument that validity involves confidence in our knowledge but not certainty". To increase our "confidence" in the researchers' storyline, Walsh suggests that it is good practice, in ethnographic research, to try to create the opportunity of asking those involved in the study to comment on the way the researchers have developed their interpretations. Aphane et al. did not appear to follow this practice. While this practice is not obligatory for ethnographic researchers, Aphane et al. could at least have indicated how else they have provided for self- and collegial criticism of their interpretations of material gathered.

6.3.2.5 Instrumental Knowledge: Providing Opportunities for Better-informed Action

Hammersley and Atkinson issue a caution to ethnographic researchers to be aware that the charge of "exploitation" can easily become leveled against them as they become involved with participants in collecting data from them. They note how, in a particular case of a researcher (Beynon, 1983) studying teachers in a school, one of the teachers commented: "When you first arrived we thought 'Here's another bloke getting a degree on our backs!' We resented the idea that we were just fodder for research" (1995, p. 273). Hammersley and Atkinson suggest that one way of addressing this ethical issue, is for researchers to try to create conditions for those being studied to derive some benefit for themselves springing out of the research. So, for instance, researchers might at times adopt the role of "citizen" or "person" deciding to help the people with whom they come into contact (1995, p. 286). As Hammersley and Atkinson note:

> There is in fact always much action engaged in by ethnographers in the field that is not directly concerned with knowledge production. By its very nature, ethnography forces one into relationships with the people being studied, and one may do things because of these relationships, over and above any connection they have with the research. (1995, p. 286)

But although Hammersley and Atkinson concede the need for ethnographic researchers to bear in mind "the ethical aspects of their work" (1995, p. 286), they state that they do not as a rule endorse the idea of researchers' roles as citizens overriding their research identity (1995, p. 287). As they indicate:

> Most of the time, in our view, then, the temptation to abandon the researcher role should be resisted. Certainly, we have little sympathy with attempts to redefine that role to make the researcher into some sort of political activist. (1995, p. 287)

According to Hammersley and Atkinson, during the research process researchers should be primarily concerned with producing knowledge. If they do indeed orient themselves largely toward producing knowledge, the information that they develop can be utilized at any time — by the hosts themselves or by others — toward better-informed action. It is in this way that benefits can be derived from the contributions of researchers (as researchers).

Hence, Hammersley and Atkinson would meet with little sympathy Aphane et al.'s explicit statement that their research was "activist" in orientation. The decision to orient the research in terms of the aim to "bring change" at a personal level as well as at the formal level, would be considered as likely to have threatened the researchers' capacities to advance knowledge about the realities. It would have been better if they could have more clearly separated out their role as researchers from their role as activists in their own minds — so that they could more easily fulfil their role as researchers attempting to generate knowledge. This clarity, in turn, would allow others to better decide how they can make use of the results of the study.

6.3.3 A Scientific Realist Assessment

6.3.3.1 Scientism Qualified: Attending to the Reality of Generative Mechanisms, Including Societal Ones

From a scientific realist point of view, Aphane et al.'s study should have been organized so as to be able to advance some knowledge about the generative causality lying behind the occurrences that they observed. They could have explored, for instance, ways in which the system of addressing inheritances (within customary practice and the general law) might reveal unwritten and unobservable power structures in the society. The structures of power may not be observable as such, but would require making inferences from specific cases examined. Aphane et al. might have explored the workings of power in, say, the way in which particular role bearers such as chiefs/administrators/judges created their judgments in relation to inheritance cases with which they were dealing. The hidden structures of power operative in the culture might also explain something about how such judgments became dealt with by those affected (surviving spouses) in the community. Aphane et al.'s study could be extended also to examine the way in which traditional practices as well as the formal legal system are linked up with other ways of structuring social relationships in the society.

Pawson and Tilley aver that those being studied should not be treated as "experts" on all aspects of the research topic. While they may know best about the way in which their own reasoning and choices have been constructed, they cannot know the "broader currents" of which their views are merely a part. As Pawson and Tilley put it:

> On the realist model, data collection is thus charged — *not* with the descriptively finite task of capturing the stakeholder's ideas, beliefs, hopes, aspirations ... but with the task of demonstrating which aspects of these beliefs are relevant to the [scientific] theory under test, so that the respondent can contribute to that test. (1997, p. 164)

The actions and views of respondents/subjects investigated in the research done by Aphane et al. should thus have been excavated in terms of an attempt to aid the development of scientific theorizing about operative mechanisms in the different contexts of interaction being studied.

6.3.3.2 Phenomenalism Reviewed: Developing Acceptable Observations

From a scientific realist point of view, theoretical preconceptions cannot be excluded from the process of observing empirical occurrences. So, for instance, Aphane et al.'s observation of the submission of women in next-of-kin meetings, would be seen as being partly a function of their preconceptions regarding women's experience of male authority in Swazi society. Or again, the researchers' observations of the way in which women are made to feel guilty about their decision to draw on the law to solve their inheritance problems, would be seen as a function partly of the researchers' prior views about guilt as a mechanism of social sanction.

According to scientific realism, it is necessary to attempt to set up some dialogue within the research community around different ways in which observations might be conceptualized. Hence, for instance, Aphane et al. could consider literature on the identification of submissive behavior in human settings; and they could also examine literature around the construction of the concept of guilt. This would mean that they could tie their own constructions into more general literature (and debates therein). As it stands, they rely on certain conceptual devices to organize their observations, and they seem to assume that others will accept them. This assumption needs to be argued for in relation to wider literature on the topics in question.

6.3.3.3 Empiricism Reviewed: Working at an Analytic Level to Understand Generative Mechanisms in Context

When discussing the contribution to knowledge that may be made through case study research, Pawson and Tilley indicate that we can "move from one case [setting] to another, not because *they* are descriptively similar, but because *we* have ideas that can encompass them both (1997, p. 119). Pawson and Tilley here draw on the distinction suggested by Sayer (and others) between quantitative generalization and qualitative abstraction (Sayer, 1984) — a distinction that equates roughly with Yin's proposed one between statistical and analytic generalization (referred to in Section 6.1).

Pawson and Tilley explain that the archetypal case of a statistical mode of generalization is "population sampling" — where an attempt is made to consider "how *representative* a particular sample is of the population from which it is drawn, in order to make generalizations back to that population" (1997, p. 119). It is based on the idea of taking "a sample of subjects ... from a finite population" in order to make claims about the population (1997, p. 119). In contrast, the "analytic" or "logical" type of generalization relies more on the skill of abstraction. In developing this skill, Pawson and Tilley suggest that researchers should avoid concentrating too much on descriptive particulars of the cases being examined. Such particulars, they argue, cannot be "cashed in cumulatively" (1997, p. 119).

In terms of these concerns, the question that Pawson and Tilley would pose to Aphane et al. is whether their study concentrated too much on the particulars of the observations on which they reported, at the expense of contributing to the cumulative development of theory about the way generative mechanisms can trigger effects in different social contexts. They have left it too much to the audiences of their study to try to identify the effects of unobservable mechanisms — such as the unobservable power that can work itself into the way that, say, widows are made to feel guilty if they do not concede to traditional obligations to in-laws. Potential readers are left to consider the way in which abstract concepts relating to these mechanisms might be developed. As scientists, however, Aphane et al. should have offered more theoretical pathways for aiding such reasoning.

6.3.3.4 *Value Freedom: Considering Possible Reformist Allegiances and if Necessary Re-organizing One's Focus*

Aphane et al. organized their study according to their conception that it was worthwhile to explore experiences (largely of women) in regard to possible inheritance-related problems in the context of social relationships (including the legal system) in Swazi society. They admittedly hoped to use the study to encourage certain changes — at a personal and formal level. A scientific realist argument would not take issue with Aphane et al. for expressing some allegiance to those whom they believed might be unduly disadvantaged within the structures of the society. However, it would be emphasized that Aphane et al. should not allow this allegiance to affect the way in which they create inferences from the data so as to excavate the structuring of the society. As a research project, the study needs to be accounted for in terms of what it can offer toward the furthering of *scientific understanding;* it cannot be justified with reference to the practical "help" that Aphane et al. might have been able to render people in the course of, and as a result of, the study undertaken.

Moreover, a scientific realist argument would wish to alert Aphane et al. as well as others concerned, to the fact that the aim of helping people can become too focused around issues of immediate concern to those studied, to the detriment of deeper considerations. Aphane et al.'s study showed that use of the general law was one option available for people trying to gain access to inheritances. It also showed up the tensions felt by certain women in deciding whether to try to access their legal entitlements under general law. But it did not draw together the various pieces of information developed in the study, toward a consideration of what kinds of radical changes might be needed to generate fundamentally different patterns in the structuring of inheritance outcomes in the society. Aphane et al.'s undertaking could have been improved if they had focused on exploring the wider structures in which existing inheritance patterns are embedded. As it happens, Aphane et al.'s allegiance to helping people with their inheritance problems, might have affected their theoretical ability to really delve into the structural causes of what was experienced as problematic.

6.3.3.5 *Instrumental knowledge: Extending Information about (Inter Alia) Social Structures*

From a scientific realist point of view, the kinds of practical input made by Aphane et al. in the process of conducting their study cannot easily be considered as being part of the research remit as such. In line with the concern of non-foundationalism about the blurring of ethnographers' roles as researchers and as citizens, a similar argument would be applied here: researchers need to address their social accountabilities in their role as *researchers*. However, if they believe that participation in action (to aid certain participants) will help them to gain increased knowledge of social structures (including knowledge of the constraints set by these structures on people's choices of action), then the action is warranted from a scientific point of view. If not, they must be careful of incorporating an activist component into the research role. As researchers, they should respect the demand on them to *advance knowledge about the structuring of the society (to the best of their ability).*

6.3.4 An Interpretivist Assessment

6.3.4.1 Scientism Opposed: Doing Social *Science*

From an interpretivist point of view, Aphane et al.'s research efforts would be appreciated because of the way in which they built up their understanding with close reference to a consideration of the variegated meanings underpinning people's interactions. For example, from Aphane et al.'s study we can see that the meaning of, say, "dependence" on in-laws could be interpreted differently by the different parties concerned. Some people considered customary ways of in-laws' creating access to the widow's inheritance, as a sign that women were being cared for. Some women accepted that it might be understood in the community as such, but felt that they preferred more immediate access to the inheritance. Others felt the hostility of in-laws, and they saw dependence on them as a constraint that needed to be addressed.

Aphane et al. tried to develop their understanding around the topic through the development of a human tie with the involved participants. However, it is also possible that those being studied were responding to the experience of interacting with the researchers — through whatever technique was being employed at that point in time by the researcher in question. This could have occurred in any of the situations in which those being studied were aware of the presence of the researcher. Aphane et al. could have given more attention to this. That is, they could have used the various encounters with participants as an opportunity to reflect, with them, upon the social contexts of interaction that they were setting up.

6.3.4.2 Phenomenalism Re-interpreted: Creating Meaningful Categories

From an interpretivist point of view, the credibility of Aphane et al.'s interpretations can be assessed in terms of the way in which they took into account participants' own meaning-making as the basis for the creation of meaningful categories. Aphane et al. appear to have developed their stories "about" the subjects through setting up different forms of dialogue with them. They also relied on intersubjective checking between the researchers in the research team to develop their understandings. And they carried information across the various research contexts, in order to increase their depth of understanding.[45] Aphane et al. did not concentrate on showing us what they did when their own interpretation of people's activities seemed to go beyond the self-understanding of these people. For instance, Aphane et al.'s concept of "allegiance" (to custom) provides a way of conceptualizing the expressions of some of the participants, while at the same time tying these expressions to the concept of "guilt" that they invoked in their own discussion. But they did not concentrate on exploring the juxtapositions, if any, between their own and participants' understandings. They could

[45]This use of a variety of techniques is sometimes interpreted as being in the spirit of qualitative research, conceived as a process of striving to develop "richness" of understanding. Bryman notes that research can be called "predominantly qualitative" even when some research instruments in the study are used in a structured context — such as, say, structured or semi-structured interviewing contexts (1997, p. 71). This is provided that the aim in using them is not to develop "precise calculations of relationships between variables" (1997, p. 73).

have left more of a paper-trail of documentation about how they were building up their categories arising out of their listening to the various participants — so that we as readers would be in a better position to judge the relevance of these categories (cf. Henwood & Pidgeon, 1993, p. 25).

6.3.4.3 Empiricism Re-interpreted: Developing Understanding of the Social World

Aphane et al. directed their efforts at building up a rich understanding of the phenomena relating to inheritance practice and law in Swaziland. Through their utilization of a variety of techniques, and through their sampling decisions that allowed them access to a variety of social situations, they explored the dynamics of interactions of widows and widowers with administrators, the spouses' families, and others in the community.

Through their examination of this variety, Aphane et al. were able to create what Strauss and Corbin call "conceptual density", while not "overdoing attempts to develop density" (1990, p. 109). Their continuing investigations were aimed at adding variation to their use of concepts (such as the concepts of, say, "customary practice" or of "legal entitlement"), so that they could be recognized to have different meanings in different contexts of use and for different participants. Furthermore, their understandings could be argued, potentially, to offer some insight into the experience of social life in other settings. Aphane et al. made some implicit comparisons with what they suggested to be similar settings — in which people might feel some sense of allegiance to customary practices, while also experiencing the pull to act against them. How these dilemmas felt by people might work themselves out, could not be generalized in the sense of stating the conditions under which certain, rather than other, outcomes are likely to occur. The "conditions" depend on the specific ways in which participants enter into relationships with one another as they work around what they define as cultural expectations. But although the researchers could not generalize, what they could do was offer an indication of ways in which their categorizations might aid the understanding of different settings. Seen in this way, Aphane et al. could be argued to have satisfied (to some extent) Henwood and Pidgeon's proposal for researchers to develop contextually sensitive theorizing as an aid to those wishing to consider its transferability to other social situations (1993, p. 27).

6.3.4.4 Value Freedom: Paying Attention to Others' Worlds

Henwood and Pidgeon advise that researchers should reflect upon values and assumptions that might influence their research, so that they can become open to appreciate information that does not tally with these (1993, p. 25). This includes developing an appreciation of others' beliefs and values as one studies the social world. Aphane et al. tried to lay open the preconceptions that they brought to the study. But it is not clear to what extent the research process was used as a forum for challenging their preconceptions (insofar as these were influencing their findings). As indicated above (Section 6.3.4.2) they could have left a more detailed paper-trail in regard to their way of conducting themselves in the various moments of the research. This would more easily allow for what Henwood and Pidgeon call an "external audit ... by immediate colleagues and more distant peers" (1993, p. 25). As it stands, the information that they supply in regard to their way of addressing their own values, is not detailed enough for us to undertake this audit.

6.3.4.5 Instrumental Knowledge: Creating Meaningful Information Relevant to Actors

Aphane et al.'s research can be seen as offering some information of utility to the hosts of the project as well as to other actors. For instance, they brought together different widows' experiences of how the process of using the general law affected their relations with in-laws and with others in the community; and they also offered information on how precedents of previous court cases could be differentially interpreted in the courts. If we can trust the researchers' sampling decisions, then the information that they supplied is of (instrumental) value to others in developing informed action.

Nevertheless, people (in future) should not treat the information as offering a view of reality as incapable of being re-created via their own interactions. So, for instance, just because in the past in-laws may have felt threatened by widows' actions, does not mean that "similar" actions on the part of widows will always be interpreted as such. Decisions taken by people have to be taken in the light of their knowledge that meanings can, and do, shift in the course of social interaction.

Aphane et al. do appear to have accomplished the task of offering information about some of the dynamics surrounding various people's relation to customary practice and to the general law. And this information can be of use to others, as long as they do not interpret any of the information as referring to fixed patterns. An interpretivist perspective would probably wish to draw out more carefully the point that any information supplied by Aphane et al. has to be seen as resting on the way in which meaning is operative in the particular settings. This applies to the settings explored in the study, as well as to other ones.

6.3.5 A Critical Theoretical Assessment

6.3.5.1 Scientism Criticized: Avoiding the Transference of an Engineering Approach to the Study of Social Reality

Morrow comments that from the point of view of critical theory, it is crucial that the research strategies employed in the study of the social world are employed in recognition of the need to break with the "natural scientific notion of technical control" (and the attendant engineering approach to social life) (1994, p. 310). What renders social research nonscientistic, is the possibility of

> critical-emancipatory knowledge ... having decisive significance for fundamental social change because it involves the fundamental transformation of individual and collective identities through liberation from previous constraints on communication and self-understanding. (1994, p. 310)

Morrow comments that critical theoretical researchers still need to acknowledge the difficulties of fulfilling a transformative remit:

> The reconstructive project of contemporary critical social science is jeopardized at the outset because of its ambiguous relationship to the given sociohistorical context within which it seeks to speak: above all, deep-set doubts about the very historical possibility (and nature) of potential transformation. (1994, p. 311)

In terms of these considerations, Aphane et al.'s attempt at organizing social change can be characterized as being "jeopardized at the outset" because of the difficult nature of the task. Whether "reconstruction" — as rooted in revitalized forms of communication on a sustained basis — was developed via the study, is still an open question. From a critical theoretical point of view, the way in which Aphane et al. set about organizing "change" would have to be linked to facilitating the subjects themselves to explore the radical meaning of their involvement in any changes. Otherwise, as Morrow notes, "previous constraints on communication and on self-understanding" are likely to become perpetuated (1994, p. 310). Aphane et al. could have given us more information concerning ways in which subjects might have been seeing their own part in any changes being facilitated through the research process.

6.3.5.2 Phenomenalism Reconsidered: Creating Realities through the Research Process

From a critical theoretical point of view, the encouragement of participants to re-envisage (and hence experience anew) "the phenomena" being studied is not unacceptable practice on the part of researchers. It is only unacceptable if the communication set up between "researchers" and "researched" is a one-way process. However, if researchers can argue that they are initiating a dialogical process through interrogating the views held by people and taking on challenge in return, this can be regarded as acceptable practice. Views exchanged could allow the different parties (including the researchers) to reflect on their initial ways of seeing "the phenomena" being explored — such as, in this case, the nature of "customary practice", the experience of "dependence on in-laws", the meaning of "access to inheritance", etc. Aphane et al. should ideally have contributed to enabling people to review their ways of experiencing the phenomena — and their attendant actions — in the light of their engagement in the communications that were set up.

6.3.5.3 Empiricism Reconsidered: "Discovering" Social Potentialities

Critical theory would appreciate Aphane et al.'s admission that their "finding out" about the social realities investigated was at the same time a move toward creating certain possibilities. They were aware that they were trying to activate possibilities in terms of a normative intention to organize transformation on a personal and formal level. As Morrow notes, there is no methodological recipe for deciding what methods can be used or how they can be used to develop transformative potential in society. So, he notes, "choices about linking theories and methods are an ongoing process that is contextually bound, not a ... decision that can be taken for granted through reference to the 'logic of science'" (1994, p. 228). Aphane et al. pursued their own approach by trying to use various techniques (on their own and linked up with other ones) to create some kind of dialogue around questions raised through the study. But they have not given us (as audiences) an indication of the difficulties, as they experienced them, of setting up genuinely communicative encounters around issues of concern. They could have offered more of a theoretical discussion around the difficulties involved in releasing the potential for change in the society.

6.3.5.4 Value Freedom Undercut: Recognizing the Value of Human Discourse

Aphane et al. offer an indication of the specific values that they brought to the study. They valued, for instance, an increased independence of women from in-laws; and they valued the opportunity for people to make use of the general law if need be in inheritance dispensations. These values they admit, might have affected the way they went about the study. But more importantly, they also indicate that they were committed to the *general value* of people respecting, while also interrogating, the viewpoints that others might wish to express around issues of concern. Aphane et al. were committed to enabling/facilitating people in the community to listen to, and address, one another's viewpoints (including those brought to the project by the researchers). This was done partly through the way in which they carried information across various research contexts. And efforts at cultivating dialogue were also made directly through the way in which they set up focus groups to facilitate discussion in the community. As long as the researchers can be argued to have allowed for interrogation of their own specific values brought to the study (and their accompanying viewpoints), critical theorists might consider Aphane et al.'s way of handling the project as a viable way to proceed.

6.3.5.5 Instrumental Knowledge Undercut: Furthering Communicative Understanding

Morrow notes that apart from questions concerning the way in which researchers handle issues of confidentiality, invasion of privacy, etc., as they relate to participants and report on their stories, there are the "larger ideological implications of research" to consider (1994, p. 235). A question of concern is whether in their pose as "scientists" researchers submit to the ideology of science and technology, in which the idea is perpetuated that science can offer expert understanding as an informed guide to action (see also Romm, 2001).

Aphane et al. were aware of what Morrow calls the "politics of knowledge" (1994, p. 318). What they offered by way of setting up the research process and by way of providing some interpretation of "findings" was not presented in the guise of possessing some special authority. Aphane et al.'s style of interacting with participants and of presenting results in workshop discussion (and in the report) will presumably have militated against such an understanding of their project. This implies that their work can be treated as not inhibiting the possibility for people to develop choices of vision and of action through a continuing discursive encounter around issues raised.

6.3.6 An Anti-foundationalist Feminist Assessment

6.3.6.1 Scientism Challenged: Overturning Traditional Research Control

Aphane et al. operated in terms of the belief that the voices of women had traditionally been somewhat muted in discussions involving ways of working out inheritance dispensations. The researchers used the research project to highlight for attention certain aspects of the relationship between surviving spouses, families, in-laws, and others in the community — aspects that had hitherto been largely unspoken about. Mauthner notes, though, that participants (as well as researchers) may feel "reluctant and vulnerable to exposing emotional aspects of intimate relationships" (1998, p. 53). She indicates that the interview situation is a "semi-public" context for voicing what might be seen as "personal" matters. The way in which researchers handle the situation of possible tension between the "private" and "public" cannot be addressed merely by wondering whether a certain "invasion of privacy" is justified for the public good (that is, for the sake of generating knowledge for "the public" to use). It has to do with how issues previously considered too private to speak about, can become part of a public sphere of discussion and debate, so that participants can recognize the value of adding their voices to the debate (while also being open to interrogation).

Aphane et al.'s manner of proceeding can be justified from an anti-foundationalist feminist point of view if participants (and others in the community — such as those attending the dissemination workshops) did not feel exploited by the way in which "information" was being developed and used by the researchers. Yet, as Alldred notes, "ethnographic research raises questions about how much to listen and how much to interpret" (1998, p. 162). It is possible that researchers may wish to engage in cultural politics, using the research to "intervene on particular issues" (1998, p. 163). They may feel that culturally dominant meanings are being perpetuated by participants' words, and that these should not be "re-presented uncritically" (1998, p. 163). Alldred suggests that researchers can, and should, add their own voice (analysis) — guided by the attempt to ward off meanings that they believe to be oppressive. She suggests that the taboo on "speaking of politics" in the research process, needs to be broken (1998, p. 163). Researchers can legitimately confront and challenge what they take to be oppressive meanings (and practices) as part of their own political practice.

Aphane et al. intended their work to be "activist" in the sense of offering an intervention in the social world on the side of "fair" discussion around issues of concern raised during the course of the project. Their admission of their political agenda would be considered from an anti-foundationalist feminist perspective as acceptable — as long as their own analyses, and suggested interventions, did not close options for others' visions and actions.

6.3.6.2 Phenomenalism Reconsidered: Working with Ambiguity

Aphane et al.'s research was focused around appreciating people's experience of ambiguities in their social world. For instance, they highlighted ambiguities springing from people's experience of uncertainty in regard to how precedents would be interpreted in the courts. And they highlighted ambiguities that were created through, for example, widows' not knowing how continued interactions with in-laws might pan out during the time at which they were involved in utilizing the general law. Furthermore, changes induced by the money economy could be experienced both as creating opportunities for certain people, but also as detrimental to the traditional hold of the morality of "care" in the community.

The researchers organized ways of interacting with participants so that they could extend their appreciation of their felt dilemmas, in relation also to stories that were told by others (including those brought to the individual and group interviews by the researchers). Focus groups in particular became an opportunity for people to discuss the ambiguity of customary practices in purportedly offering protection for women. The researchers did not try to move people toward achieving clarity of definition of the terms used in the course of the discussion. They rather chose to see the focus groups as a site to facilitate participants to enrich their understanding of issues faced — by reconsidering what was "at issue". In this way the researchers arguably found it possible to work with people's experience of ambiguity in their social world, while helping them to recognize the different angles from which they might address what they or others raised as issues of concern.

6.3.6.3 Empiricism Reconsidered: Appreciating the Situatedness of Research

In organizing the project Aphane et al. had to rely on their personal judgments (aided by regular team discussions) concerning a variety of choices, such as: what issues to follow up with participants in different contexts; how they could carry information across different contexts; approaches to be used in questioning participants; ways of facilitating group discussions, etc. They became what Gummesson (1991, p. 153) calls the primary "instrument" in the collection of the data. (See also Wolcott, 1995, p. 160.) From an anti-foundationalist feminist point of view this, however, is not a weakness that should have been corrected by trying to standardize the "research instrument" (through asking researchers to depersonalize their involvement in the project). Rather, it can be taken as a strength — as long as the researchers' making of decisions in the various contexts of interaction can be made visible by their way of accounting for them (Denzin, 1997, p. 8).

Aphane et al. could have provided more detail (in their report) on how they were indeed choosing to engage with participants in different contexts of interaction. They could also have given some indication of how (they believe) participants were choosing to interact with them in various contexts. (They could indeed have explored with participants alternative ways in which their interactions with them could be conceptualized, and practiced.) If Aphane et al. had concentrated more on this, those involved in the project (and touched thereby) could gain a better sense of the situatedness of the findings that were being developed.

Notably, in the writing of their report, Aphane et al. seemed to shift between on the one hand speaking what Alldred calls the language of re-presentation (as if they were showing "fidelity" to existing patterns of meaning in the society and to their experienced consequences), and on the other hand shying away from re-presentational speech. Their shifting style of presentation could have been a function of their sense of differing audiences' expectations concerning how they should present "conclusions".

Aphane et al. could have offered more of an open discussion around how they conceived (and worked with, and around) what they took to be various audiences' expectations in regard to their portrayal of results.

6.3.6.4 Value Freedom Undercut: Appreciating the Value of Heterogeneity

As Aphane et al. proceeded with the project, they problematized certain aspects of "customary practice". For instance, they problematized the traditional idea that women needed to be cared for by others making life choices for them. They hoped to draw attention to, and at the same time evoke, different conceptions of the notion of "care" around issues of handling inheritances. From an anti-foundationalist point of view, their commitment to evoking different visions of care can be defended in moral terms. Moral discussion must make provision for different ways of addressing the concepts used within social discourse. At least the project was able to draw attention to (and make more visible) contention surrounding the traditional view that women needed to be cared for by in-laws or other family members after the death of their spouse.

Nevertheless, Aphane et al.'s problematizing of what appeared to be a clear-cut traditional morality might have been taken even further. Their own acceptance of gender categories as constituting one of the key differentiating factors in the society, could also be questioned. Anti-foundationalist feminists might warn that the researchers' continued differentiation of men and women into apparently fixed research categories, may contribute to a further reification of stereotypes within the community. Did the researchers unwittingly reinforce the process of stereotyping by the very way in which they raised issues as being of concern more to women than to men? One possible answer to this charge is that stereotypical consciousness in a community cannot merely be ignored, for its effects can penetrate the kinds of opportunities available to those being stereotyped. Yet, it might also be argued that for the researchers to operate reflexively, they would have to bear in mind the possible social implications of continuing to use categories of "men" and "women" as if these represent obvious social divisions. Use of these categories themselves could then be qualified by the researchers' making adjustments accordingly as the research process continued.

Ideally, in terms of anti-foundationalist feminist values, Aphane et al.'s way of dealing with the research project gave those concerned the opportunity to see themselves in a new light and to consider their continuing interactions as based on a respect for differences. These differences should not, however, be treated as necessarily gender-defined ones.

6.3.6.5 Instrumental Knowledge Undercut: Questioning the Power to Define Realities

As noted above, Aphane et al. worked in terms of certain categories of thought in orienting themselves within, and reporting upon, the research project. The question of concern to anti-foundationalist feminism is whether they can account for the way these might feed back into the society. There is no way of avoiding the use of categories. But use of them should not be accompanied by the assumption that they offer an "obvious" way of framing the realities. Aphane et al. tried to make provision for the ways in which different participants wished to frame issues of concern to them. However, as noted by

Grossberg, ethnography is a practice "in which the other is inscribed within, and explained by, the power of the ethnographer's language" (Grossberg, 1989, as cited in Alldred, 1998, p. 153). Alldred poses as a question whether it is ever possible to organize ethnographic research in such a way that the centrality of researchers (with the power of their language) can be challenged. She suggests that one way of making provision for such challenge is by making it clear that representation (that is, any way of depicting worlds) is an active, not merely reflective, practice (1998, p. 158). But Alldred also notes that research that is grant-dependent may not "lend itself to radical critique of the research enterprise" (1998, p. 159).

Research as traditionally understood is an endeavor aimed at offering an observer-derived analysis of what is "in reality". It appears that Aphane et al. were able to shift — at least to some extent — this conception of the research enterprise. The researchers considered that they had some leeway, in terms of their remit, to choose not to treat the research as an exercise of finding out information that, as such, could serve as a guide to "realistic" action (on their own and others' parts). As noted above, it appears that Aphane et al. were aware of this possible expectation (from various participants and potential audiences), and in some sense catered for it. But they also appeared to be aware of the need to exercise caution in imposing their own categories and their own "authoritative voice" (as Alldred expresses it, 1998, p. 160).

6.3.7 A Trusting Constructivist Assessment

6.3.7.1 Scientism Revisited: Developing Trusting Relationships through the Research Process

Aphane et al. created a research design that, arguably, allowed both researchers and participants to develop their understandings of issues of concern in relation to inheritance practice and the law, and that allowed for some discussion in the community around what was taken to be at issue.

Holstein and Gubrium (1995) suggest that what they call "active interviewing" is a way of doing social science that explicitly provides for interviewees to develop themselves as persons through the research process. They become enabled to see others, and themselves, in a new light, by "shifting narrative positions ... throughout the interview" (1995, p. 77). Holstein and Gubrium indicate that the "active interviewer" asks

> the respondent to address a topic from one point of view, then another [as a] ... way of activating the respondent's stock of knowledge, or exploring the various ways that the respondent attaches meaning to the phenomena under investigation. The contradictions and complexities that may emerge from positional shifts are rethought to signal alternative horizons and linkages. (1995, pp. 77-78)

In keeping with Holstein and Gubrium's conception of the utility of the active interview in re-examining and reconstituting meanings, a trusting constructivist position would welcome providing opportunities for people to learn that they do not have to adopt a univocal perspective on the issues discussed. Aphane et al.'s conduct in both individual and focus group interviewing contexts would therefore be judged in terms of its enabling interviewees (with researchers) to create additional horizons as a basis for their engagement in social life. From a trusting constructivist perspective, this would mean ideally setting up experiences where people could (re)consider the discursive character of their accountabilities.

6.3.7.2 Phenomenalism Reconsidered: Evoking and Defending Observations

Aphane et al. made use of a variety of techniques to collect and carry information (perspectives) across various social contexts. They recognized, though, that the manner in which they organized cross-fertilization of ideas, was a product of the specific way in which they carried out the project. Other researchers, they admit, might well have focused on different issues; and they might well have evoked different phenomena as they interacted with the various participants.

The credibility of the "data" produced by Aphane et al. springs from their indications of how the discussions progressed with participants (partly through their own input and direction), and of how they chose to assign analytic categories that at times went beyond the expressions of participants. However, Aphane et al. did not try to insist on offering only one interpretation of reality via their analyses. So, for instance, they concentrated on showing ambiguities in terms of people's experience of inheritance practice and the law; and they showed that their own ways of viewing, say, custom, in relation to the law, had to take account of these ambiguities. They wrote the report in such a way that readers can sense the uncertainty of participants' felt relationship to "the situation", and the impossibility of capturing this in a set of analytic tools. A trusting constructivist position would appreciate their attempts to defend their observations/interpretations (and their visions of possible workable actions) in a way that still leaves space for others to see and act otherwise.

6.3.7.3 Empiricism Reconsidered: Exploring Alternative Ways of Seeing and Using Evidence

As noted in Section 6.3.7.1, what can be called active interviewing was one way in which Aphane et al. involved themselves with participants in developing some of the empirical material. Holstein and Gubrium note that when "active interviewing" is used as a research technique, the question arises as to how one can make sense of the data developed:

> Analyzing data concerning interpretive practice is somewhat less "scientific" and somewhat more "artful" than conventional interview analysis. This does not mean, however, that analysis is any less rigorous; quite to the contrary, active interview data require attention and sensitivity to both process and substance. (1995, p. 79)

Holstein and Gubrium point out that the analysis of active interview material concentrates on the way in which interpretations are built up through the interviewing process. The way in which interviewees shift narratives and redefine realities from different angles in collaboration with the interviewer must be considered as being as important to the analysis of the interview data as the actual substance of what is said. From a trusting constructivist perspective, what is additionally important is that researchers are able to account for the way in which they add their own input into the interviewing process (by inviting participants to "shift narratives" in response to ideas raised by the researcher). Their way of offering input has to take account of the need to fairly engage with participants' expressions, while inviting them to extend their story ("narrative") by approaching issues from alternative angles. So, for instance, Aphane et al. raised questions that enabled widows in individual interviews to reconsider the character of their allegiance to their in-laws; or again, in the group interview situation, the researchers invited people to reconsider the traditional notion of "care" for women in the community. From a trusting constructivist position it is acceptable that the researchers at times offered input into the discussions. But to earn the trust of people, they would have to show that they were also engaging seriously with the viewpoints being expressed. Furthermore, their understandings offered in the dissemination workshops and in the report would have to include an account of their various engagements (and an account of how the material from different research contexts was being put together bearing these in mind).

As far as the scope of application of their research is concerned, Aphane et al. suggested that the narratives constructed through their sampling decisions could show up a variety of ways of seeing "the realities" of inheritance practice and the law in Swaziland. Their theoretical sampling can be defended in terms of the development of richness of understanding (in the sense of offering alternative narratives). Their research may also be judged as relevant to an understanding of other issues besides inheritance practice and the law in Swaziland, and as possibly relevant outside of Swaziland — depending on whether plausible comparisons can be set up by others engaged in discourse around the example.

6.3.7.4 Value Freedom Undercut: Encouraging Discursive Accountability as a Value

Aphane et al. admit that their study was directed partly in terms of their own value commitments. For example, they intervened at times in the focus groups to introduce issues that might otherwise have been left off the agenda and not been openly raised by participants. These were issues that (they believed) certain participants (for example, those feeling that their power may be threatened and/or those who were specifically traditionally-minded) may have preferred not to be explicitly raised. Aphane et al.'s presence also became part of the other research contexts in which they evoked data as they engaged with participants. Those who considered that their role as researchers was to report upon, rather than to aid the direction of, social conversation, might have felt some distrust with the way in which the researchers went about the study.

Aphane et al. could not hope that all participants would necessarily welcome their contribution. To cater for those mistrusting their way of proceeding, they may have decided to try to confine their involvement to playing the role of neutral reporters. But from a trusting constructivist point of view, any pretence of striving toward value freedom would not have constituted a resolution (and might still have played into the hands of certain participants). It is preferable, rather, for the researchers to accept the responsibility of opening up for discussion their own and others' accountabilities. A trusting constructivist position would suggest that Aphane et al. could have used the various forums of interaction with participants to express some of their own ideas about the potential role of researchers in aiding "fair" discussion around issues of concern. In this way, they could have used the research process to initiate a debate around their research role. By raising openly the question of how their own accountability is to be assessed, they might acclimatize people to the idea that it can indeed be judged in processes of discursive encounter.

6.3.7.5 Instrumental Knowledge Undercut: Tempering the Possible Self-fulfillment of "Results"

Aphane et al. tried to develop ways of organizing the study that, while directive in some sense in problematizing certain issues, also allowed for mediation between different views (and extension and enrichment of these). Despite their hope that their own contribution would be interpreted as being facilitators for discussion, it is possible that some of their constructions may have served to reinforce certain patterns of relationship in the community. For example (as noted in Section 6.3.6.4), their very use of gender categories to direct much of the study, and to express results, could have served to perpetuate a consciousness of gender differentiation. A trusting constructivist position would suggest that a way to account for themselves on this score, would have been for the researchers to raise for attention their own dilemmas in harnessing the power of language in a certain direction, while not wishing to foreclose alternative ways of seeing. Explicitness in regard to their own dilemmas is one way of tempering the force of any construction(s) advanced.

Aphane et al. seemed to switch in their write-up of the project between using realist language to portray their results and using a more constructivist-oriented genre to display the constructed character of the findings. From a trusting constructivist point of view, their writing through the latter genre could have been accompanied by an explanation of implications of this for the status of their conclusions. This might have gone contrary to some participants' expectations, and to some audiences' conceptions of the meaning of scientific reporting. Nevertheless, these expectations and conceptions could themselves have been re-examined through the way in which Aphane et al. developed explanations in their report of why they were writing in the genre(s) chosen.

6.3.8 Trusting Constructivism Summarized in Relation to Alternatives

I present below a summary of the alternative interpretations of the study as I have explored them, with a brief indication of how a trusting constructivist position might develop an argument in relation to each one.

6.3.8.1 Positivism and Trusting Constructivism

From the point of view of a *positivist* position, Aphane et al. used concepts too loosely for us to know exactly what was being investigated via the study. They should have operated more rigorously in defining terms more clearly; and they could also have been more alert to looking for patterns in the data. Had they been on the lookout for patterns, they could have then developed hypotheses to be tested either in this, or in some other, study. Their funders seem to have required of them to contribute to social change on personal and formal levels (or rather, the researchers interpreted their research remit as involving this, amongst other expectations). This could have posed a threat to the scientific character of their work insofar as they concentrated more on aiding people to explore their own life circumstances and choices involved, than on providing scientific analyses across the various cases. As researchers, they should have concentrated more on developing information that could tell us something about the patterning of inheritance dispensations across the society. Aphane et al. might have felt some conflict of expectation in terms of requirements for them to act accountably as researchers. On the one hand, there was the expectation that they should use the research to aid processes of personal and social transformation; on the other hand, they were aware of an expectation to attain some information about the realities in that society. According to the positivist perspective, their justification for themselves as researchers rests on their fulfillment of the latter expectation. In their role as researchers the task in hand was to develop information that could then be of use to anyone wishing to use it.

From a *trusting constructivist* point of view, however, Aphane et al.'s manner of addressing the experience of differing expectations concerning what the research might achieve, can be regarded as defensible. They made efforts to earn people's trust in their work during the research process by setting up encounters in which people could develop an enrichment of ways of seeing issues of concern to themselves and others. Nevertheless, as they engaged with people, and as they organized their dissemination workshops and their write-up of the report, they could have rendered more explicit their defense of the stance that they were adopting as researchers. In this way they might have contributed to developing different orientations in the attitudes of participants and other audiences toward the way in which social research can, indeed, be accounted for.

6.3.8.2 Non-foundationalism and Trusting Constructivism

From the point of view of a *non-foundationalist* position, Aphane et al. should have been clearer in their own minds concerning how they wished to approach people's commonsense knowledge, and how they wished to test their own commonsense knowledge that they brought to the project. If they had been clearer on this, they could have contributed better as researchers to sifting out what could be argued to be more accurate from less accurate conceptions of the relevant realities. Insofar as the "reactivity effect" was operative as Aphane et al. engaged with the participants, they could have used this as an opportunity to find out more about patterns of meaning construction in the community. Whether or not they could have reduced this effect, at least they could have commented better on its likely operation, and on what might reasonably be inferred from this about the responses of the participants. All in all, they did not provide sufficient indication of how their study was being conducted in view of the scientific task of developing, testing, and modifying specific hypotheses.

The *trusting constructivist* retort to this (non-foundationalist) interpretation of the study is that it assumes that the results from a well-organized study clearly should take cognitive precedence over the (untested) commonsense knowledge of participants. The trusting constructivist position challenges the idea that research must be geared to subjecting people's commonsense ideas to the test of rigorous scientific procedure, so that people's actions can be informed accordingly. Aphane et al. rightfully did not expect participants and others to place their trust in them to set about this task. The project in which Aphane et al. were engaged could just as well be conceptualized as a process of setting up a researching system for various people (including the researchers) to enrich their conceptualizations of what was at issue and how issues of concern could be approached.

6.3.8.3 Scientific Realism and Trusting Constructivism

Aphane et al. were clearly keen to pay close attention to the lifeworlds of those being studied. But as a result, according to scientific realism, they were unable to use the project toward developing knowledge about the structures that would explain the kinds of responses that they came across. Aphane et al. may have helped some participants to address issues of concern to them; and they may have helped them to appreciate others' points of view as a basis for their decision-making. Yet these actors are still left bereft of knowledge of the structures within which their own and others' actions and ideas are embedded. Aphane et al. did not set up the necessary tests to be able to advance knowledge of this sort.

The *trusting constructivist* position would remark that within scientific realist argumentation, scientists are called upon to develop their tests by way of making inferences from what *they* observe (including observations of people's meanings). Their engagement with participants thus becomes a matter of seeing their views as a factor that can be fed into tests that they set up as scientists in order to advance knowledge of "the realities". A trusting constructivist position prefers to accord trust to people (including scientists) by judging their manner of engaging with differing points of view and their manner of leaving space for alternative visions. This is not to say that Aphane et al. or other researchers should refrain from all attempts to undertake inquiries from a more structurally focused angle. But it *is* to question the way in which the status of these analyses become defined. And it *is* to suggest that even when researchers judge structural analyses as being of primary importance, provision has to be made for juxtaposing this focus with the views and experiences of others. Accounting for others' views should not be a matter of considering how they can be interpreted as contributing to some so-called "scientific" test.

6.3.8.4 Interpretivism and Trusting Constructivism

Aphane et al. did not orient their project toward studying connections between defined variables. They did not try to develop a set of definite meanings for terms that could become incorporated into hypothetical statements of such connections. They instead tried to show up the variety of meanings that might be attributed to terms such as, say, allegiance to custom, care for women, dependence on in-laws, etc. As they presented their results, they indicated that these were situation-dependent in the sense that they were grounded in the meanings created by participants in situ as they defined their relationships with one another. All this is to be appreciated. But what they should also have made clear is the way in which they managed, if at all, to test (challenge) their own values and preconceptions that they brought to the study. A paper-trail of ways in which they organized their relationship with participants, and of ways in which they made decisions about how to interpret participant responses, is a vital addition to the study, according to interpretivism.

In line with this interpretivist response, *trusting constructivism* likewise appreciates Aphane et al.'s efforts at staying "close" to the meaningful character of social life. However, the trusting constructivist position focuses more on the importance of Aphane et al.'s endeavors to use the research process to develop people's discursive capacities. For instance, the active interviewing technique that they employed can be seen as useful in itself as a way of setting up instances of communication around shifting narratives. Through this technique, participants were enabled to appreciate that narratives can shift and that with the adoption of different narratives different realities can come to the fore in their consciousness. Aphane et al.'s skills in using the technique of active interviewing in individual and group contexts, and their skills in organizing cross-fertilization of ideas across various social contexts, were important factors in contributing to the quality of their study. A defense of their project could be made by focusing on how they interacted with participants in various forums to encourage forms of meaning-making that could be regarded as discursively sensitive.

6.3.8.5 Critical Theory and Trusting Constructivism

From a critical theoretical point of view, Aphane et al.'s normative commitment to raise for discussion in the Swazi community the question of how fairness in inheritance dispensations might be approached, would be respected. Aphane et al. were concerned with investigating whether the general law can be considered as a workable route to generating fair arrangements, and how this related to certain customary practices in the community. Discussion of these issues was pursued in different forums –– facilitated by the involvement of the researchers, who brought certain (investigative) skills to the project. But to what extent their transformative remit was accomplished in the sense of having a sustainable impact in the community, is still largely an open question.

From a *trusting constructivist* point of view, it would be appreciated that Aphane et al. were oriented to developing "results" through setting up communicative channels in the setting studied. Their efforts would, however, be valued primarily for their way of aiding people (including themselves) to keep in consciousness the juxtaposition between alternative narratives for viewing and addressing issues of concern (as part of their discursive orientation). A trusting constructivist position would emphasize that Aphane et al.'s manner of organizing the research process — including their dissemination workshops and report writing — should be considered in terms of its potential contribution in cultivating such an orientation. But again, to what extent this process could be argued to have been activated, is largely an open question. Aphane et al. could have offered some detail on their view of the possibilities for developing "discursivity" in (various) people's orientations in what might be considered a sustained fashion.

6.3.8.6 Anti-foundationalist Feminism and Trusting Constructivism

Anti-foundationalist feminism would endorse Aphane et al.'s attempt to draw out ambiguities in people's conceptions of the social relationships in which they were involved. They worked with the idea that social "realities" were in the process of being made, and that both they (as researchers) and those involved in the project were contributing to their ongoing construction. Their stance in this regard is to be appreciated. But Aphane et al. could still have been more open with participants and with others (for example, those attending the dissemination workshops and those reading the report) about the status they were according the various constructions that emerged in the course of the investigations.

In similar vein, a *trusting constructivist* position would suggest that Aphane et al. could have brought to the fore in their discussions with others some differing conceptions of the place of social research in society. This could have been done, for example, during the course of the individual and group "interviewing" processes, as well as in the various workshop discussions around the findings. And the defense of their position could have then been discussed in the report in the light of the details of their engagement with different points of view on this.

6.4 CONCLUSION

Aphane et al.'s study was used in this chapter as an entry point to explore the role(s) that researchers might adopt as they involve themselves as ethnographers in the lives of participants. This was discussed from various angles.

Positivism and non-foundationalism both suggest that ethnographic researchers should be careful of reducing the possibilities for finding something out about the realities under investigation. If they become involved during the project in helping people to address issues of concern, this involvement should be kept separate in their minds from the research purpose. The point of the research is to get close enough to participants to be able to collect the relevant information from them. This information can then be used to test ideas about the patterning of the social relationships in the settings examined, and to consider their comparability with similar situations. It is herein that their accountabilities as researchers lie.

Scientific realism too suggests that insofar as researchers become involved in action to help people to address issues of concern, the point of this needs to be properly theorized. If the point is to try to gain some knowledge about the constraints on action that might be incurred as a result of structural mechanisms operative in the society, then the researchers' involvement in the arena of practical action is of scientific value. If not, it must be considered as an activity separate from scientific activity and not relevant as such to the ethnographic research remit.

From an interpretivist perspective, the advantage of ethnographic research lies in the opportunities it creates for getting closer to the natural settings of human interaction than do other styles of research. The quality of data is enhanced through the detailed involvement with participants afforded by ethnographic research, as opposed to the fleeting encounters associated with other styles of research. Avoidance of fleeting encounters (as far as possible) is important so that researchers can argue that their involvement with participants has shown due respect for their meaning-making. But again, this involvement is not to be confused with any extra activities in which they may engage — which are not part of their research remit to develop knowledge.

Within critical theoretical and anti-foundationalist feminist positions, judgments about the accountability of researchers depend on our being able to assess the quality of the relationship that they set up with participants, and potentially with others who may be touched by the project. The focus is on considering how research practices can be used as sites to exemplify "fair(er)" dealings in social life (how ever this becomes defined in processes of discourse). Following a similar line of thought, a trusting constructivist position would suggest that social research should be seen as an opportunity to explore and instantiate the development of (more) trusting relationships in social life. (Whether Aphane et al.'s project can appropriately be conceptualized in this way is hard to say, as they did not comment specifically on this in their report.)

An issue that was addressed in the chapter in relation to Aphane et al.'s project, was the question of how its funding should be conceptualized (in terms of this having a bearing on the way the study was done and results reported). From a positivist point of view, the fact that the funding body (in this case the W&LSART) might have wished the study to contribute to social change at some stage, should not have become a threat to the research task at hand — namely, to develop knowledge about the topic being investigated. Whether or not Aphane et al. felt the need or obligation to help participants in practical ways during their involvement with them, the investigation would be considered as of scientific value only insofar as it offered a way of developing knowledge of the realities being faced. The function of the dissemination workshops and of the report would then be to enable people to understand the implications of the findings in terms of their possible practical utility.

Non-foundationalists would emphasize that Aphane et al.'s mode of dissemination should have paid tribute to the tentative character of results developed. This does not mean to say, though, that Aphane et al. should have presented them as having the same status as untested "commonsense". Audiences (including presumably the funders of the project) would surely criticize the report if it merely echoed commonsense ideas steeped in untested prejudices. The researchers would need to show how their ideas in regard to topics of interest had been put to the test of evidence during the research process. (Or at least they would need to show how they have created openings for scientific testing to be undertaken through further research.)

From a scientific realist point of view, the study ran into the danger of embracing a reformist political agenda by focusing too much attention on people's immediate concerns as the basis for building up an account of life in the settings studied. A theoretical context needed to be set more clearly by the researchers through providing some indication of how evidence that they were collecting could be related back to structural forces operative in the society. Ways of people's acting to (possibly) address these forces could accordingly have become better considered.

From an interpretive point of view, however, it would be appreciated that the funders left their remit sufficiently open for Aphane et al. to define how they may approach the project so that it could generate a result that would probably be meaningful to the hosts (as well as to other audiences).

From a critical theoretical point of view, Aphane et al. confronted the difficult task of trying to aid participants to develop communicative potential in the community as a way of addressing located problems. Aphane et al. tried to facilitate discussion with a view to rendering operative some communicative testing of people's varying visions in relation to alternatives. They tried to set up some public debate around the issues being raised through the project — for example, by organizing various channels for cross-fertilization of ideas, including interview processes (on individual and group levels) that could become forums for nurturing dialogical orientations in semi-public and public contexts.

From an anti-foundationalist feminist point of view, Aphane et al.'s challenge was to define a role for themselves in which they could relate to others as facilitators of discussion, while also offering their own voices at times as part of the ongoing discussion. They could have been more open with people about their epistemological orientations underlying their interaction with them. Likewise, from a trusting constructivist point of view, it would be suggested that Aphane et al. could have raised (in different forums) their own concerns about the purpose of social research in the society, as part of an endeavor on their part toward (discursive) trust earning.

What is interesting from a trusting constructivist point of view is that, having chosen a (critical) ethnographic way of proceeding, Aphane et al. were concerned to set up processes for participants as well as themselves to engage in the interrogation of viewpoints expressed during the course of the project. The ideas that were created by Aphane et al. as they made sense of their interactions with participants were constructions that they created in relation to, but not entirely in accord with, the various expressions of participants. Their own constructions, of course, had to be accounted for. So, for instance, Aphane et al. needed to account for their view (articulated in the dissemination workshops and in the report) that some of the widows' "allegiance" to custom went hand-in-hand with a dependence that might be unnecessarily constraining for them. Aphane et al.'s account of the possibility of pitting custom against the general law, was that widows might take more charge of their affairs by operating outside of customary practice (with its associations of women being dependents who had to be cared for). Nevertheless, they had to give some acknowledgement to the possibility that their view of "taking charge" with which they were operating did not coalesce with others' views — for instance, with the views of some of the widows who conceptualized their options in terms of working within, but possibly at the same time modifying, customary practices.

Ideally, the analyses afforded by Aphane et al. should have operated at the level of considering possibilities for shifting people's sensitivities, including their own ones, through activating discussion around choices of seeing and acting (in relation to the area of concern). The range of application of the results could then be seen, from a trusting constructivist point of view, as constituted through the way people in the contexts studied or in other ones decide to treat them as reality-constructors (in the sense of enriching their conceptions of possibilities for addressing the situations of which they are part). As in the initial setting, so in others, their "relevance" has to be decided as a matter of people becoming involved in discursive engagement with emerging ideas, in order to develop a defensible position in relation to their own choices of seeing and acting.

7
Exploring Action Research

7.1 INTRODUCTION

In this chapter, I organize a discussion around a project labeled as "critically reflexive action research" (Weil, 1998). I use the example as a medium to explore different visions of what action research as a strategy of social inquiry might involve, and different conceptions of how those conducting such inquiry can be called to account. Rapoport suggests that action research can be characterized in terms of its aims

> to contribute both to the practical concerns of people in an immediate problematic situation and to the goals of social science by joint collaboration within a mutually acceptable ethical framework. (1970, p. 499)

He indicates that action research is defined through the manner in which the inquiry is organized to consider ways of addressing the "problematic situations" in which people as practitioners are immersed. Whyte, Greenwood, & Lazes indicate what is meant (for them) by involving practitioners in the inquiry process. They distinguish what they call participatory action research from "the conventional model of ... research" on the grounds that

> some of the people in the organization or community under study participate actively with the professional researcher throughout the research process from the initial design to the final presentation of results and discussion of their action implications. (1991, p. 20)

They suggest that in this form of research, the roles of those involved in the inquiry are shared out — with the contribution of "professional researchers" and other researchers/participants being negotiated in the process (see also Whyte, 1991, p. 98; Eden & Huxham, 1996, p. 81; Flood & Romm, 1996, p. 135; and Smaling, 1998, p. 2).

As pointed out by Rapoport, action research is also by definition geared to generating improvements in action in the arenas being investigated. In action research, as further elucidated by Checkland and Holwell, researchers (all those involved in the inquiry) immerse themselves "in a human situation and follow it along whatever path it takes as it unfolds through time. This means that the only certain object of research becomes the change process itself" (Checkland & Holwell, 1998, p. 11). The intention of the inquiry is, as McTaggart puts it, to "involve people in deliberately changing their own action in the light of collective reflection on the perceived shortcomings of current work [or other social] practices" (1997, p. 183). McTaggart's focus on collective learning in turn echoes Argyris and Schön's suggestion that action research encourages a way of working (in practice) against cultural climates that are prohibitive of the development of learning encounters in social life (1996, p. 29).

Issues given attention in this chapter revolve around considerations of how action research as a process of learning can be defended as a research endeavor. The discussion hereof is undertaken by entering debates around whether action research should be seen as offering a way of revisiting our inquiry practices by challenging efforts at "representing" reality. The example given in the chapter is of a form of action research that is indeed intended to evoke new possibilities for experiencing our relationship with the world. As the author of the example (Weil) indicates:

> CRAR [Critically Reflexive Action Research] does not aim to create one representation of reality, but, rather, the unraveling (and documentation) of multiple realities and rhetorics that are in mutual and simultaneous interaction. (1998, p. 58)

Nevertheless, I show — using this example as starting point (and following the format of previous chapters) — how different understandings of the proper conduct of action research, can still be forwarded.

7.2 ACTION RESEARCH WITHIN THE MANAGEMENT OF PUBLIC SECTOR SERVICES (WEIL, 1998)

7.2.1 The Organization of the Project

The co-researchers' investigations in this case were undertaken around the management of public sector services in the United Kingdom. Weil (1998) reports on her involvement with 10 senior managers from different organizations in this sector, guided by her aim to "cogenerate cultures and practices that support ongoing critically reflexive learning and inquiry" (1998, p. 39). She indicates that her own understanding that she brought to the project was that

> market-based approaches [to the management of organizations] have become the primary vehicle for so-called quality and cost control, often causing public service managers to feel driven by political pressures to achieve short-term gains at the cost of long-term improvements. (1998, p. 41)

She goes on to contend that

> "reforms" in each single service can ... divert us from addressing the wider systemic costs — social, organizational, financial, human — of valuing short-term outputs (achieving more for less) rather than working to multiple bottom lines. (1998, p. 41)

She notes that the kinds of management and organizational practices that are being put forward in some circles as "efficient and effective", are sometimes based on the claim that private-sector management (organized for profit) is "good" while public sector (not-for-profit) management is "bad". Nevertheless, she remarks that, paradoxically, the worldviews that underpin this claim "are beginning to be challenged by many successful private-sector companies" (Weil, 1998, p. 40).

She states that her initiation of critically reflexive action research (CRAR) in the case reported on, and other cases in which she has been involved, is set in a context where public sector managers are

> often under considerable pressure to treat the dilemmas they face as if they *are* problems that can be defined and solved — neatly and efficiently. At the same time, a reluctance to acknowledge dilemmas can be seen as a defensive reaction against the high levels of uncertainty and ambiguity that such a stance entails. This is all the more difficult in a cultural context where "correct" ... action, rather than reflection in and on action, is valued — and indeed rewarded. (1998, p. 57)

She expresses her concern that the introduction of top-down controlling strategies as a way of reducing complexity, may not be propitious

> to achieving social outcomes and managing multiple bottom lines. Instead, we may only escalate social costs: for example, morale and recruitment fall off, and early retirement, disaffection, ... and cynicism rise. Meanwhile, the immediate gaze remains on the short-term financial bottom line. (1998, p. 57)

Weil's instigation of CRAR in the case discussed here, was organized by making provision for what she calls "on-site" and "off-site" CRAR as "interweaving cycles of managerial and organizational learning" (1998, p. 39). She explains that on-site learning involves a process where those working in an organizational context undertake inquiries, "in collaboration with colleagues and others, such as service users, *within [their] own organizational system*" (1998, p. 46). Learning processes are set up in order to aid the consideration of possibilities for inquiry and action around people's currently felt dilemmas in their organizational settings. The learning (informed by CRAR) involves the following:

1. Appreciating starting problems/dilemmas;
2. Focusing/framing inquiry cycle in context;
3. Broadening/deepening understandings of inquiry focus/question;
4. Reframing/refocusing inquiry;
5. Planning for insightful actions (individual, group, cross group, organizational);
6. Critical reflection in/on actions;
7. Communicating/learning/checking outcomes.

From 7., the cycle of learning continues as people return to 1., 2., etc., to appreciate new emergent issues (1998, p. 48).

"On-site" cycles of reflection and action, can be supported by what Weil calls "off-site" CRAR. Weil explains that off-site CRAR "brings people together with others from a range of organizations to engage in cycles of CRAR inquiry" (1998, p. 45). In the case of the project involving the consortium of 10 senior managers, off-site meetings

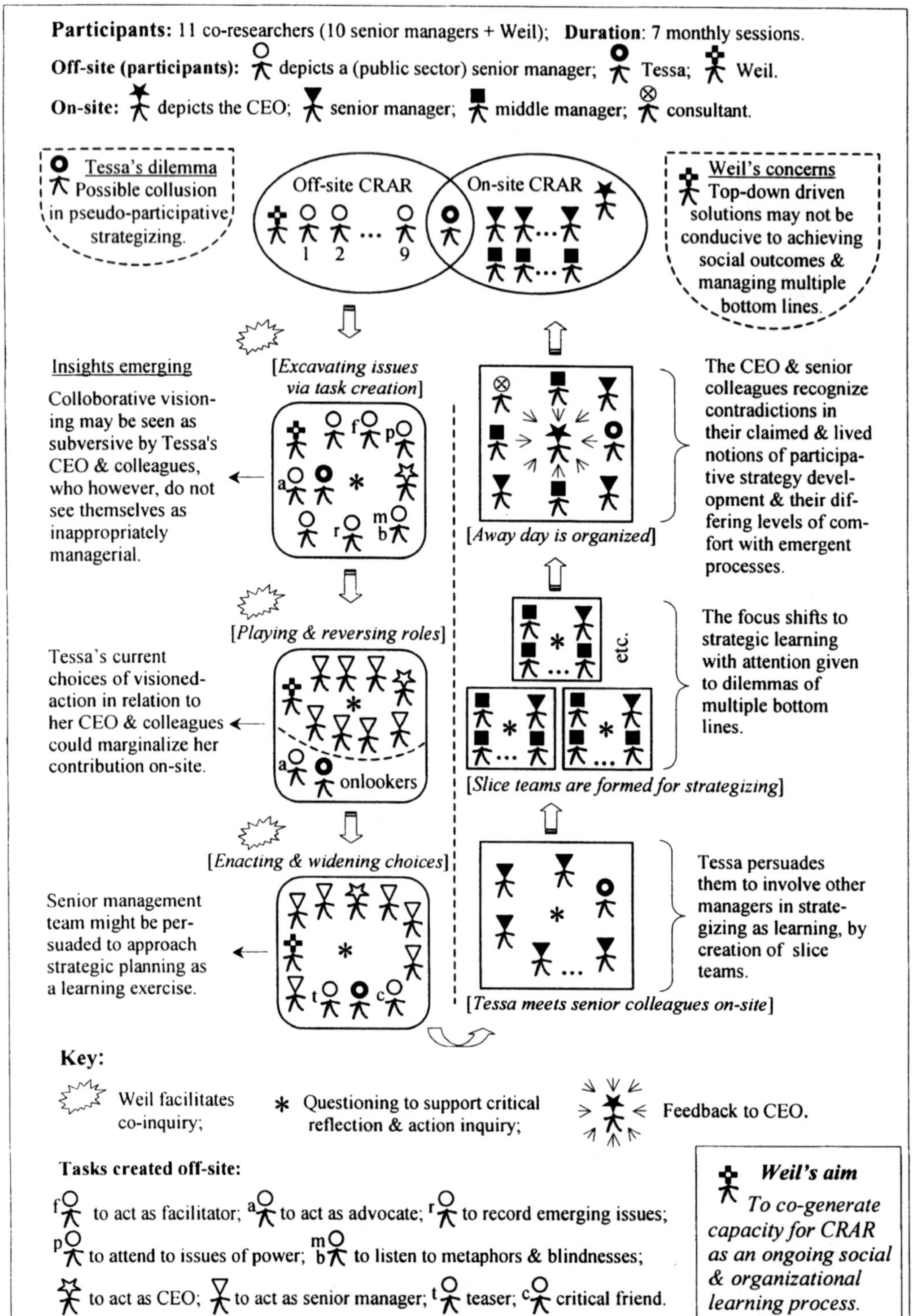

Figure 7: Weil's Critically Reflexive Action Research (CRAR) project.

were arranged monthly over seven months for a full day at a time to explore issues related to social and organizational learning and change in the context of UK public services (from the perspective of the managers). The off-site CRAR was intended to be interwoven with the on-site CRAR (see Figure 7[46]). In both cases, Weil notes that CRAR groupwork begins with people's "dilemmas of choice about responsible actions that are likely to have systemic effects" (1998, p. 46).

To explain how the co-researchers operated "off-site" to take forward certain starting problems and dilemmas, Weil hones in on a set of off-site encounters in the group — around a dilemma faced by the senior manager named "Tessa". (The name Tessa was used for purposes of reports written by Weil for different audiences.)

7.2.2 Working with "Tessa's Dilemma"

Tessa was working as a development specialist directly responsible to the Chief Executive (CEO) in a high-profile public service organization. He had recently asked her to involve middle-level managers in the strategic planning process. This was notwithstanding the fact that (in Tessa's interpretation) "he had already finalized the plan with the help of several colleagues in the senior management team" (1998, p. 47). Tessa saw her dilemma as

> being asked to collude with a pseudo-participative process, which went against the grain of what this organization would expect of other organizations whom they inspected. She knew, from middle managers' reactions in previous years, that wider participation in strategic planning was needed. But since her CEO has not offered to clarify what was open to negotiation and what was not, she felt pressured to act manipulatively. (1998, p. 47)

Tessa's dilemma was chosen by other members of the "off-site" learning network as the dilemma on which they would like to work at that point. This resulted from a participative focusing process. Each member was working on systemic dilemmas around which they wanted to do action research on-site and off-site. However, at this point in time Tessa's dilemma was chosen for off-site co-inquiry, as it represented concerns with which they could all identify. Once it had been decided to focus on Tessa's dilemma, Tessa was asked to elucidate her "perceptions and judgments about how she saw possibilities for inquiry and action in this situation" (1998, p. 48). Tessa indicated that, based on the way she felt at the time, she was planning to discuss the strategic plan with middle managers. However, Weil notes that questioning within the group soon "revealed how this could compound difficulties in the senior team, in the organization, and for her [Tessa] personally" (1998, p. 48). Weil therefore tried to

---

[46]Peter Adman constructed this figure as a way of depicting the processes involved in this CRAR project. His construction was based on his reading Weil's article (1998), as well as his reading my interpretation of Weil's account.

initiate a process of what she calls "broaden[ing] and deepen[ing] understanding of this dilemma" (1998, p. 48). Tessa was encouraged to

> brief several people in the [off-site] group to attend to things that she saw as important here. She invited one person to listen as the CEO. She invited another to attend to her metaphors and language and to "blindnesses" and contradictions between her own espoused epistemologies of practice [her views of how she could mobilize her knowledge in practice] and what was being revealed in this situation of action. ... . She asked another to attend to issues of power. Another was delegated to record key general issues throughout on the electronic white board. (1998, pp. 48-49)

To aid the process of Tessa's engagement in these conversations, Weil suggested that a lead facilitator, as well as other group members, could help Tessa to explore the situation, through their questioning of her accounts thereof. These people would at times "stop and invite reflections from those who had been specifically briefed" (to adopt the roles outlined above) (1998, p. 49). An "advocate" of Tessa's position was also appointed, to support Tessa in her deliberations. Her task was to support Tessa (and Weil) in considering ways of forwarding the inquiry along lines considered productive. (See the arrangement of this set of off-site conversations around Tessa's dilemma in Figure 7, where the various tasks of the participants are depicted.) Weil notes that the conversations that ensued, and the recordings of key issues that were located, were significant in that they could "give Tessa different data sources to take away with her to guide choices about on-site CRAR that she might initiate within her own organization" (1998, p. 49).

A number of insights were generated through the process of exploring Tessa's dilemma in this way. One insight was that "the organization had clear values about organizational learning and change, but, typically, it expressed these values and beliefs in its work with others *outside* its boundaries" (1998, p. 49). Although Tessa was in an organizational development role, "a more reflexive stance was distinctly countercultural" (1998, p. 49). This was despite Tessa's understanding that the Chief Executive and his colleagues did not see themselves as "overly controlling or inappropriately managerial" (1998, p. 49). But in Tessa's view, "her position as the only woman on the team seemed additionally marginalizing" for her in terms of her trying to institute a different style of management (1998, p. 50).

Weil notes that Tessa began to realize that her colleagues (in her organization) had

> assumed that she could just get on with involving people in strategic planning. This was her role. Dilemmas of integrity and collusion with pseudo-participation were not manifest. Her [own] preferred "epistemology of practice" was oriented toward processes of collaborative envisaging and participation ... . She wanted to generate more critical reflection on how ... "assumed good practices" [relating to participation] as applied "out there" with others, applied *within* this organization. But such forms of participation were at risk of being seen as subversive .... — despite the espoused management philosophy of the senior team. (1998, p. 50)

Based on these reflections, Weil advised that Tessa be invited to stand "outside" her situation with her advocate and act as a kind of onlooker to a conversation that was set up in the learning network — in which people took on the roles of the CEO and colleagues. (This is shown via the next arrangement of off-site conversations in Figure 7.) Weil suggests that the fruitfulness of this was that:

> Making Tessa the "silent onlooker and eavesdropper" at this point helped her to see the impact of her own choices being "played back to her", and their impact on the CEO and her colleagues. She had assumed that her lone intervention in strategic planning would have some positive effect ... . She now could become more critically reflexive about the limitations of her own constructions ... and the unintended systemic effects these might have. (1998, pp. 50-51)

The team now (still in role) was given the task of "sharing responsibility with Tessa for this strategic planning process. They spoke from a perspective of having recognized the risks of [her] marginalizing her role" (1998, p. 51). Through this set of encounters, Tessa indicated that "her entire view of the situation had been turned upside down" (1998, p. 51). She began to consider "possibilities emerging from what she had observed, helped by colleagues' notes" (from the learning network off-site) (1998, p. 51). In particular, Weil comments, she came to the conclusion that

> had she proceeded as the CEO had intended, she would have become further caught between a "rock and a hard place". Contradictions between espoused and expressed values would have intensified [in the organization] ... . Cynicism would also have increased, thereby making it more likely that "stories" unheard *inside* the organization would begin to be told *outside* the organization. This would have damaged perceptions of its capacity to fulfill its role. (1998, p. 51)

Weil remarks that despite Tessa's understanding of the disabling effects that might have been generated by her (earlier preferred) choices, and of possibilities for redefining her choices, "the desire to 'pull back' to the status quo is overwhelmingly tempting. Insight in itself does not necessarily lead to more insightful action and choice" (1998, p. 52). Weil points out that, in the light hereof, it was important to "empathize and work with Tessa's ambivalence" (1998, p. 52). It was therefore proposed that Tessa be called upon to enact a meeting with her boss and her senior colleagues (as shown through the enacted meeting in Figure 7). Weil suggested that two learning network members support Tessa in this process, with the aim of "'stretch[ing]' her repertoire of responses". One person was asked to play the role of teasing her toward alternatives and interrupting her "stuckness". Another person played "critical friend". She would notice how Tessa was using insights and translating them into action and how she was not. This was done through an "ongoing commentary and encouragement" on the part of the critical friend (1998, p. 52). Through this process Tessa also practiced discussing what had hitherto been regarded as "undiscussable" (within the organization). She also was able to redefine "the boundaries of her role with her colleagues" (1998, p. 53). She began to realize (by enacting a role reversal with her boss at one point) that "playing down her strengths ... was a far from useful strategy when experienced by him!" (1998,

p. 53).

Weil comments that learning network members then began a process of reflecting on what they had learned through their participation in aiding Tessa to come to terms with her dilemmas. Several of them spoke "about how they had realized their own complicity in some of their isolation in their organizational development roles and the more critically reflexive possibilities for action and inquiry that had been stimulated by Tessa's work" (1998, pp. 53-54). They also decided to put aside time to "share ways in which they intended to use CRAR type processes to interrupt ... 'stucknesses' within their own organizations" (1998, p. 54). Weil remarks in this regard that they had at least had the opportunity, through the process of off-site learning, to gain "experiential understanding of what it means to build collective capacity for learning" (1998, p. 54). This understanding would be relevant to their on-site engagements with others. (Weil indicates that people spoke about aspects of their continuing work in their organizations, around which they would be seeking further learning in the next off-site learning network meeting.)

Weil continues her story by noting that Tessa took insights from this off-site action based co-inquiry into her on-site longer-term action research. She "persuaded her senior team to approach the strategic planning as a learning exercise, with each leading a slice group to discuss the strategic plan" (1998, p. 55). That is, each senior manager would work with middle managers on strategizing as a learning exercise (in terms of the slice group arrangement shown in Figure 7). Weil states that this process

> released considerable energy and commitment in the organization and later led to an away day where the CEO sought feedback on the unintended effects of his behavior, style, and choices in recent months, drawing on insights generated by an external consultant. (1998, p. 55)

Springing from this, the senior managers and the CEO "began to recognize the unintended organizational effects of their differing levels of comfort with uncertainty and emergent processes, and this became a strand of continuing reflexive inquiry in their work together as a team" (Weil, 1998, p. 55).

Weil indicates that finally

> the CEO agreed that the organization should learn from this experience, enabling them to codesign a more learning oriented cycle of participative strategic learning and planning processes over the following year — rather than at a single point. (1998, p. 55)

Weil comments that what she herself would like to highlight is that

> decision making and strategic formulation/implementation [can be] approached not as static but emergent learning processes, as ongoing cycles of reflection in and on action in a complex, ever-changing system. The example indicates how CRAR can better equip managers and others to live more comfortably with discomfort, ambiguity, and uncertainty. (1998, p. 55)

She suggests that the example that she has reported upon here,

> although not dramatic in its content, and focused on senior managers' interactions, nonetheless offers a glimpse of the kind of processes that take us beyond superficial learning processes. (1998, p. 59)

Using this case as an illustrative instance, she suggests that "CRAR can keep us alert to the systemic effects, and the social and financial costs, of failing to work with and learn out of contradictions-in-action" (1998, p. 59). The case is furthermore illustrative of the point that:

> As we reduce our needs for control, we can become more capable of working with emergent processes. We can be helped to redefine the boundaries of our roles and our engagements with others. The humility and learning that can be generated by such shifts, supported by insights from multiple vantage points, can lead to more responsible choices of action and interaction. (1998, pp. 59-60)

Weil regards it as essential if CRAR is to justify its claims to aid the processes of learning to which she refers, that "documentation of CRAR journeys" are well recorded (1998, p. 55). She illustrates (1998, p. 56) that documentation can include keeping records of:

- Histories/cultures;
- Starting assumptions, values, experience, construction of dilemma/ problem?
- Possibilities for inquiry?
- Insights from literature/other sources?
- Choices/judgments/rationales?
- Nature of individual/collaborative inquiries/ data collection?
- Chronicling e.g. illuminating incidents? Revealing stories/Critical incidents;
- Reflections of experience/process of learning, "unlearning" in collaboration?
- Divergences/convergences in perspective;
- Multi-layered narratives. Stories that speak in and beyond context?
- Process of challenging, sustaining critical reflexivity?
- Revised/discarded assumptions? Noticed resistance to learning in self/others?
- New insights/outcomes;
- Shifts at individual/group/organizational levels? Local issues? Wider issues? Expected/unexpected outcomes?

Weil indicates that "these forms of documentation, to support key inquiry processes, can assist with the validation of CRAR processes and outcomes". She urges documentation in the form of "multilayered tracking of the 'messiness' of such [learning] processes" (1998, p. 55). She suggests that paying attention to documentation in this fashion, provides a way of meeting "postpositivist" criteria for acceptable research practice (1998, p. 55).

7.3 EXPLORING ALTERNATIVE ASSESSMENTS OF THE EXAMPLE

7.3.1 A Positivist Assessment

7.3.1.1 Scientism: Doing Science

The positivist tenet of scientism suggests that action research can become an acceptable strategy for social inquiry — provided that it is geared to setting up incidents of some defined "independent variable" with a view to examining effects that ensue. Via action research, these incidents can be set up in "natural" contexts (as opposed to laboratory designed situations). To explore this idea further, it is worth returning to Comte's view of what experimentation as a form of scientific inquiry consists of.

Comte suggests that experimentation in science (as explained in Chapter 4) consists of setting up "parallel [matched] cases" undisturbed by the intrusion of "irrelevant influences", and then organizing manipulations so as to observe the effects thereof (Comte, 1974, p. 254). Comte believes that the discipline of physics offers especially favorable conditions for organizing this kind of inquiry. Nevertheless, he remarks that "the science [of sociology] is not entirely deprived of this resource" (1974, p. 477). He indicates that in general, experimentation as a strategy can be used "whenever the regular course of the phenomena is interfered with in any determinate manner" (that is known to us) (1974, p. 477). In the case of sociology, he points out that the "interference" with the course of phenomena can be introduced in society by people trying to institute alterations in their social conditions. Experimentation can be used to examine the effects of any given alteration.

In the light of Comte's remarks in this regard, the attempt by Weil and others to organize alterations might be seen as classifiable as a form of experimentation — as long as it is possible to trace what has been altered and to consider what effects might have been generated by the specific alteration. For example, in organizing this project, the managers (acting as researchers) could define a variable called "style of management" and then, through action, attempt to introduce an alteration in this style of management (from, say, more top-down to more collaborative). There were 10 organizational contexts in which the various senior managers could introduce alterations in style of management, with a view to examining effects hereof on the dependent variable(s) isolated for attention, such as, say, levels of cynicism of middle managers, organizational performance, etc.

From a positivist point of view, then, this project could become an exercise in doing science. But exactly how conditions were being altered and effects followed up, would need to be more clearly specified.

7.3.1.2 Phenomenalism: Organizing Measurement Validity

The positivist tenet of phenomenalism suggests that definitions of the terms used in scientific hypotheses are a way of pointing to the phenomena under study. For instance, taking the concept of top-down management, this concept can be used to refer to definite characteristics of a particular style of managing. However, in the way in which the various researchers used it here, it seems that they were using it primarily as a way of expressing dissatisfaction at what was considered to be the inappropriate manner of organizing strategy on the part of the CEO in Tessa's organization. The term was used as an aid for Tessa to appreciate that she could legitimately confront her CEO on this. This way of using language with practical intent in everyday life is part of the process of "ordinary" interaction. But if we wish to conduct scientific investigations, we need to define more precisely the meaning we wish to give to the terms utilized in our various statements, so that we know what phenomena are being referred to.

Weil and others suggest that they were developing insights from "multiple vantage points" — for example, from the vantage point of the CEO and from Tessa's view of participatory management. They were explicitly not trying to combine these vantage points into a single view that "represents" the realities. But the vagueness of terminology employed, is not helpful in investigating the conditions (in this case, style of management) being altered through the research process, so that we can in turn trace effects back to those known alterations.

7.3.1.3 Empiricism: Seeking Causal Connections through Data Analysis

Weil's report provides certain pointers to the connection between "low levels of participation" and "high levels of cynicism" (on the part of middle managers) (1998, p. 52). This is coupled with a suggestion that "the introduction of controlling strategies that turn social actors ... into passive recipients of top-down driven solutions [is] not conducive to achieving social outcomes and managing multiple bottom lines" (1998, p. 57). Conversely, Weil suggests that it is possible for senior managers to (re)orient themselves in a manner that is thus "conducive" (1998, p. 59).

However, Weil's story ends before we can see whether in fact any of the alterations in senior management style in the on-site situations in which the managers were involved (for example, in that of Tessa's organization), do have the effects that she supposes they might have. The internal validity of the study in terms of processes created to test for causal relationships is thus rendered doubtful. Weil's beliefs (1998, p. 51) about the importance of participative learning in terms of the "far-reaching systemic consequences" that might ensue if this is not tried, still have to be tested more rigorously. Furthermore, in order to organize for the external validity of the inquiries (reaching beyond the particular senior managers involved in this case), it would be useful to try to organize comparisons across different organizations more widely. (Alternatively, if Weil believes that scientific studies have already been done on the effects of participative learning on the handling of multiple bottom lines in different organizations, she could link up her discussion more systematically with these studies.)

7.3.1.4 Value Freedom: Finding Out about Reality

The extent of the influence of Weil's prior beliefs and values on her statements about the achievements of the project, is difficult to detect through the way in which she has written it up. She clearly believes that it is worthwhile to try to institute participative learning processes in organizational life. She entered the project with the idea that decision-making and strategic formulation/implementation should be approached as a learning exercise. It seems that she used the project as a way of trying to ensure that others would appreciate the value of participative learning as a route to more effective strategy formulation. But whether she has really located any causal connection between a way of creating strategy (through, say, working with uncertainty in a particular way) and outcomes (in terms of organizational capacity to deal effectively with change) is still not clear.

Weil refers to the detailed documentation of CRAR processes that took place at various points of the project (1998, p. 55). But the problem is that these documents may themselves be a product of a value-laden way of conducting, and recording, the inquiry process. It is quite possible that the co-researchers' documentation of their learning, and of its presumed effectiveness in generating desired outcomes, refers simply to their own hopes and wishes that this is the case. It is thus difficult for us to know what cognitive status should be given to Weil and co-researchers' constructions/ narratives.

7.3.1.5 Instrumental Knowledge: Offering Useful Knowledge

Weil considers the project as being of practical use to the practitioners (senior managers) who were all involved in the investigation as co-researchers. The creation of the learning network was organized around their own felt dilemmas, and was arranged with the intention of aiding them to better address issues of concern to themselves. From their perspective, as reported by Weil, their capacity to address their management roles was improved. This was effected by harnessing their experiences in the learning network (and by participating in the process of challenging one another to extend their perceptions). In this way, they considered that they had learned, for instance, that their own collusion in pseudo-participative processes in their organizations, might lead to the consequence of creating (increased) cynicism among middle-level management, and to their defaulting into "the financial bottom line". This insight (as well as others) on the part of members of the learning network might be useful for other audiences who are similarly interested in these issues.

However, from a positivist point of view, sufficient work has not yet been done on testing hypothesized statements, for them to count as knowledge that is useful for either the participants or other audiences. We still do not know on the basis of the evidence of the project whether the institution of participative processes in organizations is indeed likely to reduce levels of cynicism and to increase capacity to handle multiple bottom lines. Further work is needed to conduct comparisons across different organizations, over time, with a view to checking any statements made in this regard. Only then can we regard the research as truly of practical relevance in helping people (those involved in the study and others) better to manage their organizational existence.

7.3.2 A Non-foundationalist Assessment

7.3.2.1 Scientism Qualified: Attending to the Effects of Human Perceptions in the Social World

In setting out his (non-foundationalist) argument, Hammersley has expressed a concern that action research is often associated (by advocates) with the idea that the role of the researcher in conducting social inquiry cannot be separated from that of "person" or "citizen". As he indicates: "We find such views among feminists, critical ethnographers and many advocates of action research" (1995, p. 80). Hammersley believes in contrast that in order to practice science it is necessary for researchers to strive to separate out their roles as scientific researchers and as citizens. In the case of Weil and co-researchers' project, this would mean that they should have been more explicit about how they were using their experiences as a basis for setting up scientific inquiries into the way that actions on their own and others' parts tend to lead to certain outcomes. For instance, it seems that one (implicit) hypothesis of theirs was that pseudo-participation leads to increased levels of cynicism in organizational existence, with attendant consequences for the management of multiple bottom lines. Conversely, it was suggested that collaborative participation might avert cynicism, as well as the tendency to address only the immediate financial bottom line. Hypothetical statements in this regard would need to be checked with detailed reference to the evidence afforded by the project.

In the process of organizing their investigations, Weil and co-researchers could also have capitalized on the opportunity to explore ways in which their perceptions of one another's expectations for "collaborative" inquiry, *might themselves have affected the process of the inquiry* (through the reactivity effect). The way in which people perceive (and react to) expectations of collaboration would be an important avenue for Weil and others to explore more systematically, in the off-site as well as on-site networks that were being set up.

As another consideration, the way in which Weil and co-researchers documented the process of their inquiries, would be looked at from a non-foundationalist position as follows. In whatever roles the participants saw themselves as assuming, they are likely to have been influenced by the fact that Weil suggested that *multiple narratives* should continually be documented in regard to their experiences. They might have thus been led to the belief that narratives cannot be adjudicated with reference to the realities of organizational life. Weil tries to account for her own stance in this regard by suggesting that it can have a salutary influence in allowing people to work with multiple realities. But a non-foundationalist position would challenge this claim, arguing that there is little, if any, evidence to support the perception (that Weil seems to be disseminating) of science as unable to make advancements in realist terms.

7.3.2.2 Phenomenalism Reviewed: Improving Instruments for Observation

Weil and co-researchers appear to have tried to explore the phenomena that they highlighted for study, while at the same time passing their own judgments thereon. This is because Weil herself, and perhaps many of the co-researchers, believe in values such as what Hammersley calls "'democracy', 'equal opportunities', 'justice'" (1995, p. 40). Guided by these values, they, for instance, judged the phenomenon of "management style", as "inappropriate" insofar as it did not match the values; and efforts were instituted to direct the phenomena in a more valued direction.

Hammersley is concerned that this way of approaching research projects, poses a threat to the research task — which is to try to explore patterns in discerned phenomena. As researchers, people could set out to examine the effects of certain styles of (senior) management on the attitudes and competencies of others; and people could experiment with altering management style with a view to exploring the effects of this alteration in various organizational contexts. But to do this, they would first need to improve the instruments for observing the phenomena in question, such as in this case, "management style", "cynicism", "capacity to address multiple bottom lines", etc. Due to the way in which Weil and co-researchers have proceeded, they have not maximized the conditions for collecting the necessary evidence to be able to either support or disconfirm proposed hypotheses.

7.3.2.3 Empiricism Reviewed: Creating Theoretical Explanations Grounded in Observations

From a non-foundationalist point of view, action research, like ethnography, could become a way for researchers to build up scientific explanations through developing an acquaintance with aspects of social life in natural settings. But the researchers involved must have a commitment to seek such explanations. When discussing ethnography, Hammersley and Atkinson suggest that researchers can orient themselves toward

> look[ing] to see whether any interesting patterns can be identified; whether anything stands out as surprising or puzzling; how the data relate to what one might have expected on the basis of common-sense knowledge, official accounts, or previous theory ... . (1995, p. 210)

Hammersley and Atkinson might regard action research as similarly affording possibilities for researchers to identify "interesting patterns" and consider "whether anything stands out as surprising". On the basis of recounting their experiences and reflecting with others on the consequences of their tried actions, people as (co)researchers could try to develop their everyday knowledge. The learning network in the case of Weil and co-researchers' project, could function as a way for people to become more scientifically oriented as they reflect upon their experiences/actions — toward developing explanations around these. They could develop their skills as scientists in using evidence to support or disconfirm statements being made. (And by linking their statements to existing theoretical literature around similar issues, they can also ensure that their work relates to arguments developed by other researchers in the scientific community.)

Weil has pointed to certain conclusions that she believes can be drawn from the study. These conclusions relate to the "local context" explored — in this case, the involvement of these senior managers in their specific public service organizations — as well as to other possible contexts. Weil states that although her work in this case was with these senior managers' interactions, it can be taken as illustrating that managers in different contexts can learn to appreciate the need for "alert[ness] to the systemic effects, and the social and financial costs, of failing to work with and learn out of contradictions-in-action" (1998, p. 59). She believes that the project undertaken serves to offer glimpses of insight about how one can try to institute (effective) learning processes in and across organizations. From a non-foundationalist point of view, the work of Weil and co-researchers does offer a source of hypotheses, as well as some empirical material that might be used in evidence to support them. But further work would need to be done to test whether the evidence can support the claims being made (both in the local context and in others that have been alluded to).

7.3.2.4 Value Freedom: Seeking knowledge through Operating Self- and Collegial Criticism

From Weil's reporting of the CRAR project it is clear that she values participative learning. On the basis of this value, she hopes that it can become a route to reducing cynicism in organizational life and to raising commitment and competence toward addressing multiple bottom lines. Her hopes and values are shot through the way in which she interacts with participants and the way in which she encourages them to challenge the "status quo" (of colluding with pseudo-participative styles of management). But the problem that she faces as a researcher, accountable to people *as a researcher*, is that her value commitments might overshadow the task of testing emerging hypotheses in relation to the evidence. Thus her hypothesis that, say, participative learning is a route to furthering the capacity to address multiple bottom lines, is in danger of becoming confirmed *no matter what the evidence of reality in fact indicates*. To increase the chances of offering a more objective account of the realities, Weil and others would have to submit their way of addressing the evidence to further scrutiny (taking into account collegial criticism as leveled by other scientists).

7.3.2.5 Instrumental Knowledge: Providing Opportunities for Better-informed Action

Hammersley indicates that debates persist in the social sciences in regard to

> how research can and should be made more relevant to practice and policy-making; and even about whether its goal should be simply the production of knowledge or whether it must be committed to more practical and political goals. (1993, p. vii)

Hammersley's position in this regard is that the goal of producing knowledge is the primary task of researchers, as researchers. However, an attempt to consider ways of making the knowledge more relevant to practice, can also be regarded as appertaining to the accountabilities of researchers. Seen in this light, action research may be one way of setting up inquiries to explore issues of particular relevance to those involved in some "practice". But action research can be classified as potentially advancing useful knowledge only as long as the researchers are *committed to the goal of knowledge-production*, and as long as they conduct their investigations along scientific lines. What is ruled out if they wish to act accountably, is to use the research process to advocate "some particular set of values" (Hammersley, 1995, p. 40). For as soon as they try to use the research process in this way, their chances of "discovering the truth" are jeopardized (Hammersley, 1995, p. 40). Weil and co-researchers' stretching of their research remit in this case, might indeed have impacted on their ability to conduct themselves as researchers. And this would have diminished their chances of developing well-informed evaluations of their own and others' activities.

7.3.3 A Scientific Realist Assessment

7.3.3.1 Scientism Qualified: Attending to the Reality of Generative Mechanisms, Including Societal Ones

Pawson and Tilley indicate that it is consistent with the ideas of scientific realism that social science be used for the purpose of creating evaluations of the effects of the implementation of social policies in society. Initial theories can be framed concerning "what is thought to work for whom in what circumstances". But these theories, to be constructed in (scientific) realist terms, must be framed in the form of "context-mechanism-outcome pattern configurations" (1997, p. 207). That is, an indication of what mechanisms might be responsible for what discerned outcomes as they appear in different contexts, needs to be offered. Using this as a basis, it is possible to go about "test[ing], arbitrat[ing] and, above all, refin[ing] the theories" (1997, p. 207). "The improved and substantiated realist theory" can then be used as an aid to "further cycles of improvement in policy-making and practice" (1997, p. 207).

Pawson and Tilley suggest that while researchers as evaluators should feed back to policy makers what they have learned, they should also at times operate by "seeking mutual enlightenment with each set of stakeholders" (that is, policy makers and practitioners) (1997, p. 207). In other words, stakeholders' views should be taken into account — as they embody theories that "often lie dormant and half-articulated and it is the evaluator's task to bring them vibrantly to life" (1997, p. 203). Thus, for instance, policy makers should be encouraged to "articulate, refine, formalize and reformulate their notions of the program theory" (the theory underlying the development of some social program) (1997, p. 203). As they involve themselves in action, they can consider the effects of implementing programs in different conditions, and can consider the differential impact "of a program across differing settings" (1997, p. 204). Their considerations in this regard can be examined by researchers/evaluators as they in turn try to "weave the decision-makers' theories into the process of realistic evaluation" (1997, p. 202). Pawson and Tilley cite a case where an evaluation

> revealed a gap in the policy makers' original organizational theory of the program; they knew much but were not know-alls. Unacknowledged contextual conditions had led Safer Cities [the program under consideration] to trigger unexpected causal mechanisms generating unexpected outcome patterns. (1997, p. 205)

From a scientific realist point of view, it is acceptable practice for researchers to aid people to learn more about the effects of implementing programs — by helping them to articulate, but also extend, their theories.

From this point of view, Weil's work with senior managers in exploring the effects of different styles of management and different ways of strategizing could be considered as one way of conducting "evaluative" research. This was effected in this case by members in the learning network acting in specific ways in their organizations and then utilizing the learning network as a forum to consider how their actions could be conceptualized. The co-researchers would, however, need to be clearer about how the theories that were emerging (or being refined) took the form of what Pawson and Tilley call realist explanations. Explanations proffered would need to be placed in the framework of the context-mechanism-outcome scheme that characterizes scientific theorizing.

7.3.3.2 Phenomenalism Reviewed: Developing Acceptable Observations

From a scientific realist point of view, theoretical preconceptions cannot be excluded from the process of observing empirical occurrences. So, for instance, Tessa's conception of the "top-down" management style of the CEO in her organization was a way of observing what she thought to be his non-participatory behavior (that is, his disinclination to invite participation). But, as Weil notes, others may not necessarily define as an indication of "top-down" management a CEO's envisaging of strategies. It was therefore important for the investigators to set up a dialogue in relation to the subject of, say, envisaging of strategies — so that observations could be conducted by taking into account different possible ways of defining the behavior of the CEO in this case. References to literature and other sources of ideas could also be used to aid this process.

But it is worth noting that although Weil points to the need to develop insightful observations, she does not appear to wish to define a way of seeing "the facts" that can be agreed upon. She seems comfortable with the notion that there may be multiple vantage points from which to see phenomena, and with the notion that doing research means encouraging people (and herself) to work with these differences. Setting up dialogues with the purpose of agreeing on observational evidence is thus not part of the CRAR agenda. From a scientific realist perspective, however, this mitigates against the chance of developing the necessary convergence across ways of seeing — necessary for scientific development of theorizing about social structures in the light of (accepted) empirical observations.

7.3.3.3 Empiricism Reviewed: Working at an Analytic Level to Understand Generative Mechanisms in Context

In order for the investigations of Weil and co-researchers to fit into the realist scheme of context-mechanism-outcome, it would be preferable if Weil had indicated how she and others were drawing inferences about the real causality operative in the contexts in which they were acting. Weil's conception of "causality" has not been explicitly spelled out in her report. It seems that proffering an account of how people's actions might be enmeshed in wider structures and of how this might influence their decisions to act, was not the focus of the study. It is therefore not clear how the actors were meant to attain knowledge of these structures, so that their actions could become better informed by such knowledge.

Weil's suggested mode of documentation — which includes chronicling illuminating incidents — is thus far insufficiently geared to operate at a level that can be "cashed in cumulatively" to develop knowledge of generative mechanisms — as Pawson and Tilley (1997, p. 119) put it. In any case, Weil and others would need to show what they were learning about these mechanisms and about how the contexts of their operation can explain the different outcomes that were being documented.

7.3.3.4 Value Freedom: Considering Possible Reformist Allegiances and if Necessary Re-Organizing One's Focus

Weil indicates that she is concerned that often managers seem to be placed under pressure not to acknowledge dilemmas of integrity as they act to meet the short-term "financial bottom line". These pressures, she believes, should be reformulated, so that people can be encouraged to work to "multiple bottom lines" in addressing their social responsibilities.

A scientific realist argument would be concerned that, in her enthusiasm to contribute to processes of social transformation, Weil herself might not be operating at the necessary analytic level to understand what lies behind managers' felt pressures to submit to financial bottom lines. Is it really likely that helping managers to try to address their dilemmas of integrity in their organizational existence can lead to genuine social transformation?

A scientific realist argument would advise drawing together the various experiences of people trying to manage bottom lines in different organizational contexts — with a view to indicating what kinds of radical changes might be needed to generate fundamentally different patterns in the structuring of outcomes in the society. For "action research" to make a contribution on this score, researchers need to direct the research process toward making explicit inferences in regard to the operation of the structures at play.

7.3.3.5 Instrumental Knowledge: Extending Information about (Inter Alia) Social Structures

The participants in the learning network focused on issues of concern to them in terms of their felt problems/dilemmas. As they proceeded, they created ways of reframing these, and ways of extending their considerations of options for action. For example, Tessa began to realize that she could exercise her own strengths as she interacted with her CEO, and that she could discuss with him what had previously been considered (by both of them) undiscussable. Together with the CEO (and others in the organization) new ways of addressing strategy formulation in the organization were tried; and new options were developed for what Weil calls "insightful action" (1998, p. 51). People (including the CEO) began to realize that management style might have effects of which they were previously unaware; and this knowledge led to their interrupting their own "stuckness", according to Weil's account.

But from a scientific realist point of view, the ways of dealing with practical issues that were tried in the course of the study, would need to be recorded more systematically than has been done by Weil and co-researchers. Lessons learned would need to be framed at a sufficiently deep analytic level so that they can be of real use to these practitioners in their continued work involvements, as well as to others. Also, the utility of the project for particular ways of organizing political practice directly in political arenas of society could be explained. To this end, theoretical inferences around the import of the co-researchers' different attempts to open up their choices of action would have to include a more detailed account of the structures in which they were enmeshed.

7.3.4 An Interpretivist Assessment

7.3.4.1 Scientism Opposed: Doing Social *Science*

From an interpretivist perspective, Weil and co-researchers' project might be viewed as one way of building up understanding of the social world with close reference to a consideration of its meaningfulness to actors. The project operated in keeping with the interpretivist claim that social inquiry should be directed toward trying to ascertain ways in which outcomes in social life are predicated upon human meaning-making. Weil did not concentrate on trying to organize predictions from hypothesized independent variables toward dependent ones, but rather on allowing people to recognize how their own (and others') interpretations of situations might make a difference to the continuing unfolding of possibilities.

The interpretivist position would appreciate Weil's efforts to display (and work with) the way in which terms were employed differently by people in different contexts of use. For example, the researchers (those involved in the learning network) excavated the richness of the notion of "top-down" management. When used by Tessa in the support group (off-site), it related to her concerns that she was being asked to collude in what she considered to be pseudo-participative management. When used by her in interaction with her CEO (on-site), her intention was to suggest to him that he could ask for feedback on his style of management. The meaning of the term thus shifted depending on its context of use.

What is especially important about Weil and co-researchers' investigations is that they created a way of examining and extending people's participation in defining their social realities as they interact with others. This is not, however, to say that the interpretivist position would see action research in general, and Weil's proposed CRAR in particular, as the ideal way of exploring social realities. From an interpretivist position, the danger of researchers' involving themselves as "actors" in social settings studied, is that they might be tempted to use this as an opportunity to progress their own values. As shown in Section 7.3.4.4 the project might be questioned on this score from an interpretivist point of view.

7.3.4.2 Phenomenalism Re-interpreted: Creating Meaningful Categories

The co-researchers' investigations involved an exploration of what was meaningful to people (in this case they themselves) in their "everyday" existence. But Weil offers an indication that she did not expect her own way of making sense of people's activities (on- and off-site) to mesh entirely with that of all of the other members of the learning network. Nor did she expect them all to agree on the meaning of "the phenomena" that were being called up for discussion. Hence, for instance, her own conception of Tessa's "risk-taking" might have differed from Tessa's conceptualization of her activities. Tessa may regard her display of strength in relation to her CEO as being more risky than Weil and some of the other participants appreciate. They in turn might believe that in action Tessa can assess the risks (for herself as well as for others to whom she feels responsible, such as middle-level managers) and on the basis hereof, make her own decisions for "insightful action". Weil in any case provides us with an indication of the processes in the learning network of aiding Tessa through her felt dilemma, by leaving what she calls a trail of documentation around this. This is in line with interpretivist requirements for acceptable social research. Those with access to the documents (and readers of Weil's reports) are invited to judge whether the categories created in Weil's storytelling can be said to be sufficiently grounded in the various participants' meaning-making.

7.3.4.3 Empiricism Re-interpreted: Developing Understanding of the Social World

The route that Weil took to understanding the social world, was to participate with a set of practitioners in working through their felt dilemmas. In this way, she (and they) were able to enrich their understanding of the pressures placed upon them in their work situations; and at the same time they could come to appreciate how their interpretations of these pressures might make a difference to their consideration of options for action. In the process of the inquiry, they recognized some of the ways in which they were party to the constitution of their social world.

The display of ways in which the social world becomes constituted through human activity, can be seen as fitting in with the interpretivist view of the manner in which theorizing should be organized (in order to pass as valid). The co-researchers operated with Tessa's dilemma as a focus point to consider how Tessa, and they, could conceive options for action in their on-site situations. Weil notes that at one point Tessa herself might have decided to slip back into the status quo of colluding in processes of what she considered to be pseudo-participative strategizing. This, Weil notes, would have been "understandable". Hence, even from Tessa's (single) example, we are able to gather that people's responses are not predictable. These responses depend upon the way people decide to interpret the situations in which they are involved with others (such as in this case, the CEO and others). Weil indicates, by implication, that had Tessa decided to operate differently, a different pattern of events might well have ensued in her organization, and indeed many of the other senior managers in the learning network too might have altered their conceptions of options for responsible action.

But further work could still be done by Weil and co-researchers to elucidate ways in which they believe the social features of Tessa's situation were built up, and how this might be compared with their own experiences (in their different on-site involvements). If this had been done more explicitly, then we would be in a better position, as audiences, to consider how to approach the possible "transferability" of the different narratives that were developed in the course of the project.

In considering what is called "generalizability" or "transferability" of qualitative inquiry, Schofield discusses the question of how one can generalize from an unusual site to more typical ones. Using an example of what she calls an atypical case she proposes that:

> Some of the findings from the study I have been discussing were clearly linked to unusual aspects of the school and hence have very limited generalizability to other situations, although they may none the less be important in demonstrating what is possible, even if not what is generally likely. (1993, p. 219).

Schofield's comments in this regard are made in the context of studying what she calls "desegregating school systems" (1993, p. 217). She indicates that the study in question could be used to "shed light on what can be accomplished in a serious and sophisticated effort to structure an environment conducive to fostering positive relations between students" (1993, p. 218). She indicates that she decided to concentrate on the task of "seeing whether and how difficulties [concerning their relations] can be overcome" (1993, p. 218). She organized her study so that she could "see what happens under conditions that might be expected to foster relatively positive outcomes" (1993, p. 218). As she argues, "if things went well at such a site, the study would then provide an opportunity to gain some insight into how and why they go well and into what the still-intractable problems are" (1993, p. 218).

Schofield's suggestions relate to qualitative inquiry organized with an "action" component — in that she wished to "see what happens" when efforts were made to foster positive intergroup relations (as defined, presumably, by the different participants). Her study, as she explains it, can be compared with Tessa's on-site involvements in trying to re-organize the "environment" of management decision-making in her organization. Schofield's point is that one can use the findings generated in specific instances to *consider possibilities* that people can bear in mind as options that might be tried. Whether or not the findings can indeed be "applied" in other situations is then a matter of others comparing the study with their own situation and making judgments on this basis. What is important, from an interpretivist point of view, is that the understanding offered by detailed reference to case material (as in Weil and co-researchers' project), is illustrative of possibilities that are recognized to be contingent on people's organization of their interactions.

7.3.4.4 Value Freedom: Paying Attention to Others' Worlds

Henwood and Pidgeon advise that social researchers should develop an appreciation of others' beliefs and values (1993, p. 25). It is important that the values of researchers do not function to occlude their vision of what is valuable and meaningful to others in their everyday experience. From this (interpretivist) point of view, the suggestion would be that Weil, in particular, should not allow her own way of assigning meaning to — and valuing — participative learning, to overshadow others' possible ways of assigning meaning. From Weil's report, it seems that most of the participants in the learning network, and those in the on-site situations (as they appear in Weil's report hereon) came to appreciate the value of participative learning guided by CRAR. From an interpretivist point of view, though, it is possible that Weil overstepped her role as researcher by trying to encourage others to accept the importance of this way of learning. If we knew that the actors in the social world (for example, the senior managers in the learning network and those with whom these senior managers were working on-site) themselves consider the desirability of the values propounded by Weil, then it is acceptable conduct to explore ways in which such values might be implemented in different contexts. And one can use the inquiry process to evaluate whether or not they are being well implemented. But from Weil's own account it is not clear to what extent the different actors do, indeed, subscribe to such values. It is possible that the "cynicism" of which Weil speaks, implies that the cynics might be less interested in pursuing these values than is hoped by Weil (and by some of the other participants). Weil could supply more detail in regard to how she and co-researchers dealt with their own values as they interacted with others — so as not to impose these onto others. This would make it easier for us to consider whether their behavior can be accounted for from an interpretivist perspective.

7.3.4.5 Instrumental Knowledge: Creating Meaningful Information Relevant to Actors

Setting up a "learning network" was the approach used in this case to gain insight into pressures felt by (senior) managers, and possible ways of addressing them. These insights took the form of revealing stories, chronicles of incidents, etc., that were developed by tapping into various people's experiences and by extending their reflections on them (in the learning network). Arising from this, Weil constructed what can be argued to be a plausible account of how senior managers and others might benefit from processes of learning set up to aid critically reflexive inquiry. She believes that stories told by her and co-researchers on the basis of the experiences in the learning network, point to *possibilities* for them, and others, to consider as options for better addressing their social responsibilities.

To offer information that is more usable, though, it would be preferable for Weil's discussion of such possibilities to continue, instead of her account hereof having been closed where it has been left off thus far.

7.3.5 A Critical Theoretical Assessment

7.3.5.1 Scientism Criticized: Avoiding the Transference of an Engineering Approach to the Study of Social Reality

Morrow notes that from a critical theoretical point of view, one cannot define in advance of considering the context in which researchers are operating, how "critical theorizing" should be forwarded through the research process (1994, p. 228). He suggests, though, that what is important is that the critical theoretical methodological strategy is based on the "agency-structure dialectic and the interpretive structural approach to historical explanation" (1994, p. 228). That is, research should be organized in such a way that due account is taken of how agents can and do make choices, while recognizing that they do so in conditions that are not simply of their own making. A sense of the histories that have shaped the patterning of choices and outcomes, is part of any critical research agenda.

Morrow suggests that action research can be seen as one strategy that can, in principle, be used to forward the concerns of critical theory. He cites Harvey who suggests that action research can be seen as one amongst the "analytic techniques" that might be utilized by "critical social researchers" (Harvey, 1990, as cited in Morrow, 1994, p. 227). Morrow argues that action research may be classed as a "special case of critical ethnography" (1994, p. 258). As in critical ethnography, the researcher's participation in people's everyday lives is coupled with an attempt to "erode the expert/subject distinction" (1994, p. 258). In the case of action research, people's propensities to reflect on their own and others' experiences in concrete contexts of action, are highlighted (and nurtured).

The project reported upon by Weil can be appreciated as an attempt on her part to work with people in setting up a process whereby they could re-examine the way they, and others, might be affected by their choices of action. Weil's style of action research (CRAR) fits in with the critical theoretical preference for using the research process to oppose conventional "engineering" metaphors of control. Weil hoped that through the project people might learn to treat with suspicion the "control" mode of operating in the world, where it is believed that outcomes can be engineered on the basis of (supposed) knowledge of reality. She notes, though, that cultural pressures to operate in a guise of controlling the situation, are still ever-present. So Weil notes that her endeavors (and those of others committed to a different style of operation in the world) are made all the more difficult in a cultural context where "'correct' ... action, rather than reflection in and on action, is valued — and indeed rewarded" (1998, p. 57). In critical theoretical terms the worthiness of her endeavors can be judged by considering the contribution that was made (or that could be argued to have been made) to enabling people to revisit this mentality.

7.3.5.2 Phenomenalism Reconsidered: Creating Realities through the Research Process

Weil indicates that her aim in the project was to set up an inquiry process to enable people to recognize how their previous way of envisaging realities might have been blocking their own and others' potential to create more responsible choices therein. For instance, when Tessa perceived the CEO's management style as top-down, she was encouraged through the group work to recognize that the meaning that she assigned to the CEO's management approach, could be subjected to dialogue. In the process of dialogue (and especially through the enacted meeting), Tessa was aided to reconsider her vision of his management style, by recognizing that he might appreciate a display of strength on her part. This allowed her to consider alternative, more communicatively oriented, actions in relation to the CEO. From a critical theoretical point of view, the point of setting up the learning network should indeed have been to increase people's capacities to work toward furthering their communicative action both off- and on-site — insofar as this was conceived possible.

7.5.3.3 Empiricism Reconsidered: "Discovering" Social Potentialities

Critical theory challenges a research orientation that strives to locate empirical facts as a basis for theorizing about reality. From this perspective, Weil and co-researchers' project would be appreciated for its intention to aid participants — through the learning network — to avoid stuckness in what appears to be factual givenness. Hence, for instance, Tessa could "discover" — in action in relation to her CEO — that her acting in keeping with the value of participation that she espoused, could be made more workable than what she had previously assumed.

From a critical theoretical point of view, one of the main advantages of the project as described by Weil was that it was geared toward aiding actors to find ways of developing the fluidity of the social world so that they could avoid acting solely under the apparently factual pressure of the financial bottom line. In endorsing processes of self- and collective transformation through social inquiry, critical theory does not argue "abstractly" for what Morrow calls an impossible exclusion of normative questions (1994, p. 239). The normative question of how ways of life can be improved, directs the organization of the inquiry. Weil's advice to use facilitation, role-play, etc., to draw out people's propensities to reflect on their social responsibilities in intersubjective encounter with others, would thus be appreciated as consistent with the normative intentions of critical theory.

7.3.5.4 Value Freedom Undercut: Recognizing the Value of Human Discourse

Critical theory endorses discourse as a form of social interaction geared to communicative co-ordination of action, as a way of opposing the unreflected-upon infiltration of standards of "efficiency" into all spheres of human existence. Weil and co-researchers' network — where they set up an explicit forum for considering how they could review these standards — would thus be seen as an acceptable route for social inquiry.

Critical theorists might argue, however, that Weil and co-researchers could have made more explicit to one another (and to others interested) how they were trying to activate communicative logic both off-site and on-site in their relations with others. And critical theorists would also be interested in knowing about the processes of developing more convergence of perspective amidst the continued divergence of which Weil speaks.

But critical theorists would appreciate that those involved in the project were in practice providing certain instances of how communication could proceed in relation to their felt dilemmas. And their documentation of their communicative processes serves as an exemplar to others too of ways in which discourse can become a route to creating what Weil calls more "insightful action" (1998, p. 51).

7.3.5.5 Instrumental Knowledge Undercut: Furthering Communicative Understanding

Morrow indicates that although action research may appropriately be drawn upon by (critical theoretical) researchers, we should be careful of glorifying the contributions that can be made by such research (1994, p. 317). He suggests that "a multiplicity of different relationships to theory, research, and practice" needs to be incorporated in the critical research tradition (1994, p. 317). A possible weakness of action research lies in the temptation that it presents to "retreat from 'knowledge' to immediate 'practice'" (1994, p. 316). Morrow notes in this context that Habermas believes in the importance of still preserving some distinction between the institution of "science" and the process of "political organization" (1994, p. 317). He cites Habermas as suggesting that:

> I don't think that we can ever again, or even that we should ever again, bridge the institutional differentiation between the science system and political agitation and political organization and political action. ... . So there are just bridges between us as participants in some sort of political action and as members of the science community. (Habermas, 1992, as cited in Morrow, 1994, p. 317)

Taking part as researchers in trying to set in motion more communicative processes in different spheres of society, may be one way of creating a connection between themselves as political actors and as members of a science community. But if they are to use action research with this purpose, researchers would need to be committed to acting as "members of the science community" (as an institution), at the same time as they engage in "political" transformative action. This is in order that they can take some distance from immediate practical pressures.

The question of concern to critical theory is whether the route chosen in action research to connect social inquiry and (political) action, should be regarded as a workable way of challenging what Morrow calls "existing forms of administration and centralization" (1994, p. 320). Weil highlights in her report that she saw her aim as "cogenerating cultures and practices that support ongoing critically reflexive learning and inquiry" (1998, p. 39). Morrow's argument is that the undertaking of action research within (and across) organizations in society is one potential option for organizing social transformation — although it should be regarded as a "long march" (1994, p. 320).

Weil's co-researching efforts can be seen as resonating with the critical theoretical tradition of trying to connect "science" (as critically reflexive inquiry) and "social practice", without submitting to narrow strategic demands set by "existing forms of administration" — as Morrow (1994, p. 320) puts it. In this project, the co-researchers were senior managers who had funded their participation from their own budgets and were not formally answerable to others for any predetermined notions of outcomes or progress. Weil and co-researchers thus were in a (favorable) position of not having to justify the practical "success" of their involvement in the learning network with reference to clear performance measures.

7.3.6 An Anti-foundationalist Feminist Assessment

7.3.6.1 Scientism Challenged: Overturning Traditional Research Control

Weil did not try to present herself as "professional" researcher who, as such, knew how to organize the inquiry process. At times, as part of her involvement in the learning network, she added advice/suggestions as to how the learning process might proceed. And she also added her own concerns regarding the possible social consequences of managers' strategizing around short-term focus points. But her inputs in this regard could have all been challenged, given the climate of challenge that she was attempting to foster. In this sense her involvement in the inquiries was not meant to exercise a professional voice over and above that of the other participants.

As far as recommendations for action arising out of the inquiry process are concerned, Weil notes that options for action emerged for Tessa and other participants as they began to recognize that there was not one way of seeing their situations that could be "reasoned" to be more or less accurate reflections thereof. Weil highlights with reference to the example of Tessa's dilemma that Tessa, as others, were experiencing contradictions in action. She suggests that an appeal on any of their parts to some source of authority to pass judgment on "the realities" would not be an appropriate way to deal with the experienced uncertainties. In line with anti-foundationalist feminist concerns, Weil herself favors ways of "seeing the world as emergent, learning from complexity, diversity, the unknown" (1998, p. 44). Through her recounting of Tessa's story — as part of a wider story about developing learning potential — she hopes to display the strengths of working with this (anti-foundationalist) position in action. She also opens the story(ies) to the public sphere of discussion and debate — in keeping with Mauthner's comments that anti-foundationalist feminist inspired research re-opens for discussion certain issues by transferring them from the semi-private research situation to the more public sphere (Mauthner, 1998, p. 53).

As Mauthner notes, researchers can excavate "hidden facets of a relationship in the construction of unique narratives" (1998, p. 53). In this case, Weil renders public (to other audiences) how the co-researchers excavated Tessa's personal dilemma of integrity concerning her relationship with her CEO and colleagues. In this way Weil creates a forum for public discussion in regard to the handling of Tessa's experience of dilemma, by developing a narrative around the processes involved in exploring it in the learning network.

7.3.6.2 Phenomenalism Reconsidered: Working with Ambiguity

According to anti-foundationalist feminism, it is important for people to recognize that their ways of viewing situations can be re-examined with a view to extending their repertoires for choice-making. So, for example, Weil indicates that in the course of the project Tessa began to experience the possibility of showing her strength to her CEO. Whether he really was (before engaging with her in her new adopted role) displaying signs of top-down management, was not a point that needed clarity. What was important for Tessa was to create some scope to move beyond a univocal, unambiguous way of seeing his "real" behavior. Once she could do this, it seems that both he, and she, could find a way of working together.

Likewise, phenomena such as "capacity to address multiple bottom lines", "organizational performance", etc., did not need to be assigned a clear-cut definition for inquiries and attendant actions to proceed fruitfully. Of course, Weil hoped that as a result of the setting up of the CRAR learning network, members hereof and others with whom they were working on-site might improve their insightful action in their organizational contexts. But it is still not specified what time frame we (those concerned) should be operating with here, as a basis for considering the fruitfulness of the project in these terms. For instance, how long is needed before it should be decided that people's capacity to address "multiple bottom lines" was improved? And what would be needed in evidence hereof?

From an anti-foundationalist feminist point of view, the learning network was geared, inter alia, to enabling people to appreciate that trying to close the definitions of these, as other, phenomena could (adversely) affect finding ways of working with diverse experiences of the issues being raised by use of the terms.

7.3.6.3 Empiricism Reconsidered: Appreciating the Situatedness of Research

Weil focuses in her report on the manner in which the specific interactions in the learning network proceeded, and on how she chose to interject in the process. From an anti-foundationalist feminist point of view, she could perhaps also have provided (further) documentation around how she felt that participants were responding to her enthusiasm toward the possibilities of CRAR. However, Weil does suggest that researchers when working in this mode of co-inquiry need to attend to the documentation of starting assumptions and whether and how these shift. This would include how taken-for-granted practices and strategies have been challenged and changed. She suggests that this should entail multiple forms of knowing — not merely analytic approaches and rationalist reasoning (1998, p. 56).

Weil's report on the inquiry process is written in a genre that allows us to realize that what she writes about possibilities for CRAR's effectiveness, is partly a product of her own way of making sense of the different incidents that emerged in the course of the learning process. She indicates that her "story" points to what *she* sees as the potential for CRAR learning processes to make a difference to people's organizational existence — by their confronting with more comfort the ambiguity of their experienced world. But again, Weil's story could have been juxtaposed with other ones pointing, for instance, to people's continuing "discomfort" with trying to manage in the light of ambiguity and uncertainty. Although she does allude to the idea that learning "from contradictions at play" in organizational life is easier said than done (1998, p. 54), she could have brought this more openly into juxtaposition with her — at times optimistic — claims in regard to the potential for CRAR as a learning process to help people to address their management role(s).

7.3.6.4 Value Freedom Undercut: Appreciating the Value of Heterogeneity

Weil indicates that the commitment informing what she calls her "postpositivist" orientation adopted in the project, was to foster "collaborative learning in and from action, based on multiple perspectives" (1998, p. 44). From an anti-foundationalist feminist point of view, Weil's explicit commitment to nurture people's propensities to work with multiple narratives as they conduct themselves in the world, can be defended in moral terms. This is especially because Weil indicates that, as she sees it, values also need to be "cogenerated" in practice (1998, p. 44); her own sense of what is valuable can therefore itself be challenged in the process of engaging with others. But what could not be given a warrant within an anti-foundationalist feminist perspective would be to insist on developing an inquiry process in order to make (apparently reasonable) predictions from conditions to outcomes as a way of reducing uncertainty in people's experienced world. For this insistence itself might threaten the possibility of people's coming to terms with a variety of (differing) ways of treating problematic situations (experienced by them as dilemmas in action). Weil therefore rightfully expresses her value commitment to averting what she sees as a threat to the appreciation of heterogeneous ways of operating in the world.

7.3.6.5 Instrumental Knowledge Undercut: Questioning the Power to Define Realities

Weil indicates that as she sees it, the processes involved in the learning network have provided some "experiential understanding of what it means to build collective capacity for learning" (1998, p. 54). Weil considers it as important that participants seem to have furthered their capacities to engage in processes of inquiry that can help them to better address their "multiple accountabilities" (1998, p. 55). To be able to address these accountabilities, she suggests that they need to "let go" of the urge to command and control (1998, p. 55). And she suggests that the processes of CRAR can offer one way of helping people thus to "let go". As she sees it, then, the processes involved in the learning network are of "practical" import for people primarily in the sense of helping them to "let go" (from attempting better to control the world). Other members of the learning network may well offer other indications of what they believe they have learned through participation in the network. An anti-foundationalist feminist perspective would appreciate that in any case Weil does not wish her own account of the case (through the report) to be treated as offering a set of findings about what participative learning "really" can achieve, independently of the experience of those involved.

7.3.7 A Trusting Constructivist Assessment

7.3.7.1 Scientism Revisited: Developing Trusting Relationships through the Research Process

From a trusting constructivist perspective, Weil's involvement in the learning network could be seen as a way of using the inquiry process to enable participants to experience a form of trust building based on supporting but also challenging their different visions of issues of concern. In order for the participants to accept that their own constructions around their felt problems/dilemmas could be reworked, they needed to recognize that they were engaging with one another with a view to enriching their conceptualizations of their work situations. Arising from this process, each of them in turn could deal more accountably with the complexities of judgment facing them "on-site". At the end of the day (or rather, of the different days set aside monthly for the off-site learning), people in the group were (presumably) becoming more equipped to self-evaluate their ways of responding in their on-site situations, taking into account their own possible complicity in perpetuating a culture of submitting to the pressure of short-termism.

McKay notes that the advantage of action research — as a form of inquiry enabling people to become (more) reflective practitioners — is that people learn to devise ways of developing their roles as self-evaluators (1997, p. 13). In other words, through participating in coming together with others to "address particular problems or insights they might encounter in their [practical] situation" they develop skills in evaluating others (by both supporting and challenging them) as well as in evaluating themselves (by adopting the same attitude vis-à-vis their own experiences) (McKay, 1997, p. 12). What is important is that they utilize their learning encounters to extend their vision of possibilities for action. For this to occur they need what McKay calls some theoretical literacy in regard to different ways of conceptualizing the unfolding events in which they are involved. Hence McKay suggests that "practitioners of all varieties and persuasions need to have a degree of theoretical literacy if they are to play a role in social development" (1997, p. 11).

In the case of the CRAR project, the people in the learning network tapped into a variety of sources of ideas, such as literature and other sources (along with the pooling of insights around the possible meaning of their own experiences) — as a way of extending their "literacy". From a trusting constructivist point of view, this display of literacy is a way of accounting for choices of vision and of action as a process of engagement with ongoing controversies. Shotter indicates in this respect that provision of "good reasons" is the stuff of everyday discourse — wherein people "live [their] daily lives in an ambiance of conversation, discussion, argumentation, negotiation, criticism and justification; much of it to do with problems of intelligibility and the legitimation of claims to truth" (1993b, p. 29). Shotter argues, in line with constructivist argumentation, that to give good reasons (as a way of accounting for action) is not to base one's reasoning on "unwarranted claims to ... access an independent, extralinguistic reality" (1993b, p. 28). Likewise Weil would argue that this unwarranted way of creating reasons for action is not conducive to what she calls "finding responsible pathways through multiple realities" (1998, p. 44). Weil could nevertheless have offered more detail in her report on how, both off-site and on-site, the assigning of trust to others' judgments was being conceptualized by the participants (including by herself).

7.3.7.2 Phenomenalism Reconsidered: Evoking and Defending Observations

Weil used the action research process as an opportunity for people (including herself) to experience the notion that, as Doctorow puts it, "the world in which we live is still to be formed" (Doctorow, 1977, as cited in Gummesson, 1991, p. 122). Her aim was to enable people to consider how their ways of looking at "the phenomena" might influence their own and others' conceptions of options for action in their situated activity. Hence, if, for instance, Tessa viewed the CEO as intransigent in his management approach, this might affect her way of addressing him; while if she viewed him as able to appreciate some of her felt dilemmas, this would allow her to interact with him in a different way. If she viewed him in the former fashion (as intransigent), she would not simply be "observing" the facts of his style, but could well contribute to creating a certain form of interaction between herself and him (and also between him and others in the on-site situations). The group off-site allowed Tessa to extend her initial repertoire of interaction in relation to the CEO by allowing her to appreciate that he could be observed differently. The additional (and alternative) observations, however, do not mean that Tessa thereby developed an appreciation of what management style the CEO "really" was operating. This, as other phenomena, were recognized by Tessa and others in the learning network to be ambiguous. But her recognition that, as Gummesson puts it, "facts are not that unambiguous" (1991, p. 123), could be seen as indeed helping her to increase her range of options for action. From a trusting constructivist point of view, as long as Tessa can defend her position adopted in relation to these options, she can be considered as operating in an accountable fashion. (The defense hereof in this case would involve Tessa taking seriously others' views and engaging discursively with them, both off- and on-site, as a basis for her continuing actions.)

A trusting constructivist position would concur with Weil's argument that traditional notions of value-free observation might encourage people to try to minimize uncertainty with reference to "the facts". This, as Weil argues, can run counter to their developing the capacity to work with continuing contradictions-in-action. It is in view hereof that Weil opts to work with an alternative (epistemological) stance. As she sees it, either way she is unavoidably complicit — in the stance she adopts — in "making a difference" to the manner in which people might experience their world.

7.3.7.3 Empiricism Reconsidered: Exploring Alternative Ways of Seeing and Using Evidence

Weil hoped that the CRAR network could aid the participants to develop their repertoire of responses in the situations in which they were involved, by seeing them from multiple vantage points. Thus, for example, Weil notes that Tessa wanted to "understand better how [her initial stuckness] ... may be restricting choices of action and inquiry" (1998, p. 49). The learning network provided her with "different data sources to take away with her to guide choices about on-site CRAR that she might initiate in her own organization" (1998, p. 49). The choices that Tessa might discern arising out of her addressal of the "data sources" were left to Tessa to make. But she was accountable to others on-site for the way she chose to operate her role as a development specialist. And she would also have to account for her activities in following meetings of the learning network (off-site). In further meetings Tessa would need to explain how, in the light of varying understandings that were emerging within the learning network, she could defend whatever judgments she was making in her situated activity "on-site". The co-researchers thus expected Tessa to act in a way that would warrant people's trust in her — without the group expecting that she would need to act as if there was a single operative reality to which she would need to appeal.

Just as Weil encouraged the learning network to be used as a forum where multiple narratives could be formulated, so she styles her report accordingly. She notes that, from one perspective (the one that she favors in her report) "the example indicates how CRAR can better equip managers and others to live more comfortably with discomfort, ambiguity, and uncertainty" (1998, p. 55). She expresses this statement as part of a judgment on her part about what might possibly be learned through the example. But she has indicated at various points in the report that there may be other stories to be told out of the evidence of the project. The overall story that she presents is thus — on her own admission — an indication of her judgment that it is *worthwhile to continue to use CRAR* in certain contexts — as a possible way of better equipping managers and others to live with uncertainty.

A trusting constructivist position would suggest that Weil's way of drawing on the evidence of the case can be defended in terms of her explicit location of how her personal choices were brought to bear in the making of her "conclusions". Weil's defense of her conclusions is on the basis of her indicating that it is one of the stories that can be told, taking into account that it is not the only one that can emerge in relation to the evidence. Readers operating in terms of a trusting constructivist position would indeed not wish her to "make her case" by trying to reason in some abstract (impersonal) manner from evidence to conclusions. They would appreciate that her choice of reasoning springs from a way of making sense of continuing controversies in relation to the issues addressed through the project.

7.3.7.4 Value Freedom Undercut: Encouraging Discursive Accountability as a Value

Weil's involvement in the CRAR project in this case, was guided by her commitment to exploring and extending the participants' capacities (including her own) to develop "critically reflexive choices of responsible action and inquiry in complex and ever shifting situations" (1998, p. 58). Her involvement in the learning processes, is from the point of view of her belief that managers should try to work out of "contradictions-in-action" — rather than glossing over these contradictions (such as by focusing, for example, only on the "financial bottom line"). As far as Weil sees it, whether or not CRAR can indeed aid people to address their multiple accountabilities (in discursive engagement with others), depends, partly at least, on whether people themselves decide to embrace CRAR-type social and organizational learning processes.

From a trusting constructivist position, it is acceptable that Weil's value commitments (toward the installation of CRAR-type processes) guide her involvement in the learning network. This does not, however, release her from the need to continue to reflect upon her own values. So, for instance, she could have indicated to us what, if any, challenges may have been presented to her through the learning network; and she could have commented upon what this might have meant to her in terms of re-envisaging her values. She might have offered a section in her report indicating to us how her starting values shifted, if at all, in any way through the project. As it stands, Weil does, nonetheless, indicate to us where her current value commitments lie. But she is still faced with a dilemma as to how to present her creation of her story about the case. She cannot (with honesty) portray herself as having tried to put aside for the purpose of the research her own desire to "make real" the potential effectiveness of CRAR. She admits that she operated by urging people to consider the potential effectiveness of CRAR in terms of aiding organizational capacity to address ambiguity and uncertainty. She was at the same time trying to mobilize a willingness on others' parts to invest time in trying to make participation in such processes effective, while recognizing that criteria for considering effectiveness are themselves necessarily contentious.[47] From a trusting constructivist perspective, Weil was placed in the difficult position of having to try to defend CRAR in a culture where people might view it in terms of a "quick fix solution" and assess it on this basis. This renders Weil's accounting for herself (and the portrayal of herself) a difficult accomplishment!

7.3.7.5 Instrumental Knowledge Undercut: Tempering the Possible Self-fulfillment of "Results"

Weil recognizes that the manner in which claims about the world are presented, may have a self-fulfilling effect as they become part of the way people envisage, and enact, their social practices. In order to counterbalance the perpetuation of a culture of storytelling about the hazards of trying to work with contradictions-in-action, Weil encouraged stories (and critical incidents) to be excavated in the learning network around the experience of working with contradictions. Weil's account (to audiences outside the learning network) of the CRAR processes involved in this case, is likewise an attempt to offer a somewhat countercultural narrative — alongside what she sees as culturally dominant pressures to diminish ambiguity and uncertainty (1998, p. 44). Through her script she tries to counter stories that focus on the need to develop clear performance indicators so as to ensure "so-called quality and cost control" (1998, p. 41). Stories in regard to the need to control performance mean (for Weil) that those whose performance is being "measured" may find it difficult to live with the dilemmas that they face in action.

A trusting constructivist position appreciates that Weil's manner of dealing with what she sees as the self-fulfillment of organizing univocal narration, is by trying to encourage a different way of life to be experienced. This is a way of life where people's accountabilities can be assessed (and, by implication, trust earned) through their active engagement in critically reflexive co-inquiry processes as a basis for finding "responsible pathways through multiple realities" (1998, p. 44).

7.3.8 Trusting Constructivism Summarized in Relation to Alternatives

I present below a summary of the alternative visions of the project as I have explored them, with a brief indication of how a trusting constructivist position might develop an argument in relation to each one.

---

[47] Adman and Warren point out that the term "effectiveness" when invoked in organizational discourse is used in a "soft" fashion to call forth discussion about what it might mean to "do a better job" (1996, p. 63). They suggest that to appreciate the softness of the concept, we need to appreciate the way in which "measurements" of it arise (and are contested) in processes of social interaction. Bell and Morse (1999, p. 155) apply a similar argument in relation to sustainability indicators for the effectiveness of environmental policies. They suggest that just because we can never arrive at a definite set of indicators of this, does not mean that it is impossible to talk around the concept (and create responsible actions ensuing from the discussions).

7.3.8.1 Positivism and Trusting Constructivism

From the point of view of a *positivist* position, the project described by Weil can be regarded as scientifically defensible only if it can be said to embrace a research design allowing for reasoned argument from evidence gathered toward the drawing of some conclusions. If those involved in the learning network in this case wish the project to be placed under the banner of scientific inquiry, they would need to specify what hypotheses were being developed in the project and how the "critical incidents" to which Weil refers, have a bearing on proving or disproving them. As it stands, it is not possible to decide whether a CRAR-type learning process instituted in an organization by senior managers does indeed have the effect of reducing cynicism of middle-level managers, as Weil believes (and hopes) it might have. Likewise it is impossible to tell whether a CRAR-type learning process instituted in an organization by senior managers indeed allows them, and others, to better address what Weil calls "multiple bottom lines". We would need clearer indicators of the variables under study in order to be able to argue (with good reason) that the effects that Weil hopes might be instituted through CRAR, do in fact ensue.

However, from a *trusting constructivist* point of view, we need to recognize that what positivism might interpret as being a scientific study, relies to a large extent on our placing trust in the "scientific" possibility of observing the effects of introducing clearly definable alterations in situations under study. Yet it is by no means clear that one can ever set up processes of social inquiry in this way. People's trust in scientists who present this as being their aim, may thus be misplaced. In this case, Weil preferred to organize the inquiry as a process of extending deliberations around what the researchers/participants took to be revealing experiences relevant to the inquiry. The point of the inquiry was not first and foremost to develop statements in regard to the existence of links between certain independent and dependent variables that can be argued to hold across different social contexts. It was more to offer narratives that can be taken into account by actors making their own judgments about possibilities for insightful action in the contexts in which they are situated. Members of the learning network were thus invited through their involvement in the network to extend their repertoire for insightful action; and other audiences hearing the stories told too can decide what they might mean for them in their own contexts of action.

7.3.8.2 Non-foundationalism and Trusting Constructivism

A *non-foundationalist* position would forward the complaint/criticism that Weil and co-researchers' study was not sufficiently well designed to set up scientific tests that would allow us to adjudicate statements made with reference to their grounding in the evidence. If the study had been thus designed, then at least we would be more likely able to consider to what extent the commonsense knowledge brought to the project by the participants, matches up with the realities under consideration. It is quite possible that due to their immersion in their own management practice, they do not have the sufficient detachment to judge the evidence of their own and others' experience in terms of its bearing on hypotheses being developed. If Weil and others had availed themselves of mechanisms for introducing more objectivity into the process of drawing conclusions, statements made would be more likely able to offer a better appreciation of the realities. And then, if this appreciation is fed back into the social network, people can make use of the improved information to aid their decisions for action. From a non-foundationalist point of view, there seems little point in considering the learning endeavors undertaken as being "scientific" — unless the aim was to advance knowledge of the realities being investigated.

The *trusting constructivist* response to this (non-foundationalist) assessment of the study is that it assumes that the knowing process must be directed at realities that exist irrespective of the knowers involved. The trusting constructivist position challenges this assumption. Weil's and others' decision in this case to utilize a variety of avenues for creating stories, and to allow for a variety of stories to emerge, was a way of setting up a forum where people could engage with alternative narratives as a basis for developing options for insightful action. In order to earn the trust of those assessing the research, they are called upon to defend their way of proceeding with the inquiries. They can do this by indicating in what sense they were aimed at generating insights that could be translated into responsible action and change (as indeed Weil has done). If the defense of such learning processes rested solely on conceding to non-foundationalist criteria for judging the worth of science, reasoning would never be able to move beyond what is regarded as acceptable by the scientific "colleagues" to which non-foundationalist argumentation appeals.

7.3.8.3 Scientific Realism and Trusting Constructivism

According to a *scientific realist* position, Weil's keenness to set up processes of collaborative learning, might have been at the expense of her developing inferences about the causal mechanisms underlying the various reported experiences in the group. Although Weil refers to literature and other sources that could be used as an aid in the group's reflections, and although she refers to the need to contextualize experiences with reference to an understanding of histories and cultures, it is unclear whether participants were facilitated toward gaining an understanding of the structures of the society in which their actions were embedded. Weil should have been more specific about this.

The *trusting constructivist* response to this (scientific realist) argument is that it suggests that participants in the learning network would have had to gear themselves to thinking that their aim was to advance knowledge of "the realities" in which their activities were enmeshed. A trusting constructivist position prefers to accord trust to people by judging their manner of engaging with differing points of view and their manner of leaving space for alternative visions. In this case, the various members in the learning network seem indeed to have allowed one another to appreciate that there would always be multiple operative narratives, and that it was out of the engagement with this multiplicity that they could learn to work with the dilemmas faced by them in their situated activity. When there was disagreement (or what Weil calls divergence) they did not try to address this by appealing to one another to make reasonable inferences about what "really" is.

7.3.8.4 Interpretivism and Trusting Constructivism

Weil and co-researchers organized their inquiries with the intention of extending their repertoires for their situated activity in everyday life, based on the pooling of their various experiences and their reflections thereon. From an interpretivist point of view, this approach to investigating the social world, may be seen as one way of "getting close" to an understanding of everyday social settings. However, Weil could have been more attuned to the way in which people in the learning network may have been giving meaning to mutual interactions. Also, she should have been more alert to how her own value commitments may have been influencing her involvement with others in the project. Nonetheless, the advantage of Weil's approach is that it does not try to create generalizations about human organizational life in the form of "if-then" causal patterns, but concentrates on portraying possibilities for ways of interacting that might be relevant to actors in different social contexts.

As does interpretivism, a *trusting constructivist* position can appreciate Weil's efforts at organizing investigations by working with people's "everyday" felt dilemmas. However, a trusting constructivist position is less wary than is interpretivism of the values that Weil admittedly brought to the project. Just because Weil admits that her views on CRAR's effectiveness are a function partly of her own value commitment to make it effective, people are in a better position to create their own choices of vision and of action in relation to possibilities presented by Weil about CRAR. Weil could defend her drawing on the values that she brought to the study, by arguing that it is in terms of these very values, that she allows others to recognize the situatedness of her own ideas. This means that people need not to be awed by them as if they present some "authority". But Weil cannot defend herself in a culture where people are expecting that others *must* operate by striving toward value freedom in their role as researchers. A trusting constructivist position is sympathetic to Weil's suggestion that her accountability as a researcher does not rest on her trying to act in line with such expectations. Her commitment to critical reflexivity through a project based, inter alia, on the use of critical friends as integral to the inquiry process, can be respected.

7.3.8.5 Critical Theory and Trusting Constructivism

Weil's conscious and explicit intent to explore, with others, a way of inquiring that could allow them to develop their capacities for intersubjectivity, resonates with the normative commitments of *critical theory*. This is especially if the co-researchers' intentions were to have been directed toward developing mutual understanding in the form of consensus around the issues brought to the fore. How they were managing to work together around different understandings that were being developed, could have been given more attention in Weil's write-up. Also, how they managed to work with their emerging understandings in the face of cultural pressures for measuring the success of their organizational actions against "factual" performance indicators, could have been given more attention.

From a *trusting constructivist* point of view, it would be appreciated that Weil tried to operate the "organizational route" (in this case via the management of public sector organizations) as a way of contributing to social transformation. The project was set within a framework of critical concerns about dominant approaches to public sector reforms in the United Kingdom, for example, concerns about the excessive use of market-based approaches to service design and delivery. Weil's co-researching efforts with senior managers in this sector (as part of a process of social development) could, however, have been accounted for in a more concerted fashion for the benefit of other audiences. Had Weil set up a more detailed discussion of her felt accountabilities, those judging her said commitment to social transformation could be helped to appreciate the epistemological stance that she brought to bear in the project.

7.3.8.6 Anti-foundationalist Feminism and Trusting Constructivism

Weil's way of accounting for her involvement in the learning network can be considered as fitting in with the epistemological (and ethical) arguments of *anti-foundationalist feminism*. Weil indicates that she hoped to instantiate an epistemology of practice that would enable people to organize inquiry toward an excavation of multiple realities, as a way of broadening the conceptualization of felt dilemmas. She opposed in practice the notion of research as an endeavor leading to the development of a picture of social reality argued (using some form of reasoning) to provide a more or less accurate "representation" thereof. She expressed a commitment to processes of learning oriented toward appreciating multiple realities as a basis for addressing multiple accountabilities.

From a *trusting constructivist* point of view, insofar as Weil feels that she is operating in a cultural milieu where people expect a certain form of "scientific" inquiry to lead to knowledge of reality, her ways of justifying her endeavors become indeed a more difficult accomplishment. A trusting constructivist position recognizes that researching efforts in society need to be accounted for; but following anti-foundationalist feminism on this score, it is preferred that they be assessed in terms of the skills utilized to develop, and engage with, multiple narratives. A trusting constructivist position appreciates that Weil's task of defending her decision to help people in the learning network (including herself) to "find pathways through multiple realities" is rendered the more difficult if the cultural climate (as she sees it) is not conducive to this kind of research activity.

7.4 CONCLUSION

Weil's co-researching project was used in this chapter as an example of an attempt to set up social inquiry as a learning encounter involving practitioners engaged in some social setting. In considering this, as other, action research projects, positivism and non-foundationalism both would suggest that action researchers must be careful not to interpret all observational material coming forth, as pointing to the need to continue with certain lines of action that they happen to favor. In this case, although Weil refers to the validation of inquiries in terms of standards of rigor, she has not defined clearly enough the hypotheses that were being tested concerning the connection between actions and outcomes within the project. This needs to be more clearly specified, so that information of relevance to decisions regarding continuing action can be accrued.

Scientific realism too suggests that although it is in principle possible to organize action research as one way of evaluating social programs, the role of the researchers should be clarified so that they recognize that in *this role* they are seeking to develop an understanding of the operative realities. According to scientific realism, the research task in action research, as in all forms of research, is to try to gain some knowledge about the real structural mechanisms that are at play in the society (and their manifestation in the contexts that are being investigated).

From an interpretivist perspective, the advantage of action research lies in the opportunities it can create for paying close attention to the experiences of people in terms of their ways of operating in meaningful reality. Weil's working closely with senior managers in exploring and excavating their dilemmas, could well have enhanced the research capacity to generate a plausible account of how they address the pressures of their management role(s). This is provided that Weil and co-researchers were able to adopt a research role somewhat separate from their role as value-committed actors (wishing to progress certain values in the course of the inquiry).

Within critical theoretical and anti-foundationalist feminist positions, attention is given to ways in which research can become a process of expanding people's sensitivities to others, in relationships of mutuality (in whatever way this is considered). Weil's CRAR initiatives, underpinned by the value of enabling people to experience (new) possibilities for mutual learning, may therefore be regarded as one viable research route, amongst others available.

From a trusting constructivist point of view, what is important about Weil's employment of CRAR is her use of the research process as a way of building up trust relationships between those involved in the learning network. Her participation in the learning network can be seen as involving an effort on her part to organize an experience of trust in the group based on people supporting as well as challenging others as "critical friends". They thus could learn that they do not need to try to work toward developing a representation of "the realities" in order to excavate options for insightful, responsible action on each of their parts. At least they will have been exposed — through this action research project — to a different epistemology of practice (and attendant conception of trust).

As far as the funding of the research effort is concerned, Weil reports that she and her co-inquirers did not feel compelled to try to show that instituting CRAR in the learning network could produce definite tangible results. Nevertheless, from positivist and non-foundationalist points of view, Weil's "story" (presented in her report) should have been an attempt to draw together more systematically the evidence from the various critical incidents and other documentation produced during the project. This would have allowed her to make some statements about whether it is reasonable to suggest that participative learning is really likely to be effective in, say, aiding managers to address multiple bottom lines. If she believes that more evidence toward the testing of such statements is needed, she should indicate more clearly that this is her position.

From a non-foundationalist point of view, it would be emphasized that at least she should indicate to us how well tested the results are and therefore how tentatively she believes they should be understood.

From an interpretivist point of view, she might have ventured to offer more clarity on the way in which she envisaged outcomes being created through different meaning-making activities on the parts of those involved. This might have rendered the results more meaningful to the senior managers funding the project. Other audiences could also have contextualized accordingly the information developed.

However, seen from a scientific realist point of view, the research process could all too easily embrace a reformist political agenda by focusing too much attention on people's immediate concerns — in this case, the concerns of managers having to address their specific felt dilemmas. More attention should have been paid to understanding how the pressures that they were experiencing could be related back to the structures of the market system gripping both private and public service organizations.

From a critical theoretical point of view, Weil could be seen as taking up the challenge of trying to work with people on practical issues of concern, while bearing in mind a sense of her wider social responsibilities in terms of excavating possibilities for transformative action. Of course, due to systems' pressures for efficiency operating in contradiction to lifeworld potentialities toward co-ordinating action along different lines, the participants (as Weil herself) were bound to experience dilemmas in action within the organizational contexts in which they were operating. Trying to set up some debate through developing participative learning off- and on-site is at least a way of recognizing these, and, on this basis, exploring possibilities for extending repertoires of action. The fact that the senior managers allowed Weil access into their world of action is an indication that they felt the need on some level to subvert the "immediate" pressure to act in accordance with the financial bottom line. It was this felt need that Weil tried to work with as she chose the "organizational" route to transformative action.

From an anti-foundationalist feminist point of view, it would be appreciated that Weil was intent on defining a role for herself in which she could relate to others without posing as having the skills to better explore their worlds. This did not prevent her from participating in the learning network by introducing ideas in regard to possible ways of proceeding with the inquiry, and in regard to possible ways of interpreting "critical incidents" emerging through the inquiry. People could recognize the skills and concerns that she brought to the study from her experience. But they were also encouraged to draw on their own experience — both in considering ways of conducting the inquiry process and in considering ways of interpreting unfolding events (off-site and on-site).

From a trusting constructivist point of view, it would be noted that Weil displayed an awareness of multiple expectations that might have been brought to bear on the co-researchers in terms of their research remit. Weil bore these in mind while at the same time trying to develop a way of operating that meant overturning traditional ideas about the relationship between "knowing" and "acting". She felt it worthwhile to operate with the understanding that people need not first try to "know" as a basis for sound action; for in action they might experience dilemmas, that, when reflected upon, can enable the development of "responsible pathways through multiple realities". But she recognizes that this suggested relationship between knowing and acting that she herself tried to enact, is socially respectable only if people are prepared to place more trust in others to act in accordance with felt contradictions (in relation to an engagement with ongoing controversies).

Weil presents the research approach (CRAR) and the specific details of the inquiry process in this case (see Figure 7), as one way of responding to what she takes to be her responsibilities as an inquirer/person. She regards herself as contributing to setting up an inquiry process in which she and others could extend themselves as persons in working with alternative "realities" that became presented. *She does not regard herself as accountable primarily to a scientific community of colleagues operating with a traditional epistemology of practice with the aim of finding out about reality.* Nor does she see herself as enskilling practitioners in the learning network to operate in a fashion that might be seen by such "colleagues" as doing science. Rather, the aim was to introduce an alternative approach to inquiry and an alternative way of addressing the ambiguity of (factual) evidence to that offered by traditional forms of inquiry.

Of course Weil, with others, might have chosen to organize the inquiry strategy differently. Weil might indeed herself have chosen not only to become involved with senior managers primarily "off-site", but also with some of them "on-site".[48] She might also have chosen to consider other ways of extending her own and others' accounts of the difficulties of handling multiple bottom lines, by inviting people (such as middle managers, other employees, customers, etc.) into person-to-person interviews (along the lines of, say, the active interviewing discussed in Chapter 6). But from a trusting constructivist point of view, Weil's discussion of processes that were adopted does not need to be defended as the "best" way of proceeding in the circumstances (or in similar ones). It need be defended only as *a* way to proceed — in view of concerns that might be raised in relation to the purpose of the inquiry.

Considering the research "products" (as insights) that were generated in the process of the inquiry, Weil indicates that these had the status (for her) of aiding people to conceive possibilities for developing their choices of vision and action. They are not, for her, an expression of an attempt to "represent" reality. This means that alternative visions might well be presented. To whatever audiences she directs her story(ies), Weil invites them to come to their own conclusions and generate their own readings of the documents/critical incidents reported. To aid continued discourse around her write-up, she offers other narratives in juxtaposition to her optimism about the effectiveness of CRAR — by, for instance, acknowledging the possibility of continuing cynicism that might run counter to making CRAR effective. The glimpses she provides of the possible effectiveness of CRAR can be read as glimpses indeed of a manner of organizing social relationships where people can earn the trust of others through their way of working out of contradictions-in-action. But, as she notes, the appeal for trust via this route also depends on the development of a cultural climate where accounting for oneself in recognition of the experience of dilemma, is rewarded — rather than devalued.

---

[48] The approach that Weil opted for (by operating "off-site" in this case) can be seen as bearing similarity with what Moggridge and Reason (1996, p. 162) call setting up cooperative inquiry groups.

8
Conclusion: Accounting for Different Conceptions of Accountability in Social Research

8.1 INTRODUCTION

This chapter is organized to pinpoint my argument for my preference for a trusting constructivist conception of accountability in social research in view of what I consider to be alternatives. Certain alternatives have been elucidated in the book by drawing on various categories under which different arguments were presented. At the end of this chapter (Section 8.8), I offer more detail on how I regard the use of these categories as part of the process of organizing the debate. At this point, though, using the categories, I iterate my defense of the trusting constructivist position as has been explored in the book, by summarizing the main arguments.

8.2 SOCIAL RESEARCH AS DOING SCIENCE

The tenet of *scientism*, which I isolated in Chapter 1 (following Delanty) as being one of the tenets associated with positivist argumentation, suggests that it is crucial that the processes involved in the doing of social research can be regarded as being scientific. If not, it is difficult to account for them, and for the amount of social costs afforded to "professional" social inquiry in society. Non-foundationalism and scientific realism qualify the positivist tenet of scientism somewhat. (While accepting the unity of the sciences, they consider how an understanding of social science needs to recognize the human element in the social world.) But they still follow the idea that the scientific community can rightfully be expected to define the limits of acceptable scientific research in order to advance knowledge of realities investigated — whether natural or social reality. The interpretivist position in turn stipulates that the study of social reality must take specific account of the meaningful character thereof. It is suggested that researchers should be able to defend the plausibility of accounts offered by showing how results can reasonably be argued to relate to people's meaningful experiences.

Non-foundationalism, scientific realism, and interpretivism all admit that whatever results are offered by social researchers as advancements toward our knowledge of reality, cannot be taken as certain knowledge. In all of these positions, the positivist tenets of both *phenomenalism* and *empiricism* are revised — albeit in different ways — to take account of the impossibility of finally grounding results in "the evidence". Notwithstanding the revisions made to the positivist tenets of phenomenalism and empiricism, these positions still incorporate both the tenet of *value freedom* (as something to be striven for) and of *instrumental knowledge*. In terms of these last two tenets, it is suggested that as long as social researchers act accountably as researchers, and as long as mechanisms exist for judging the way in which they create findings, results achieved can be regarded as likely to be better than those that would be achieved if scientific processes were not followed. People can then base their actions on the informed understandings that represent the state of knowledge (to date) developed via scientific inquiry.

However, I have suggested in this book — following critical theory and anti-foundationalist feminism on this score — that the positivist tenets of value freedom and of instrumental knowledge need radical revision in the light of the recognition of the impossibility of accessing extra-linguistic reality. My argument is that as soon as the positivist tenets of phenomenalism and empiricism become revised to cater for the impossibility of our checking statements against (extra-linguistic) reality, the tenets of value freedom and instrumental knowledge also need to be revisited. The four examples of research that were used in Chapters 4-7 were meant to offer an indication of how research processes can be argued (within critical theoretical, anti-foundationalist feminist, and trusting constructivist argumentation) to *evoke*, rather than to represent, realities being investigated.

I suggested in Chapters 4 through to 7, that positivist, non-foundationalist, scientific realist, and interpretive forms of argumentation all in their own ways try to turn research strategies into processes for developing knowledge of social reality. However, I argued that there is no need to conceptualize research processes in terms of their contribution to attaining knowledge of realities posited to exist outside of the knowing process. The important requirement for social researchers earning others' (colleagues', participants' or other audiences') trust is that they can defend the choice of research strategy(ies) adopted, while simultaneously pointing toward a recognition of alternative ways of creating/evoking research material.

According to trusting constructivism, the products (stories) arising through processes of theorizing should not be presented as enabling humans to get closer to "the truth", defined as representation of reality. They should be presented as an option for making sense — which thereby opens the way for readers/audiences to enrich their own possibilities for sense-making as they engage with the proffered accounts. It is in this way that I have suggested that social researchers can deal (more) accountably with the potential self-fulfilling effect that their story telling might create in society — if people come to act in terms of the visions that they advance (even when they are presented as somewhat uncertain).

The belief that is disseminated in society that by striving to get closer to the truth the scientific community can manage to organize this, is based, as I see it, on an act of faith. But this self-understanding of science allows scientists to free themselves from the responsibility of having to consider the possible social consequences of their own acts of "knowing" for others in the social world. The presumption that following canons of scientific inquiry *is* likely to lead to better understanding, allows researchers to consider their sole accountability as consisting in striving for such understanding and presenting their state of knowledge to others. They need not concern themselves with the way in which their proposed insights might impact on others' enactments of meaning — because, as far as they are concerned, this impact is likely to be for the better (in that it allows people to obtain better understanding of their realities).

The longstanding dispute about whether or not science should be seen as advancing our knowledge of realities existing independently of discourse, cannot be resolved with reference to "the evidence" about what science "really" can achieve. But my argument is that just because we can never know whether so-called scientific investigation is leading to advancements of knowledge defined as reflection of external realities, researchers do need to be particularly wary of presenting it as a process of developing such knowledge. Working with a conception of knowing as a process of developing constructions might be considered preferable not because it can be proved that theorizing constitutes, rather than reflects, "realities", but because it provides more scope for inquirers to engage in a discursively accountable way with others.

The notion of discursive accountability suggests that as we engage with "the world", we acknowledge that we have to take responsibility for the possibility that our own understandings (and ways of expressing and working with our understandings) might affect outcomes (or at least the experience of these) for others. Our understandings cannot be isolated from the (self-fulfilling) impact that they might have. When knowing is seen as a process that is aimed at accessing some independent reality, people in society might easily come to rely on the knowledge-producing machinery of science — just because of its presumption to bring us closer to "the truth". Although it is recognized within, say, non-foundationalist, scientific realist, and interpretivist argumentation, that the statements afforded by scientists can never be regarded as certain, the statements are still presented as being of some instrumental utility. Their utility springs from the (presumed) fact that as long as they arise out of acceptable ways of doing science, they can rightfully be accorded some authority in relation to other ways of becoming informed about reality. They can thus be used to aid people in organizing more informed action in relation to the world.

It is at this point that trusting constructivist argumentation takes issue with these realist-oriented positions. The trusting constructivist concern is that when knowledge-production is conceptualized in this way, "reason" is oriented toward trying to claim recourse to the evidence (to date) of reality. I have suggested that the "reasonable" reference to reality, however, might run anathema to people's trying to build trust by defending their choices of vision and action as part of their *engagement with alternatives in discourse*.

The view of trust that I am proposing as a way of people orienting themselves to others in processes of discourse might be seen as relevant more to so-called high trust cultural climates — where people associate trust awarding with respect for others as persons. Maclagan points out that in such cultural settings, people work with the idea that "to trust someone is to recognize their abilities and potential" (1998, p. 55). He notes that "in high trust societies people expect to be trusted, and in the organizational context this suggests that violations of that expectation are likely to be interpreted by people as a sign of others' lack of respect for them" (1998, p. 58). Maclagan raises the question as to whether some allegiance to the notion of trust is implicit in the codes of all societies, albeit "ultimately bound up with social customs and codes" (Trusted, 1987, as cited in Maclagan, 1998, p. 58). Maclagan suggests that the ethics of keeping trust (once this is awarded on some basis) could be argued to be a "universal norm" (1998, p. 59). As he sees it, no matter on what basis trust is placed (and it may be placed on different bases in different cultural contexts and also be placed with lesser or greater degrees in different contexts, depending on how much discretion is granted to those with whom it is placed) there appears to be some normative requirement to keep trust once placed. (See also Fukuyama, 1995, p. 26; and Jones, 2000, p. 197.)

My argument in this book is that one can award trust in people's capacities to make judgments about ways of seeing and acting, without expecting them to have to justify themselves with reference to the soundness of their vision in relation to "the realities". Requirements on them to "keep trust" could be requirements on them to be *discursively oriented to defending choices of vision and action in serious engagement with alternatives*. I am not proposing that this basis for placing trust and for judging people's trustworthiness should be operative in every society and every cultural context, without people deciding its relevance for them. Nevertheless, I believe that it should be seriously considered as affording a style of human interaction that can take into account the human impossibility of our accessing independently existing realities. I have tried to defend my suggestion that this is an option to be considered, by showing how I have developed it in engagement with other ways of approaching the question of what "knowing" as an enterprise in society might involve. I believe, along with critical theoretical and anti-foundationalist feminist argumentation, that any answer to the question of what knowing involves, at the same time endorses some value in society. I evoke the value of discursive accountability (as potentially operative in contexts of "professional" social inquiry and in everyday discourse) as soon as I defend the trusting constructivist position. I present this value as an option (way of life) to be given consideration.

I now proceed to offer a summary of how a trusting constructivist position can be built up in relation to debates about the employment of the research strategies that I discussed in Chapters 4-7.

8.3 ACCOUNTING FOR EXPERIMENTATION

In Chapter 4, I used the example of Dovidio et al.'s (self-labeled) experiment to open a discussion on differing ways of treating the accountabilities of inquirers involved in setting up experiments as a way of organizing social research. Employing the categories discussed and developed in Chapters 1-3, I suggested the following.

From a *positivist* point of view, Dovidio et al.'s accountability as scientists would be judged by focusing on the way in which they used the scientific process to create conclusions about causal relationships operative in social reality based on the evidence that they collected. The (argued-for) internal validity of the experiment (that is, the ability of it to establish whether causal connections exist between the independent variable of group identification and certain patterns of response) rests on Dovidio et al.'s convincing readers that scientific procedures were well invoked in the experiment. For instance, procedures for operationalizing variables of concern have to be accounted for, and reasons for regarding observed evidence as pointing to a relationship between the isolated variables would have to be spelled out (more fully than is done by Dovidio et al. thus far). The external validity of the experiment would be seen as resting on their reasonable way of comparing the results of this experiment and other studies (in which they or other scientists were involved) in order to justify general inferences being made (about existing causal connections). The reasoning involved in organizing these comparisons would have to be rendered (more) explicit. The accountability of Dovidio et al. is thus a matter of their ability to defend statements regarding the likely validity (internal and external) of the results, by pointing to the processes undergone in order to come to the conclusions that they did.

From a *non-foundationalist* point of view, attention would be given to the possible sources of error that might have been operative in the research process, so that critique (and self-critique) could become a way of locating and trying to minimize these errors. Also, the product (the results) would not be portrayed as ever providing true statements, but rather as presenting statements that have become developed through the setting up of many studies, in efforts to disprove the results. If the results still stand up to the test of the evidence in a variety of research situations, then there is a stronger chance that they tell us something about the realities being investigated. Insofar as Dovidio et al. could be said to have taken into account possible critique of their scientific process, and also to have been sufficiently critical of other studies undertaken by themselves and others, they might be said to be acting accountably as scientists. But as it stands, they presented their report around the claim that they were (probably) able to collect data via their instruments of observation, and that they could use these data as evidence toward the conclusions that they reached. They could, at least, have been more circumspect about the way in which conclusions were developed.

From a *scientific realist* point of view, similar observations could be made about the research process and the need to operate self- and collegial criticism in order to act accountably. However, "colleagues" should not be confined to those working with a successionist view of causality such as the view upheld by Dovidio et al. Rather, colleagues who recognize the need to create experiments aimed at uncovering generative mechanisms in social reality should also be consulted, with a view to considering how such experimentation might be designed and accounted for.

From an *interpretivist* point of view, the experiment can be accounted for only if "variables" isolated provide openings for generating categories that connect with people's meaningful experience of social reality. The variable of "group identification" for instance, could become an opportunity to open a discussion on the meaning hereof for participants (in this or some other research investigation). Only in this way, can the efforts of Dovidio et al. become defended as contributing to an understanding of the constitution of social reality.

From a *critical theoretical* point of view, because Dovidio et al. were too much focused on the effort of finding out about causal relationships operative in social reality, they were unable to recognize that as they interacted with participants in the experimental process and as they presented results, they could all too easily be perpetuating (in a self-fulfilling way) the "found" relationships. They have not accounted sufficiently for this possible consequence of their work.

From an *anti-foundationalist feminist* point of view, similarly, accounting for the worth of research undertaken is not a matter of trying to defend conclusions regarding causal relationships holding in reality. The focus in this case would be on how researchers might justify the setting up of the inquiry process so as not to endorse patterns of (traditional) researcher control. The selection process of using "subjects" to take part in experiments has to be more carefully arranged so that they can voice opinions and engage in argument in regard to ways of conceptualizing their involvement in the study. Efforts also need to be made for setting up arenas for discussion within the process of conducting the research, so that on an individual and/or a group basis people can reflect on this process. This is one way, in any case, of building up a defense of experimentation from an anti-foundationalist feminist point of view.

From a *trusting constructivist* point of view, following critical theoretical and anti-foundationalist feminist perspectives, the process of the research should not be accounted for in terms of its argued-for ability to extract information relevant to the testing of hypotheses. The defense of experimentation has to take into account the kinds of objections to it that have been raised from various quarters, including critical theoretical and anti-foundationalist feminist ones. From a trusting constructivist point of view, participants and other audiences would need to be made aware of the evoked character of "phenomena" being investigated and of any statements made in relation thereto. When researchers interact with participants, and as they record the process, their accountability to participants and other audiences rests on their orientation toward developing their own and others' capacities to engage discursively with alternative interpretations of the achievements of the inquiry.

8.4 ACCOUNTING FOR SURVEY RESEARCH

In Chapter 5, I used the research project organized by Romm et al. to open a discussion on different ways of treating the accountabilities of researchers involved in organizing surveys and reporting on ensuing findings. I detailed various arguments in regard to Romm et al.'s endeavors as follows.

From a *positivist* point of view, Romm et al.'s accountability as scientists would be judged by focusing on the way in which they used the survey to collect information in regard to the variables under investigation. Although the survey was largely descriptive, it could in principle (in another study, for instance) be extended so as to put to the test some hypotheses about possible relationships holding between variables. So, for example, one could put to the test a hypothesis suggesting that tutorial contact has some effect on the way students learn (as long as terms used in the hypotheses can be well defined). Provided that Romm et al.'s survey (as it stands) is recognized in this case to be an exploratory attempt to develop some hypotheses in regard to possible connections between variables (so that correlations or causal relationships can be tested rigorously at some future date), they can be considered as contributing to the doing of science. But Romm et al. still could have offered an explicit account of why they believe that their own prejudices did not unduly influence their way of organizing their observations through their questionnaires, interviews and discussions (as techniques of observation). Putting aside these concerns, they have made a good case for why the data collected via the samples chosen, can be argued to be representative of the populations of interest. So the population validity of the survey (the generalizability of results to the populations of past ABET students at UNISA, tutors, and regional co-ordinators) need not be called into question. But more analytic work still needs to be done in trying to discern whether relationships of correlation or causation hold between the variables isolated.

From a *non-foundationalist* point of view, it would also be remarked that thus far it is difficult to consider whether the evidence collected by Romm et al. offers us any advancements of knowledge into patterns of correlation or causation that might hold between variables studied. Nonetheless, their survey does offer a starting point to develop hypotheses in this regard. Romm et al. could have, however, at this stage provided more information on why they believe that the answers given by respondents in the various contexts in which techniques of observation were administered, should be taken as telling us about realities (rather than simply about the way in which interactions with respondents were organized). Their report could have benefited from such commentary. They could have based their commentary on indeed asking some of the participants about this (where possible). This would have allowed Romm et al. to report better on the (potential) reactivity effect. They would also have benefited from conducting more of a debate with colleagues in the scientific community regarding their way of defining the terms used in the various questionnaires and the interviews. To act accountably as scientists they would need to orient themselves to taking on board possible and actual collegial criticism. As it stands, they have not shown sufficiently clearly how they have used their involvement in the scientific community as a way of minimizing their own possible errors of judgment.

From a *scientific realist* point of view, again, the need to operate self- and collegial criticism in order to act accountably would be focused upon. However, in line with the scientific realist insistence on the need to be aware of generative causality that may be operative in social reality, colleagues who work solely with a successionist view of causality would have to be criticized as part of the process of dialogue in the scientific community. Romm et al.'s work would be criticized from a scientific realist perspective with a view to inviting them (or others reading their work) to widen the scope of their considerations to include inferences about the operative causal mechanisms in this case. Adding these considerations would increase the possible utility of the information for funders of the report and other audiences (including those involved in and touched by ABET concerns). These audiences could become better informed of the depth of the changes that may be needed in order to achieve desired educational objectives in South Africa.

From an *interpretivist* point of view, the survey organized by Romm et al. was an interesting attempt to account for the character of "variables" as constructs relating to people's meaningful experience of social reality. Romm et al. tried to explore chosen variables by considering the meaning that might be attributed to the ticking of the closed response categories in questionnaires, to the comments made by respondents via open-ended ones, and to their answers given through interactions in (face-to-face) interviews. Although Romm et al. did not explore meanings with participants directly in the case of their collection of data from students, they did this indirectly by involving tutors who worked closely with them. They were also able in their report to account for their interpretations of the data collected by showing how they were created in engagement with others' interpretations of what might have been involved in students' (and others') interactions with the various devices for "data collection". In this way, Romm et al. made efforts to improve the quality of the data that was gleaned from the fleeting encounters organized through the questionnaires and the interview guide administration. They are thus more easily able to defend their suggestions that they achieved some closeness to people's everyday experiences — and hence arguably were able to develop categories grounded in meaningful experience. Their account probably would be appreciated by those interested in understanding the processes through which the data were collected in interaction with different respondents/participants. And the utility of the results achieved is greater than had they not tried to compensate in the way that they did for the fleeting character of most of the interactions with respondents (the majority of whom were past students in this case).

From a *critical theoretical* point of view, Romm et al.'s way of organizing the study might be defended as an effort to open up argumentation around the "facts" generated through the process of the research. The research process became used as an opportunity to organize intersubjective encounters around the discerned data. The validity of what Romm et al. discerned through the questionnaires and the interview guide lies in whether they can argue that both during the research process and thereafter in the presentation of findings, they have provided space for communicative encounter as the basis for people working with the visions that arose. Romm et al.'s conduct of discussions with various participants and their way of explaining the status of "the findings" in the report, can be appreciated as at least embracing the value of intersubjective communication as a way of organizing social inquiry.

From an *anti-foundationalist feminist* point of view, Romm et al.'s study might be seen as contributing to the value of organizing dialogue across heterogeneous experiences of reality. They tried to prevent from becoming impositional any specific conception of relevant categories in terms of which to set questions and code responses of participants. Although they still felt that an argument could be made for defining their work as a form of survey, they made efforts to set up arenas for discussion within the process of the research, and the report was written in a genre that presumably would discourage univocal readings of "the findings".

From a *trusting constructivist* point of view, Romm et al.'s work could be assessed in terms of considerations of whether they seemed to be taking responsibility for their ways of seeing realities and at the same time providing leeway for others to work with different visions. The making of a statement suggesting that, say, contact with tutors aids student learning, can be properly accounted for only by the researchers' recognizing that such a statement itself might contribute to the creation of certain effects. For example, people (both tutors and students) might become oriented to embracing the kinds of learning focused upon through the way they are questioned about this and/or through the way their responses are interpreted. Romm et al. did not regard as being irrelevant to their research remit this potential self-fulfilling effect of ways of organizing and reporting on the collection of the survey data. While some of the audiences (for example, the funders) might at first have been surprised by this manner of conducting and reporting on the findings of a survey, Romm et al. seem to have created a possibility for their appreciating that the meaning of research and its purpose in society is itself not clear-cut. This meant that Romm et al. could account for their inquiries by simultaneously calling on a definition of accountability that appealed to others to consider awarding trust on a basis with which they might not at first have been familiar. But Romm et al.'s attempt to earn trust while at the same time engaging in a discourse on the meaning of their own accountability, implies that they did have to take seriously into account differing expectations for their accountability that they considered to be at play (or rather, to be played around) as they organized their interactions with others.

8.5 ACCOUNTING FOR ETHNOGRAPHIC RESEARCH

In Chapter 6, I used the example of Aphane et al.'s research project to open a discussion on differing ways of treating the accountabilities of those involved in inquiry classed as ethnographic. I pointed out that the (proper) involvement of ethnographic researchers with participants (as hosts of their studies), as well as their ways of reporting "about" them, could be conceptualized, and judged, differently.

From a *positivist* point of view, Aphane et al.'s accountability as scientists would be judged by focusing on the way in which they used the ethnographic process to develop ideas about possible correlations or relationships of causality that may exist between variables that have been shown to be relevant in people's everyday existence. The way people experience their existence provides one source of hypotheses about correlations (and causal relationships) that might obtain in social reality. So, for example, if women in the study in question believe that they are disadvantaged relative to men in obtaining access to inheritance in the society, a hypothesis can emerge that can be tested in this and in further studies. The internal validity of ethnographic work derives from the fact that small samples are utilized where one can make initial explorations into patterns of relationship holding between what is deemed as relevant to people's lives in their natural settings. Although the variables may not yet be clearly defined, and may rely more on the (sometimes ambiguous) descriptions used in everyday life, researchers can try to convert the ambiguity into more rigorous definitions of terms, so as to seek patterns in the discerned data. The external validity (generalizability) of their research is a matter of their linking to other investigations the findings gleaned from the small samples studied, so as to generate further information in regard to the variables studied. In Aphane et al.'s study they did not really offer a clear enough account of why they believe the information that they collected from the samples studied, and their (implicit) comparison of this with information derived from other sources, was done in a way that can be termed scientific.

In the case of Aphane et al.'s research, the likelihood is that their own initial prejudices in regard to the status of women in this and other societies influenced the way in which they both collected and interpreted the relevant data. Any attempt at defending the validity of their results, would have to take into account their own admission that they came to the study with prejudices that might have affected their conclusions. But in any case, they have provided an exploratory investigation into possible patterns that may exist in reality. Further studies (including not only ethnographic ones) can be used to test hypotheses regarding such patterns.

From a *non-foundationalist* point of view, attention would be given to the measures that could be introduced to organize the study in a systematic manner as a scientific one. In ethnographic research as in other forms, researchers need to direct their research activity to the effort of finding out about the patterning of reality. Insofar as they allow themselves to be caught up in people's (and their own initial) commonsense conceptions, or to bring theoretical views to the study that are immune to criticism (by colleagues and others), they cannot account for their work under the umbrella of science. Conversely, if they are able to distance themselves from the idea that commonsense perceptions are "valid in their own terms", they can operate better as scientists. So, for instance, Aphane et al. could have organized their study so as to test the commonsense perception that gender affects access to resources (in this and other societies). On the basis of the investigations undertaken of inheritance access in the samples studied in this case (using the variety of techniques of observation that they drew upon), we can consider what evidence there is for holding onto the claim that gender does indeed appear to affect access (and in what contexts). We cannot decide with any certainty to what extent the samples studied allow us to generalize to the wider society; nor can we decide with certainty that in other societies similar social patterns around gender and access to different kinds of resource exist. But the study — if organized properly — could have increased our chances of providing conceptions that are more firmly grounded than a simple allegiance to commonsense conceptions.

However, Aphane et al. have not given us sufficient evidence that they were committed to undertaking the study as a scientific one. Their indication of their political allegiances that they brought to the study, could well have affected their way of speaking to participants and their way of interpreting their responses. And participants' recognition (or perception) of the researchers' expected results, might have affected the way they interacted with the researchers. Aphane et al. could have made more effort to consider the import of these intrusions into the process of doing their ethnography. They could have given more attention to the possible reactivity effect that their own presence as researchers introduced into the study as they interacted with others in various contexts of interaction. They could have studied this effect more systematically as they conducted the research, and they could have included their observations on this in their report.

From a *scientific realist* point of view, the discussion on Aphane et al.'s accountability would be directed at considering their use of the skill of abstraction to develop insights that can be considered as advancing knowledge of social realities. Judged in these terms, comments that they made about, say, the way women are made to feel guilty when they try to gain access to their inheritance, would be seen as too descriptive at present to be "cashed in cumulatively". They should have concentrated more on trying to convert the particularities of the information gained from the detailed examination of (small) samples into information relevant to our knowledge of "deeper" realities in this, and other, cases. Because they made scant attempts to design their study so as to be able to gather information with this in mind, their accountability as scientists can be questioned. However, others reading their report, such as colleagues in the scientific community, may be able to make use of some of the observations developed in the study, toward further theorizing. And they themselves, when faced with this possible criticism of their work, might likewise rethink how they can do more analytic work to find theoretical pathways toward a deeper understanding of the realities investigated.

From an *interpretivist* point of view, the study organized by Aphane et al., as with all (well-organized) ethnographic studies, has the advantage of being able to build up an understanding based on people's meaningful experience of social reality. However, Aphane et al. arguably were not sufficiently sensitive to exploring in the process of the research the way in which their interaction with respondents might have operated to call forth certain meanings (as part of the researcher-participant interaction process). Also, Aphane et al. could have offered readers of their report more detail on how they built up their understandings through their engagement with various participants, including instances when participants disagreed with their own ways of making sense of their situations. This they could have done by leaving behind a more detailed paper-trail of their way of dealing with others during the research process. As regards their efforts to offer analyses that cast light on issues of relevance not only to the participants studied (as hosts of the project), but also to others, they could have been clearer on how they were organizing their comparisons across other contexts (social settings). Readers (including the hosts of the project and other audiences, academic and others) would then be in a better position to decide on how relevant their work is for their own understanding of different social settings.

From a *critical theoretical* point of view, Aphane et al. were admirably trying to use the research process to experiment with a style of critical ethnographic research whose aim was to contribute to social change in the human relationships studied. The study of inheritance issues and ways of treating these in the society, were taken as opportunities at times to raise larger concerns — for instance, the concern that inheritance matters in the society would be reduced to an effort to find administrative solutions to problems identified. Aphane et al. set up processes for participants to exchange views with the researchers, and with others, around different ways of treating inheritance phenomena. Aphane et al.'s efforts to set up these communicative processes can be considered as part of their accountabilities as researchers. Seen from a critical theoretical point of view though, these efforts, while undertaken in good faith, might have been organized better. Aphane et al. could have given more thought to, and allowed participants also to reflect more carefully upon, what is involved in communicative encounter as a basis for co-ordinating their actions around issues of concern raised in the study.

From an *anti-foundationalist feminist* point of view, Aphane et al.'s style of setting up processes for the cultivation of dialogue around the experience of realities is what would be focused upon in judging their style of research into others' lives. Their efforts at mediating between different people's perceptions by carrying them across different contexts of data "collection", and their setting up of individual and group discussions so that people could recognize and work with alternative ways of treating their own and others' experiences, would be judged favorably. Indeed, unless the researchers could be argued to have made such attempts, it is difficult to consider them as having acted accountably from an anti-foundationalist feminist point of view. From this point of view, they could not justifiably have used the participants to extract information from them, without organizing a discussion in terms of which participants could benefit by extending their appreciation of different perceptions of the realities and different ways of treating options for action. Because such discussions were organized, and multiple ways of seeing and acting were evoked through the discussions, it is also easier for other audiences to engage in a dialogical fashion with what is presented in the report. (And because the report shifts between realist and narrative ways of elucidating insights, it is also easier for audiences not to be awed by the researchers' presentations.)

From a *trusting constructivist* point of view, following critical theory and anti-foundationalist feminism, participants and other audiences would need to be made aware of the evoked character of "phenomena" brought to the fore through the particular interactions taking place between self-appointed researchers and others. The active interviewing technique — whereby both researchers and participants are called upon to interrogate their views in relation to challenge leveled by others — is one way of highlighting the manner in which people's visions can develop as part of a discursive process. But to organize interviewing in this fashion does depend on making efforts to build up a human relationship in the research process with which the parties concerned may not be familiar. A trusting constructivist position would suggest that Aphane et al.'s manner of conducting themselves with participants (in individual and group settings) might be conceptualized as defensible as an effort to arrange forums where people could extend their experience of trust building through processes of discourse. A trusting constructivist position would also appreciate Aphane et al.'s way of organizing the write-up of the study in the light of their recognition of the need to take into account different possible conceptions of what kind of "product(s)" needed to be created via the research process. It would be emphasized that they did not need to concede to any of these conceptions, but that they did need to act responsibly in finding a way of presenting their report.

8.6 ACCOUNTING FOR ACTION RESEARCH

In Chapter 7, I used the example of Weil's report on a case of her (action) research to open a discussion on differing ways of viewing the accountabilities of inquirers involved in treating research as an opportunity for learning-in-action. I suggested the following.

From a *positivist* point of view, Weil and co-researchers' accountability as scientists would be judged by focusing on the way in which they used the setting of their everyday actions to test hypotheses concerning relationships of causality operative in social reality. The internal validity of the research needs to be argued for in terms of its workability in establishing the existence (or not) of causal connections. Hence, in this case, the question is whether causal connections exist between, say, the independent variable of "participative learning" and certain patterns of response on the part of senior and middle managers. Insofar as the researchers' involvement in, and reflection upon, change processes instituted in social situations can be monitored so that we can (reasonably) trace observed outcomes back to the instituted changes, the internal validity of the research can be defended. And insofar as the researchers can reasonably make comparisons across different situations, the external validity of the research is increased. Weil's particular way of organizing the action research with co-researchers can be classed as an attempt to organize experimentation in situ with participants who did have access to various organizations, and who could in principle monitor effects of instituting certain changes. But Weil could have offered more explanation in her report as to why she believes that the processes followed in order to test hypotheses regarding participative learning, were not unduly influenced by her own and others' prior beliefs about its effectiveness. To operate accountably as researchers, the co-inquirers could not allow their own agendas (for example, their desire to see participation instituted in their organizations) to affect the way they drew conclusions from the evidence at hand.

From a *non-foundationalist* point of view, although action research may conceivably be a way of finding out about patterns existing in social reality, we must be cautious of people blurring the distinction between researchers and actors who wish to pursue certain agendas. Action research can be used in a way that merits the label of professional inquiry only if the inquirers make efforts to test hypotheses with reference to the evidence that they are collecting through their involvement in change processes. But action researchers' collection of evidence may be fraught with specific difficulties arising from the way in which the co-researchers relate to one another. Weil and co-researchers could have explored these difficulties more. And Weil and co-researchers also would need to recognize challenges arising from colleagues in the scientific community who may wonder about the rigor of their inquiries. For instance, such colleagues may question the way in which terms such as "participative learning" and "management of multiple bottom lines" have been used. They seem to have been used in a loose way — so that it is difficult for us to recognize what features of reality are being taken by the researchers as indicators of these terms. Because of the loose way in which terms are used and hypotheses set up (around these terms) we do not know what kinds of evidence Weil and co-researchers would take as disproving their hypotheses.

Although Weil suggests that she and others were reflexive in regard to their own worldviews, it seems that she was not inclined to challenge the view that participative learning leads to desirable outcomes (such as effective management of multiple bottom lines). An indication of how the attitude of reflexivity referred to by Weil can be linked up with efforts at doing science (as a process of testing hypotheses) is necessary in order that we can consider the researchers as being accountable as researchers. Also, although Weil discusses with favor the idea that there exist multiple realities and multiple stories with which people in action have to engage, this idea can be challenged. Indeed, if not challenged, people in society (and the participants in the project) may well become more cynical in regard to the possibility of acting in a more informed way with reference to an understanding of reality. Weil is concerned that a stance that implies that there exists a world "out there" to be more or less accurately apprehended, does not allow managers to address effectively multiple bottom lines. But this argument of hers needs to be tested against the evidence, rather than simply propounded. Once tested, it is possible that another conclusion (other than the one she propounds) may be found to be more in line with the evidence. According to non-foundationalism, the evidence to date points to the idea that science — if properly practiced — can be used to make advancements in knowledge of reality.

From a *scientific realist* point of view, Weil's recognition that there are multiple realities with which actors may have to engage, would be seen as a sign that there exist different layers of reality, all of which are, however, "real" (in the sense that they exist outside of human knowing processes). The task of researchers is to try to access these realities (including those hidden and unobservable). Insofar as Weil works with the assumption that action research can be a way of exploring the various layers of reality, and insofar as the research becomes a kind of realist evaluation of the effectiveness of styles of strategizing in organizations, it may be accounted for in scientific terms. However, Weil and co-researchers would have to operate at the necessary analytic level to gain some understanding of what is needed to bring about transformation of the structures of society in which the organizations being investigated are embedded. For example, Weil's hope that participative learning can lead to the management of multiple bottom lines, might be overoptimistic and grounded more in her own values and hopes than in an effort to come to grips with the operative realities.

From an *interpretivist* point of view, Weil and co-researchers rightfully recognized — and made visible to themselves and to other audiences of their accounts — that patterns of events in social life depend on how people define their options for action. Situations can appear to people as admitting of many possibilities for meaningful action. Weil and co-researchers' action research project became a way of investigating how people can shift their patterns of response in engagement (through critique and collaboration) with others. The study offers insights to the participants themselves in regard to ways of organizing such shifts, and it also can potentially be of use to other audiences. This is not necessarily because it shows what is "generally likely" in terms of the link between action and its outcomes, but because it can point to what is *possible* in society. Actors in society can compare the insights generated by Weil and others (in regard to their way of working with possibilities) with their own situations to decide the utility of the project for their own actions. Therefore, the study was helpful both to participants (and its internal validity can be accounted for in these terms) and to others (if insights are regarded as "transferable" to the exploration of possibilities in other situations). Seen in this light, the plausibility of the study could probably be defended from an interpretivist point of view. However, readers need to bear in mind (in deciding on the likely validity of the project) that Weil possibly overstepped her role as researcher by being too committed to the value of participative learning to take the necessary steps to examine others' perceptions of its effectiveness (and workability). Had she indicated in her report how she engaged with others' views on this (and indeed how the co-researchers in the learning network might have engaged with others in their own organizations around the meaning of this), the report would have been improved.

From a *critical theoretical* point of view, Weil and co-researchers expressed a commitment to try to shift what might be considered as manipulative/strategic responses to situations, toward the development of a more communicative orientation. Although Weil and co-researchers were themselves not oriented to reaching toward consensus of visions, at least they were aware that the process of doing research and the presentation of findings could become forums for inviting communicative encounter as the basis for the co-ordination of action. Weil's efforts (with others) to create spaces for communication and democracy in the organizational arenas in which they were involved, can thus be appreciated. But, on Weil's own admission, trying to turn around the dominant forms of administration and control was no easy task. Weil hints at the difficulties of the challenges involved in working communicatively within systems geared toward "efficiency" of performance. She could have spelled out in more detail how she and others addressed these challenges as researchers involved in social transformation. If she could have accounted better for her manner of addressing these challenges as part of the process of social transformation, her inquiries (with others) can be defended as accountable social research in critical theoretical terms.

From an *anti-foundationalist feminist* point of view, Weil's use of the research process to organize collaborative learning could be regarded as a way of shifting patterns of (traditional) researcher control. In the process people learned, by way of experience, that it is possible to see and act by embracing an appreciation of multiple realities. This did not mean that responsible action in "the world" was threatened. On the contrary, it allowed people to find responsible pathways through multiple realities (according to Weil). Weil's and others' stories in regard to the finding of such pathways (in this case through the aid of setting up the action research project) can be helpful to wider publics. This is provided that stories about the research are not told as if they are reporting about, in this case, participative learning and its (general) effects. If the report offered by Weil, and the stories told by her and the other co-researchers in various contexts, are given the status of, indeed, narratives, then they can encapsulate the (feminist) value of respecting the heterogeneous ways in which people may come to terms with the(ir) worlds. Weil puts up front her own epistemological preference for cultivating the propensity of people to find responsible pathways through multiple realities. She expresses her preference in view of what she sees as the dominant cultural expectation that people try to find a way of coming to grips with the external world by striving for "objectivity". She accepts that her own research is not value free, but she argues that to pretend that research can ever be value free is itself to become complicit in endorsing a specific way of relating to/in the world with which she does wish to identify and for which she cannot responsibly account (to herself or to others).

From a *trusting constructivist* point of view, Weil has tried to offer a countercultural narrative to the dominant epistemology as part of an effort to allow people to experience their worlds anew. Weil has defended her approach by suggesting that it seems to offer a way (though of course not the only way) of organizing research with others around issues of concern (or issues raised for concern). Those judging the value of the project can decide for themselves what "lessons" might be learned from it; and Weil accepts that this is the case. The detailed documentation of learning processes that she provides and the multiple narratives that she also provides, mean that we (as readers) can see how she has developed some view of the value of participative learning, without seeing this view as one that excludes others' experiences. Weil's way of working with her own recognition of multiple realities is a way of working that would be appreciated from a trusting constructivist perspective. She recognizes that in speaking about participative learning in a particular way, she evokes experiences of "it" for consideration by herself and others. Her statements regarding the effectiveness of such learning likewise are not constructions about the (likely) "reality" of its effectiveness (given the evidence to date), but statements that might become real if people decide to act in a certain way in regard to them. Because she indicates that we are embedded in the realities that we create, she leaves openings for people to define ways of, indeed, creating realities.

From a trusting constructivist point of view, Weil's accountability can be assessed in terms of a discussion around her portrayal of her way of working (responsibly) with a view of reality as constructed.

8.7 DEFENDING RESEARCH PROCESSES AND REPORTING ON THEIR PRODUCTS

Four actual examples of research were used in the book to point to the manner in which research processes might be judged using alternative understandings of the value of social research in society. In this book I could necessarily not cover all possible research strategies and techniques that might be used by social researchers. I chose to concentrate on what could be classed, respectively, as experimentation, survey research, ethnography, and action research strategies. (The classing of the examples does not need to be seen as a once-and-for-all classification. But I tried to defend my current suggested classifications, in dialogue with authors' remarks and with others' views, as a way of seeing the research projects in question.)

From a trusting constructivist point of view, the defense of research strategies that come to be adopted at any point in time, needs to be made on the basis of indicating how the choice of approach (and the details of the techniques employed along the way) has emerged in recognition of the possibility of employing alternatives. Decisions to pursue certain courses of inquiry at points in time cannot be justified on the basis that the chosen ones are the most appropriate, given the realities that researchers are setting out to investigate. For "realities" can indeed admit of different forms of investigation as they become evoked differently via these different forms. In defense of their chosen strategy(ies) and techniques, researchers therefore need to provide an indication of what kinds of data (phenomena) they are choosing to collect/evoke via the process. This means that findings created (as research "products") can be contextualised (by those confronted by them) in view of the strategies and techniques adopted, and the particular way of adopting them in the case in question. This also means that researchers leave open the space for both themselves and others to engage in another way with the "research material" if they choose to organize studies differently.

As indicated through the four examples discussed in this book, there is a variety of ways in which findings arising out of research processes can be reported. I believe that it is incumbent on social researchers to take seriously the objections raised by those who resist the authoritative voice of traditional scientific reporting. I have tried to show that researchers who regard their accountability as researchers as being solely to advance knowledge, see their relationship to others in society as being that of knowledge-providers. Despite their suggestions that whatever knowledge provisions they offer should be treated circumspectly in society because these are not certain, their genre of writing (in realist style) sets them up as offering (some) knowledge of "the realities". My argument is that once they recognize the fallibility (and contentious character) of human reason, they could just as well decide to write in a way that admits the fundamental fragility of any claims to know.

I have suggested that it is important that social researchers and others are at least aware of the value choices they are making when they operate in terms of the expectation that science — taken as a whole — can serve to advance knowledge of independently existing realities. I have argued that this is not the only feasible way for scientists to define their role as (accountable) inquirers. It is noteworthy in this respect that those adhering to more realist-oriented conceptions of the purpose of science in society suggest that the search for knowledge of realities existing outside of human consciousness is unavoidable for humans. Instead of regarding this claim as an expression of their own way of defining the human condition, they suggest that, as humans, people simply cannot live by relinquishing the quest for truth (defined as representation of reality). Nevertheless, despite the strength of the assertion that the search for knowledge (defined in realist terms) is a human necessity, I have suggested that we can reconsider it by referring to proponents' own admissions that we can never be sure of the cognitive status of any specific statement or set of statements made in the search for knowledge. Admitting that we can never be sure of this, and that scientists (as professional inquirers) cannot claim to offer such assurance, we recognize that people still do manage to live. Taking an alternative starting position, then, it may be reasonably argued that there is no necessity to insist that humans strive (for example, through the aid of scientific activity) to seek "the truth". Varying stances regarding ways of defining both human knowing and human living are choices that have to be accounted for. I have argued in this book that researchers therefore need to account for their own complicity in endorsing certain ways of knowing and living in society.

8.8 ADDRESSING OTHERS' ARGUMENTS

A concern that I wish to consider — in conclusion — relates to the question of whether in categorizing arguments regarding accountability in social research, I have caricatured positions with which I have less sympathy, and treated more fairly those with which I am more in sympathy. My response to this question is presented in this section — by offering some remarks on different ways in which the terms caricature and categorization may be utilized. (See also Romm, 1998c.)

Caricaturing of another's argument is normally regarded as presenting a threat to organizing debate between arguments in reasonable terms — more of a threat than, say, attempting to categorize aspects of another's position in order to highlight its opposition to, or difference from, one's own. But the question with which I wish to begin the discussion here is whether there is any way of grounding the claim that a person is "really" engaged in caricaturing (rather than making an effort to categorize) some argument.

My suggestion in this regard is that the very orientation adopted by people when evoking the term "caricature", sets the tone for the way in which they might address one another's textual interpretations. If it is taken (following some form of realist epistemology) that texts "have meaning" that can be *misrepresented* (by interpretations), then the act of accusing the other of misrepresentation (caricature) calls on the other to admit that s/he was wrong in the interpretation. If, however, the use of the term "caricature" is taken as an indication that the offended party feels that the other has not tried to come to terms with his or her account of a possible meaning that can be generated from the text (and from various encounters with it), then this becomes a call for both to revisit the expressed arguments and the way in which they have been expressed.

Let us consider an example. Those who (self-categorize) their position as realist-oriented might argue that I have misrepresented their argument regarding the fact that the search for knowledge of reality is a human necessity. They might argue that they have clearly stated their point that people need to act in the world as if there are certain realities that have to be taken into account; and "reality" for people means (by definition) that which has to be treated as external to themselves. Such authors might therefore argue that I have in this book misrepresented their claims regarding the human endeavor to know about reality. In other words, I have engaged in caricaturing what are called realist-oriented positions by not accurately reporting on their substance. However, it is possible to suggest that the term caricature — if and when invoked — can be treated not as a plea to report more accurately on the substance of an argument presented by some author(s), but rather to consider ways in which the terms of the discussion between arguments can become enriched. So, for instance, a discussion could be opened around terms used in expressions claiming that in action people need to orient themselves to considering things as existing outside of human consciousness. And one can then revisit, accordingly, arguments concerning the human experience of "reality".

Speaking (admittedly) from a constructivist perspective, I have suggested that there is no manner in which we can finally ground a complaint that another person has "really" been *caricaturing* as opposed to *categorizing* an argument. The charge of caricature implies that there is an experience that the other has not respected, or tried to come to grips with, the rationale of a position taken. We can never know whether this experience is justified or not — but what we can do is try to set up an alternative style of relating, so that our relationship can become experienced as more constructive (for furthering the arguments).

But now another issue arises. When a category has been employed to make sense of a person's argument, does this in itself imply that the argument becomes oversimplified by being categorized? I answer this question largely in the negative. While categorization does involve some selectivity in the way people come to terms with authors' expressed arguments, it also at the same time gives people the opportunity to consider the extent to which they feel the category does (or does not) do justice to points being made. In processes of argument, people may decide that their reasons for wanting to identify themselves with some aspect of a category (as advanced by themselves at some point in time or by others) can be modified/extended. They may reconsider the way in which they wish (or whether they wish) to be thus identified. By arguing around the relevance of some suggested category to (aspects of) their work, they can at the same time refine/develop their argument.

Let us consider again by way of example the categorization of "realism" and "constructivism" as epistemological positions. A realist-oriented epistemological stance has been associated in this book with some adherence to the idea that knowledge-seeking involves trying to get to grips with realities existing outside the knowing process. This implies knowers' acting accountably by making efforts to avoid the danger of clinging to their own prejudices when confronted by realities. It implies their making efforts to minimize possibilities of bias in the understanding of reality. It does not, however, imply that those who try to minimize bias will always be successful in doing so (as noted by Hammersley, 1997, paragraph 1.10). But, the argument goes, they should at least try to do so. The research community can act as a mechanism to socialize its members so that they can maintain an allegiance to the principle of striving for "objectivity" in their scientific work. In addition, the community can serve to iron out certain biases through the fact that researchers are required to criticize one another's work in terms of its likelihood of displaying bias. (See the discussions of this in Chapter 2, Sections 2.3.1, 2.4.1, and 2.5.1.)

Those rejecting positivist accounts of what it means to seek unbiased knowledge of reality, express identification with what is called a "realist" epistemology only insofar as it is recognized that this is not tantamount to assuming that scientists (and the scientific community) will always be able to operate to remove/minimize bias. They regard it as unjustified to associate their support of realist definitions of truth-seeking with what they would see as positivist/foundationalist forms of realism. The term "foundationalist realism" is for them not a relevant category in terms of which to advance debate about the issues at stake. For the issue is how we might take into account the fallibility of all efforts of human knowing, while still accepting that, as Hammersley and Gomm put it, humans cannot interpret reality "just however we like" (1997b, paragraph 1.9). It is this issue, they contend, that now has to be engaged with when we consider the character of the scientific enterprise aimed at advancing knowledge of reality.

Just as the meaning of "realism" can be qualified so that it does not imply a positivist phenomenalist or empiricist argument, so too can the category of "constructivism" be qualified. It need not be associated with the so-called "relativistic" belief that all constructions are to be regarded as acceptable ways of constructing realities. The epistemological category of "constructivism" as elaborated upon in this book encapsulates the suggestion that constructions offered in processes of discourse should be regarded as openings for people to engage in interrogating statements in order that their situatedness can be reflected upon. This implies in turn that people should be prepared to adopt an attitude of reflexivity in regard to their own knowing endeavors. To ascribe to an attitude of reflexivity in constructivist terms is to suggest, as Jacobson and Jacques put it, that we need to "become more reflexive about the ways that situated knowers [inquirers] and knowns [participants in social inquiry processes] influence the production of inevitably perspective-dependent knowledge" (1997, p. 56). As Weil argues, this implies the adoption of "a worldview that sees us as implicated and embedded in the realities we are creating, including through our rhetorics" (1998, p. 43). From a constructivist point of view, then, the issue that needs to be addressed is how we can take into account our embeddedness in realities that we create as we engage in "knowing".

Whether we continue to find the categories of, say, "realism" versus "constructivism" as useful ones in terms of which to conduct debates about human knowing, depends on whether the parties engaged in debate feel that these terms still help to highlight differences between arguments being advanced. What labels might be used to express differing arguments, can be decided in processes of argument. But labeling (of arguments) is part of the process of people engaging with others in processes of discourse.

Of course, it is still possible in the course of discourse that people will feel that their arguments run the risk of being caricatured as others position (or categorize) different arguments in relation thereto. My suggestion here is that the decision to level the charge of caricature, should depend on whether the "offended party" feels that the other's attempted categorizations at the same time occlude the possibility to explore alternative ways of furthering the debate about issues that have been raised for concern. The charge of caricature — leveled with the intention to continue discursive engagement between alternatives — then becomes an invitation (and plea) for the continuation of the discussion in terms that allow the respective parties to take into account the different concerns raised.

I have tried in this book to organize my discussion of the different arguments on accountability by not denuding the rationality of positions that I see as differing from my own. It must be remembered that (as far as I see it) categories are devices for positioning different arguments in relation to one another, rather than devices for "representing" a particular author's views. Readers can consider whether they believe that arguments (as articulated by themselves or by others) regarding researcher accountability can become better appreciated by placing aspects of them under any of the categories that I have discussed. And they can consider whether my discussions of the categories (and my extrapolations of views on accountability linked to these categories) are helpful in extending their visions of some possible ways of practicing and/or assessing accountability in social research.

References

Adman, P., & Warren, L. (1996). *Information systems management: Perspectives for higher education*. Lancashire: Mimir Press.

Alldred, P. (1998). Ethnography and discourse analysis: Dilemmas in representing the voices of children. In J. Ribbens & R. Edwards (Eds.), *Feminist dilemmas in qualitative research: Public knowledge and private lives*. London: Sage.

Amir, Y. (1969). Contact hypothesis in ethnic relations. *Psychological Bulletin, 71*, 319-342.

Aphane, M.-J. D., Dlamini, T., Magwaza, M., Manzini, N., Mkhonta, F., Mthembu, L., & Vilikazi, M.-L. (1993). *Inheritance in Swaziland: Law and practice*. Report coordinated by M-J. D. Aphane. Swaziland: Social Science Research Unit, University of Swaziland.

Arber, S. (1993). Designing samples. In N. Gilbert (Ed.), *Researching social life*. London: Sage.

Argyris, C., & Schön, D. A. (1996). *Organizational learning II: Theory, method, and practice*. Reading: Addison-Wesley Publishing Company.

Armstrong, A. K. (1992). *Struggling over scarce resources: Women and maintenance in Southern Africa*. Harare: University of Zimbabwe Publications.

Bailey, K. D. (1987). *Methods of social research*. London: Collier Macmillan Publishers.

Barry, D., & Elmes, M. (1997a). On paradigms and narratives: Barry and Elmes' response. *Academy of Management Review, 22*, 847-849.

Barry, D., & Elmes, M. (1997b). Strategy retold: Toward a narrative view of strategic discourse. *Academy of Management Review, 22*, 429-452.

Bell. S., & Morse, S. (1999). *Sustainability indicators: Measuring the immeasurable*. London: Earthscan Publications.

Beynon, J. (1983). Ways-in and staying-in: Fieldwork as problem solving. In M. Hammersley (Ed.), *The ethnography of schooling: Methodological issues*. Driffield: Nafferton.

Bhaskar, R. (1975). *A realist theory of science*. Leeds: Leeds Books Ltd.

Bhaskar, R. (1979). *The possibility of naturalism*. Brighton: Harvester.

Brand, A. (1990). *The force of reason*. London: Allen & Unwin.

Brewer, M. B., & Miller, N. (1984). Beyond the contact hypothesis: Theoretical perspectives on desegregation. In N. Miller & M. B. Brewer (Eds.), *Groups in contact: The psychology of desegregation*. Orlando: Academic Press.

Bridgman, P. W. (1952). *The nature of some of our physical concepts*. New York: Philosophical Library.

Bryman, A. (1992). *Quantity and quality in social research*. London: Routledge.

Bryman, A. (1997). Quantitative and qualitative research: Further reflections on their integration. In J. Brannen (Ed.), *Mixing methods: Qualitative and quantitative research*. Aldershot: Avebury.

Callaway, H. (1992). Ethnography and experience: Gender implications in fieldwork and texts. In J. Okely & H. Callaway (Eds.), *Anthropology and auto-biography*. London: Routledge.

Checkland, P., & Holwell, S. (1988). Action research: Its nature and validity. *Systemic Practice and Action Research, 11*, 9-21.

Comte, A. (1855). *The positive philosophy*. Freely translated and condensed by H. Martineau. New York: C. Blanchard.

Comte, A. (1970). *Introduction to positive philosophy*. Edited and with an introduction by F. Ferré. Indianapolis: The Bobbs-Merrill Co.

Comte, A. (1974). *Positive philosophy*. Translated and condensed by H. Martineau. New York: Ams Press.

Connerton, P. (Ed.) (1976). *Critical sociology*. Harmondsworth: Penguin.

Cook, S. W. (1985). Experimenting on social issues: The case of school desegregation. *American Psychologist, 40*, 452-460.

Cook, T. D. (2000). Toward a practical theory of external validity. In L. Bickman (Ed.), *Validity and social experimentation*. London: Sage.

Cooperrider, D., Barrett, F., & Srivastva, S. (1995). Social construction and appreciative inquiry: A journey in organizational theory. In D-M. Hosking, H. P. Dachler, & K. J. Gergen (Eds.), *Management and organization: Relational alternatives to individualism*. Aldershot: Avebury.

Crotty, M. (1998). *The foundations of social research: Meaning and perspective in the research process*. London: Sage.

Dachler, H. P., & Hosking, D-M. (1995). The primacy of relations in socially constructing organizational realities. In D-M. Hosking, H. P. Dachler, & K. J. Gergen (Eds.), *Management and organization: Relational alternatives to individualism*. Aldershot: Avebury.

Davidson, J. O'C., & Layder, D. (1994). *Methods, sex and madness*. London: Routledge.

Davis, J. (1997). *Alternate realities: How science shapes our vision of the world*. New York: Plenum.

Davis, J. H., Schoorman, F. D., & Donaldson, L. (1997). Toward a stewardship theory of management. *Academy of Management Review, 22*, 20-47.

Deetz, S. (1996). Describing differences in approaches to organization science: Rethinking Burrell and Morgan and their legacy. *Organization Science, 7*, 191-207.

Delanty, G. (1997). *Social science: Beyond constructivism and realism*. Buckingham: Open University Press.

Delanty, G. (1998). Social theory and European transformation: Is there a European society? *Sociological Research Online, 3*, <http://www.socresonline.org.uk/socresonline/3/1/1html>

Denzin, N. K. (1970). *The research act*. Chicago: Aldine.

Denzin, N. K. (1997). *Interpretive ethnography: Ethnographic practices for the 21st century*. London: Sage.

De Vaus, D. A. (1996). *Surveys in social research*. London: UCL Press.

Doctorow, E. L. (1977). False documents. *American Review, 5*, 215-232.

Donaldson, L. (1997). A positivist alternative to the structure-action approach. *Organization Studies, 18*, 77-92.

Dovidio, J. F., Gaertner, S. L., Validzic, A., Matoka, K., Johnson, B., & Frazier, S. (1997). Extending the benefits of recategorization: Evaluations, self-disclosure, and helping. *Journal of Experimental Social Psychology, 33*, 401-420.

Eden, C., & Huxham, C. (1996). Action research for management research. *British Journal of Management, 7*, 75-86.

Edwards, D. (1997). *Discourse and cognition*. London: Sage.

Fals-Borda, O. (1991). Some basic ingredients. In O. Fals-Borda & M. A. Rahman (Eds.), *Action and knowledge*. New York: The Apex Press.

Fay, B. (1996). *Contemporary philosophy of social science: A multicultural approach*. Oxford: Blackwell Publishers.

Feyerabend, P. (1993). *Against method*. London: Verso.

Fielding, N. (1993a). Ethnography. In N. Gilbert (Ed.), *Researching social life*. London: Sage.

Fielding, N. (1993b). Qualitative interviewing. In N. Gilbert (Ed.), *Researching social life*. London: Sage.

Fisher, F. G. (1996). Transdisciplinary teaching: Overcoming the inevitable dislocation in applying science without regard to its social context. In W. W. Gasparski, M. K. Mlicki, & B. M. Banathy (Eds.), *Social agency: Praxiology (vol. 4)*. New Brunswick: Transaction.

Flood, R. L., & Romm, N. R. A. (1996). *Diversity management: Triple loop learning*. Chichester: John Wiley & Sons.

Foddy, W. (1995). *Constructing questions for interviews and questionnaires: Theory and practice in social research*. Cambridge: Cambridge University Press.

Fritzman, J. M. (1990). Lyotard's paralogy and Rorty's pluralism: Their differences and pedagogical implications. *Educational Theory, 40*, 371-380.

Fukuyama, F. (1995). *Trust: The social virtues and the creation of prosperity*. London: Hamish Hamilton.

Gaertner, S. L., Mann, J. A., Dovidio, J. F., Murrell, A. J., & Pomare, M. (1990). How does cooperation reduce intergroup bias? *Journal of Personality and Social Psychology, 59*, 692-704.

Gergen, K. J., & Whitney, D. (1996). Technologies of representation in the global corporation: Power and polyphony. In D. M. Boje, R. P. Gephart Jr., & T. J. Thatchenkery (Eds.), *Postmodern management and organization theory*. London: Sage.

Gill, J., & Johnson, P. (1991). *Research methods for managers*. London: Paul Chapman.

Glaser, B. G., & Strauss, A. L. (1967). *The discovery of grounded theory: Strategies for qualitative research*. New York: Aldine.

Gouldner, A. W. (1975). *The dark side of the dialectic*. Amsterdam: Sociology Institute, University of Amsterdam.

Gouldner, A. W. (1976). *The dialectic of ideology and technology*. New York: Seabury Press.

Gouldner, A. W. (1980). *The two Marxisms: Contradictions and anomalies in the development of theory*. London: Macmillan.

Gregory, W. J., & Romm, N. R. A. (1996). Toward multi-agency dialogue: Facilitation as fair education. In W. W. Gasparski, M. K. Mlicki, & B. N. Banathy (Eds.), *Social agency: Praxiology (vol. 4)*. New Brunswick: Transaction.

Gregory, W. J., & Romm, N. R. A. (2001). Critical facilitation: Learning through intervention in group processes. *Management Learning* (in press).

Grossberg, L. (1989). On the road with three ethnographers. *Journal of Communication Inquiry, 13*, 23-26.

Guba, E. G., & Lincoln, Y. S. (1994). Competing paradigms in qualitative research. In N. K. Denzin & Y. S. Lincoln (Eds.), *Handbook of qualitative research*. London: Sage.

Gummesson, E. (1991). *Qualitative methods in management research*. London: Sage.

Habermas, J. (1972). *Knowledge and human interests*. London: Heinemann.

Habermas, J. (1974). *Theory and practice*. Boston: Beacon Press.

Habermas, J. (1976a). The analytical theory of science and dialectics. In T. W. Adorno, H. Albert, R. Dahrendorf, J. Habermas, H. Pilot, & K. R. Popper (Eds.), *The positivist dispute in German sociology*. London: Heinemann.

Habermas, J. (1976b). A positivistically bisected rationalism. In T. W. Adorno, H. Albert, R. Dahrendorf, J. Habermas, H. Pilot, & K. R. Popper (Eds.), *The positivist dispute in German sociology*. London: Heinemann.

Habermas, J. (1978). Rationalism divided in two. In A. Giddens (Ed.), *Positivism and sociology*. London: Heinemann.

Habermas, J. (1979). An interview with Jürgen Habermas, by Angelo Bolaffi. *Telos, 39*, 163-172.

Habermas, J. (1981). Modernity versus postmodernity. *New German Critique, 22*, 3-14.

Habermas, J. (1982). A reply to my critics. In J. B. Thompson & D. Held (Eds.), *Habermas: Critical debates*. London: Macmillan.

Habermas, J. (1984). *The theory of communicative action (vol. 1): Reason and the rationalization of society*. Boston: Beacon Press.

Habermas, J. (1992). Concluding remarks. In C. Calhoun (Ed.), *Habermas and the public sphere*. Cambridge: MIT Press.

Habermas, J. (1993). *Justification and application: Remarks on discourse ethics*. Cambridge: Polity Press.

Habermas, J. (1996). *Between facts and norms*. Cambridge: Polity Press.

Hall, J. R. (1999). *Cultures of inquiry: From epistemology to discourse in sociohistorical research*. Cambridge: Cambridge University Press.

Hammersley, M. (1985). From ethnography to theory: A program and paradigm in the sociology of education. *Sociology, 19*, 244-59.

Hammersley, M. (1993). Introduction. In M. Hammersley (Ed.), *Social research: Philosophy, politics and practice*. London: Sage.

Hammersley, M. (1995). *The politics of social research*. London: Sage.

Hammersley, M. (1997). A reply to Humphries. *Sociological Research Online, 2*, <http://www.socresonline.org.uk/socresonline/2/4/6.html>

Hammersley, M., & Atkinson, P. (1995). *Ethnography: Principles in practice*. London: Routledge.

Hammersley, M., & Gomm, R. (1997a). Bias in social research. *Sociological Research Online, 2*, <http://www.socresonline.org.uk/socresonline/2/1/2.html>

Hammersley, M., & Gomm, R. (1997b). A response to Romm. *Sociological Research Online, 2*, <http://www.socresonline.org.uk/socresonline/2/4/7.html>

Harrison, C. W., & Hood-Williams, J. (1998). More varieties than Heinz: Social categories and sociality in Humphries, Hammersley and beyond. *Sociological Research Online, 3*, <http://www.socresonline.org.uk/socresonline/3/1/8.html>

Hartsock, N. (1990). Foucault on power: A theory for women? In L. J. Nicholson (Ed.), *Feminism/postmodernism*. New York: Routledge.

Harvey, L. (1990). *Critical social research*. London: Unwin Hyman.

Held, D. (1980). *Introduction to critical theory: Horkheimer to Habermas*. London: Hutchinson.

Henwood, K. L., & Pidgeon, N. F. (1993). Qualitative research and psychological theorizing. In M. Hammersley (Ed.), *Social research: Philosophy, politics and practice*. London: Sage.

Heritage, J. (1984). *Garfinkel and ethnomethodology*. Cambridge: Polity Press.

Holstein, J. A., & Gubrium, J. F. (1995). *The active interview*. London: Sage.

Hoy, D. C. (1986). Power, repression, progress: Foucault, Lukes, and the Frankfurt School. In D. C. Hoy (Ed.), *Foucault: A critical reader*. Oxford: Basil Blackwell.

Jackson, M. C. (1991). *Systems methodology for the management sciences*. New York: Plenum.

Jacobson, S. W., & Jacques, R. (1997). Destabilizing the field: Poststructuralist knowledge-making strategies in a postindustrial era. *Journal of Management Inquiry, 6*, 42-59.

Jervis, R. (1997). *System effects: Complexity in political and social life*. Princeton: Princeton University Press.

Jones, B. (2000). The money or the principle. *Work, Employment and Society, 14*, 191-200.

Josselson, R. (1996). On writing other people's lives: Self-analytic reflections of a narrative researcher. In R. Josselson (Ed.), *Ethics and process in the narrative study of lives*. London: Sage.

Keat, R. (1981). *The politics of social theory: Habermas, Freud and the critique of positivism*. Oxford: Basil Blackwell.

Keat, R., & Urry, J. (1982). *Social theory as science*. London: Routledge & Kegan Paul.

Keys, P. (1997). Approaches to understanding the process of Operational Research: Review, critique and extension. *Omega, 25*, 1-13.

Kilduff, M., & Mehra, A. (1997). Postmodernism and organizational research. *Academy of Management Review, 22*, 453-481.

Knorr-Cetina, K. D. (1981). *The manufacture of knowledge*. Oxford: Pergamon.

Kolakowski, L. (1993). An overall view of positivism. In M. Hammersley (Ed.), *Social research: Philosophy, politics and practice*. London: Sage.

Kramer, R. M., & Brewer, M. B. (1984). Effects of group identity on resource use in a simulated commons dilemma. *Journal of Personality and Social Psychology, 46*, 1044-1057.

Kuhn, T. (1970). *The structure of scientific revolutions*. Chicago: University of Chicago Press.

Lather, P. (1991). *Getting smart: Feminist research and pedagogy with/in the postmodern*. New York: Routledge.

Lather, P. (1993). Fertile obsession: Validity after poststructuralism. *The Sociological Quarterly, 34*, 673-693.

Lather, P. (1995). The validity of angels: Interpretive and textual strategies in researching the lives of women with HIV/Aids. *Qualitative Inquiry, 1*, 42-68.

Layder, D. (1993). *New strategies in social research: An introduction and guide*. Cambridge: Polity Press.

Lincoln, Y. S. (1995a). Emerging criteria for qualitative and interpretive research. *Qualitative Inquiry, 1*, 275-289.

Lincoln, Y. S. (1995b). The sixth moment: Emerging problems in qualitative research. *Studies in Symbolic Interaction, 19*, 37-55.

Lincoln, Y. S., & Guba, E. G. (1985). *Naturalistic inquiry*. Beverly Hills, CA: Sage.

Lofland, J. (1974). Styles of reporting qualitative field research. *American Sociologist, 9*, 101-111.

Lyotard, J. F. (1984). *The postmodern condition: A report on knowledge*. Manchester: Manchester University Press.

Maclagan, P. (1998). *Management and morality: A developmental perspective*. London: Sage.

Marshall, C. (1985). Appropriate criteria of trustworthiness and goodness for qualitative research on education organizations. *Quality and Quantity, 19*, 353-373.

Mauthner, M. (1998). Bringing silent voices into a public discourse: Researching accounts of sister relationships. In J. Ribbens & R. Edwards (Eds.), *Feminist dilemmas in qualitative research: Public knowledge and private lives*. London: Sage.

Maxwell, J. A. (1996). Qualitative research design: An interactive approach. *Applied Social Research Methods Series, vol. 41*. London: Sage.

Maynard, M. (1994). Methods, practice and epistemology: The debate about feminism and research. In M. Maynard & J. Purvis (Eds.), *Researching women's lives from a feminist perspective*. London: Taylor and Francis.

McGowan, J. (1991). *Postmodernism and its critics*. Ithaca: Cornell University Press.

McKay, V. I. (1997). The importance of reflection and action research in the training of community educators: A sociological contribution. *Educare, 26*, 4-14.

McKay, V. I., & Romm, N. R. A. (1992). *People's education in theoretical perspective*. Cape Town: Longman.

McTaggart, R. (1997). Revitalizing management as a scientific activity. *Management Learning, 28*, 177-195.

Merton, R. K. (1964). *Social theory and social structure*. London: The Free Press.

Midgley, G. (1996a). The ideal of unity and the practice of pluralism in systems science. In R. L. Flood & N. R. A. Romm (Eds.), *Critical systems thinking: Current research and practice*. New York: Plenum.

Midgley, G. (1996b). What is this thing called CST? In R. F. Flood & N. R. A. Romm (Eds.), *Critical systems thinking: Current research and practice*. New York: Plenum.

Midgley, G. (2000). *Systemic intervention: Philosophy, methodology and practice*. New York: Kluwer Academic/Plenum publishers.

Mill, J. S. (1961). *Auguste Comte and positivism*. Ann Arbor: University of Michigan Press.

Moggridge, A., & Reason, P. (1996). Human inquiry: Steps towards emancipatory practice. *Systems Practice, 9*, 159-175.

Morrow, R. A. (1994). *Critical theory and methodology*. London: Sage.

Mykhalovskiy, E. (1997). Reconsidering "Table Talk": Critical thoughts on the relationship between sociology, autobiography, and self-indulgence. In R. Hertz (Ed.), *Reflexivity and voice*. London: Sage.

Oakley, A. (1998). Gender, methodology and people's ways of knowing: Some problems with feminism and the paradigm debate in social science. *Sociology, 32*, 707-731.

O'Neill, J. (1995). *The poverty of postmodernism*. London: Routledge.

Page, R. N. (2000). The turn inward in qualitative research. *Harvard Educational Review, 70*, 23-38.

Palmer, P. J. (1987). Community, conflict, and ways of knowing: Ways to deepen our educational agenda. *Change, 19*, 20-25.

Pawson, R., & Tilley, N. (1997). *Realistic evaluation*. London: Sage.

Popper, K. R. (1959). *The logic of scientific discovery*. London: Hutchinson.

Popper, K. R. (1961). *The poverty of historicism*. London: Routledge & Kegan Paul.

Popper, K. R. (1966). *The open society and its enemies, vol. 2*. London: Routledge & Kegan Paul.

Popper, K. R. (1969). *Conjectures and refutations*. London: Routledge & Kegan Paul.

Popper, K. R. (1976). The logic of the social sciences. In T. W. Adorno, H. Albert, R. Dahrendorf, J. Habermas, H. Pilot, & K. R. Popper (Eds.), *The positivist dispute in German sociology*. London: Heinemann.

Popper, K. R. (1978). *Unended quest: An intellectual autobiography*. London: Fontana/Collins.

Popper, K. R. (1994). *The myth of the framework: In defence of science and rationality*. London: Routledge.

Rapoport, R. N. (1970). Three dilemmas in action research. *Human Relations, 23*, 499-513.

Readings, B. (1991). *Introducing Lyotard*. London: Routledge & Kegan Paul.

Reason, P. (1994). Inquiry and alienation. In P. Reason (Ed.), *Participation in human inquiry*. London: Sage.

Rehg, W. (1996). Translator's introduction. In J. Habermas (Ed.), *Between facts and norms.* Cambridge: Polity Press.

Remenyi, D., Williams, B., Money, A., & Swartz, E. (1998). *Doing research in business and management: An introduction to process and method.* London: Sage.

Rennie, D. L., Phillips, J. R., & Quartaro, G. K. (1988). Grounded theory: A promising approach to conceptualization in psychology? *Canadian Psychology, 29*, 139-150.

Rhode, D. L. (1989). *Justice and gender: Sex discrimination and the law.* Cambridge, Mass.: Massachusetts Institute of Technology.

Richardson, L. (1992). The consequences of poetic representation: Writing the other, rewriting the self. In C. Ellis & M. Flaherty (Eds.), *Windows on lived experience.* Newbury Park, CA: Sage.

Romm, N. R. A. (1990). Gouldner's reflexive methodological approach. In C. J. Alant (Ed.), *Sociology and society.* Johannesburg: Southern Book Publishers.

Romm, N. R. A. (1991). *The methodologies of positivism and Marxism: A sociological debate.* London: Macmillan.

Romm, N. R. A. (1994). Symbolic theory. In N. R. A. Romm & M. Sarakinsky (Eds.), *Social theory.* Johannesburg: Heinemann.

Romm, N. R. A. (1995). Knowing as intervention. *Systems Practice, 8*, 137-167.

Romm, N. R. A. (1996a). Critical theory as a way to understand development. In J. K. Coetzee & J. Graaff (Eds.), *Development and reconstruction in the new South Africa.* Halfway House (South Africa): International Thomson Publishing.

Romm, N. R. A. (1996b). Inquiry-and-intervention in systems planning: Probing methodological rationalities. *World Futures, 47*, 25-36.

Romm, N. R. A. (1996c). Systems methodologies and intervention: The issue of researcher responsibility. In R. L. Flood & N. R. A. Romm (Eds.), *Critical systems thinking: Current research and practice.* New York: Plenum.

Romm, N. R. A. (1997a). Becoming more accountable: A comment on Hammersley and Gomm. *Sociological Research Online, 2*, <http://www.socresonline.org.uk/socresonline/2/3/2.html>

Romm, N. R. A. (1997b), Implications of regarding information as meaningful rather than factual. In R. Winder, S. Probert, & I. Beeson (Eds.), *Philosophical aspects of information systems.* London: Taylor and Francis.

Romm, N. R. A. (1998a). Interdisciplinary practice as reflexivity. *Systemic Practice and Action Research, 11*, 63-77.

Romm, N. R. A. (1998b). Nurturing sustainability through triple loop learning: Discursive accountability as responsible action. *Systemist, 20*, 40-52.

Romm, N. R. A. (1998c). Caricaturing and categorizing in processes of argument. *Sociological Research Online, 3*, <http://www.socresonline.org.uk/socresonline/3/2/10.html>

Romm, N. R. A. (2001). Critical theoretical concerns in relation to development. In J. K. Coetzee, J. Graaff, F. Hendricks, & G. Wood (Eds.), *Development: Theory, policy and practice.* Cape Town: Oxford University Press.

Romm, N. R. A. (with support from Flood, R. L., & Adman, P.) (Romm et al., 1998). *An evaluation of ABET at UNISA.* Pretoria: University of South Africa Press.

Salipante, P., & Bouwen, R. (1995). The social construction of grievances: Organizational conflict as multiple perspectives. In D-M. Hosking, H. P. Dachler, & K. J. Gergen (Eds.), *Management and organization: Relational alternatives to individualism.* Aldershot: Avebury.

Sapsford, R. (1999). *Survey research.* London: Sage.

Sayer, A. (1984). *Method in social science: A realist approach.* London: Hutchinson.

Sayer, D. (1979). Science as critique: Marx vs Althusser. In J. Mepham & D-H. Ruben (Eds.), *Issues in Marxist philosophy, vol. 3.* Brighton: Harvester Press.

Sayer, D. (1983). *Marx's method: Ideology, science and critique in 'Capital'.* Brighton: Harvester Press.

Schofield, J. W. (1993). Increasing the generalizability of qualitative research. In M. Hammersley (Ed.), *Social research.* London: Sage.

Scholte, B. (1972). Toward a reflexive and critical anthropology. In D. Hymes (Ed.), *Reinventing anthropology.* New York: Pantheon Books.

Seale, C., & Filmer, P. (1998). Doing social surveys. In C. Seale (Ed.), *Researching society and culture*. London: Sage.

Sharma, A. (1997). Professional as agent: Knowledge asymmetry in agency exchange. *Academy of Management Review, 22*, 758-798.

Shipman, M. (1982). *The limitations of social research*. London: Longman.

Shotter, J. (1993a). *Cultural politics of everyday life*. Buckingham: Open University Press.

Shotter, J. (1993b). *Conversational realities: Constructing life through language*. London: Sage.

Sidell, M. (1993). Interpreting. In P. Shakespeare, D. Atkinson, & S. French (Eds,), *Reflecting on research practice*. Buckingham: Open University Press.

Simons, J. (1995). *Foucault and the political*. London: Routledge.

Slater, D. (1998). Analyzing cultural objects: Content analysis and semiotics. In C. Seale (Ed.), *Researching society and culture*. London: Sage.

Smaling, A. (1998). Dialogical partnership: The relationship between the researcher and the researched in action research. In B. Boog, H. Coenen, L. Keune, & R. Lammers (Eds.), *The complexity of relationships in action research*. Tilburg: Tilburg University Press.

Smelser, N. (1984). *Sociology*. Englewood Cliffs: Prentice-Hall.

Stanley, L., & Wise, S. (1983). *Breaking out: Feminist consciousness and feminist research*. London: Routledge & Kegan Paul.

Stanley, L., & Wise, S. (1991). Feminist research, feminist consciousness, and experiences of sexism. In M. M. Fonow & J. A. Cook (Eds.), *Beyond methodology*. Bloomington: Indiana University Press.

Stanley, L., & Wise, S. (1993). *Breaking out again: Feminist epistemology and ontology*. London: Routledge & Kegan Paul.

Stewart, D., & Shamdasani, P. (1990). *Focus groups: Theory and practice*. London: Sage.

Strauss, A. L., & Corbin, J. (1990). *Basics of qualitative research: Grounded theory, procedures and techniques*. London: Sage.

Swingewood, A. (1984). *A short history of sociological thought*. London: Macmillan.

Temple, B. (1997). Collegial accountability and bias: The solution or the problem? *Sociological Research Online, 2*, <http://www/socresonline.org.uk/socresonline/2/4/8.html>

Trusted, J. (1987). *Moral principles and social values*. London: Routledge & Kegan Paul.

Van Maanen, J. (1988). *Tales of the field: On writing ethnography*. Chicago: University of Chicago Press.

Walsh, D. (1998). Doing ethnography. In C. Seale (Ed.), *Researching society and culture*. London: Sage.

Weaver, J. H. (1997). *Conquering statistics: Numbers without the crunch*. New York: Plenum Trade.

Weber, M. (1949). *The methodology of the social sciences*. Translated and edited by E. A. Shils & H. A. Finch. New York: The Free Press.

Weber, M. (1970). *Essays in sociology*. Translated, edited and with an introduction by H. H. Gerth & C. Wright Mills. London: Routledge & Kegan Paul.

Weber, M. (1973). The definitions of sociology, social action and social relationship. In K. Thompson & J. Tunstall (Eds.), *Sociological perspectives*. Harmondsworth: Penguin Education.

Weil, S. (1996). From the other side of silence: New possibilities for dialogue in academic writing. *Changes, 14*, 223-231.

Weil, S. (1997). Social and organizational learning and unlearning in a different key: An introduction to the principles of critical learning theatre (CLT) and dialectical inquiry (DI). In F. A. Stowell, R. L. Ison, R. Armson, J. Holloway, S. Jackson, & S. McRobb (Eds.), *Systems for sustainability: People, organizations and environments*. New York: Plenum.

Weil, S. (1998). Rhetorics and realities in public service organizations: Systemic practice and organizational learning as critically reflexive action research (CRAR). *Systemic Practice and Action Research, 11*, 37-62.

Whyte, W. F., Greenwood, D. J., & Lazes, P. (1991). Participatory action research: Through practice to science in social research. In W. F. Whyte (Ed.), *Participatory action research*. London: Sage.

Wicks, A. C., Berman, S. L., & Jones, T. M. (1999). The structure of optimal trust: Moral and strategic implications. *Academy of Management Review, 24*, 99-116.

Wilder, D. A. (1984). Predictions of belief: Homogeneity and similarity following social categorization. *British Journal of Social Psychology, 23*, 323-333.

Wolcott, H. F. (1995). *The art of fieldwork.* London: Altamira Press.

Woodbrooks, C. (1991). *The construction of identity through the presentation of self: Black women candidates interviewing for administrative positions at a research university.* Unpublished dissertation, Ohio State University.

Yin, R. K. (1994). *Case study research: Design and methods.* London: Sage.

Yin, R. K. (2000). Rival explanations as an alternative to reforms as "experiments". In L. Bickman (Ed.), *Validity and social experimentation.* London: Sage.

Young, I. M. (1990a). *Justice and the politics of difference.* Princeton: Princeton University Press.

Young, I. M. (1990b). The ideal of community and the politics of difference. In L. J. Nicholson (Ed.), *Feminism/postmodernism.* New York: Routledge.

Author Index

Subject Index

Printed in the United Kingdom
by Lightning Source UK Ltd.
133731UK00010B/5/A

9 780306 465642